Ecological Studies, Vol. 191

Analysis and Synthesis

Ecological Studies

Further volumes can be found at springer.com

Volume 186
Biological Invasions in New Zealand (2006)
R.B. Allen and W.G. Lee (Eds.)

Volume 187
Managed Ecosystems and CO_2: Case Studies, Processes, and Perspectives (2006)
J. Nösberger, S.P. Long, R.J. Norby, M. Stitt, G.R. Hendrey, and H. Blum (Eds.)

Volume 188
Boreal Peatland Ecosystem (2006)
R.K. Wieder and D.H. Vitt (Eds.)

Volume 189
Ecology of Harmful Algae (2006)
E. Granéli and J.T. Turner (Eds.)

Volume 190
Wetlands and Natural Resource Management (2006)
J.T.A. Verhoeven, B. Beltman, R. Bobbink, and D.F. Whigham (Eds.)

Volume 191
Wetlands: Functioning, Biodiversity Conservation, and Restoration (2006)
R. Bobbink, B. Beltman, J.T.A. Verhoeven, and D.F. Whigham (Eds.)

Volume 192
Geological Approaches to Coral Reef Ecology (2007)
R.B. Aronson (Ed.)

Volume 193
Biological Invasions (2007)
W. Nentwig (Ed.)

Volume 194
***Clusia*: A Woody Neotropical Genus of Remarkable Plasticity and Diversity** (2007)
U. Lüttge (Ed.)

Volume 195
The Ecology of Browsing and Grazing (2008)
I.J. Gordon and H.H.T. Prins (Eds.)

Volume 196
Western North American *Juniperus* Communites: A Dynamic Vegetation Type (2008)
O. Van Auken (Ed.)

Volume 197
Ecology of Baltic Coastal Waters (2008)
U. Schiewer (Ed.)

Volume 198
Gradients in a Tropical Mountain Ecosystem of Ecuador (2008)
E. Beck, J. Bendix, I. Kottke, F. Makeschin, R. Mosandl (Eds.)

Volume 199
Hydrological and Biological Responses to Forest Practices: The Alsea Watershed Study (2008)
J.D. Stednick (Ed.)

Volume 200
Arid Dune Ecosystems: The Nizzana Sands in the Negev Desert (2008)
S.-W. Breckle, A. Yair, and M. Veste (Eds.)

Volume 201
The Everglades Experiments: Lessons for Ecosystem Restoration (2008)
C. Richardson (Ed.)

Volume 202
Ecosystem Organization of a Complex Landscape: Long-Term Research in the Bornhöved Lake District, Germany (2008)
O. Fränzle, L. Kappen, H.-P. Blume, and K. Dierssen (Eds.)

Roland Bobbink • Boudewijn Beltman •
Jos T.A. Verhoeven • Dennis F. Whigham
Editors

Wetlands: Functioning, Biodiversity Conservation, and Restoration

Springer

Dr. Roland Bobbink
Dr. Boudewijn Beltman
Prof. Dr. Jos T.A.Verhoeven
Landscape Ecology
Institute of Environmental Biology
Utrecht University
PO Box 800.84
3508 TB Utrecht
The Netherlands

Prof. Dr. Dennis F. Whigham
Smithsonian Environmental Research Center
PO Box 28
Edgewater, MD 21037
USA

Cover illustration: Large photo: Overview of an undisturbed tidal freshwater wetland system at Jug Bay Sanctuary on the Patuxent River, Maryland, USA (Photo R. Bobbink). Small photos: *Top* Dense beds of the invasive species *Crassostrea gigas*, the Pacific oyster, near Yerseke, The Netherlands, after intentional introduction of this species for cultivation (Photo D.W. Thieltges). *Middle* Restoration of a degraded rich fen by removal of the vegetation and the acidified top layer of the peat, The Netherlands (Photo B. Beltman). *Bottom* Extensive aerenchyma in the stem of pond weed (*Potamogeton pectinatus*) (Photo by Summers and Jackson 1994).

ISBN 978-3-540-77420-4

Ecological Studies ISSN 0070-8356

Library of Congress Control Number: 2007943144

Cover design: WMXDesign GmbH, Heidelberg, Germany

Printed on acid-free paper

9 8 7 6 5 4 3 2 1

springer.com

Preface

The two volumes on "Wetlands as a Natural Resource" in the book series Ecological Studies (Volumes 190, 191) are based on the highlights of the 7th INTECOL International Wetland Conference in Utrecht, 25–30 July 2004. This conference brought together about 900 participants from 61 countries, who discussed a very broad range of science-, policy- and management-oriented issues related to wetland ecology and hydrology, wetland conservation and creation, the impact of global change and wetlands as a resource in terms of food, flood protection and water quality enhancement. The participants were from different sectors of society, i.e., science and technology (scientists 45 %; PhD students 20 %), natural resource management (20 %) and policy (15 %). There were 38 symposia with invited speakers centered around the nine conference themes. We have given the organizers of these symposia the opportunity to produce one chapter for these books with the integrated content of their symposium. This has resulted in 25 chapters, of which 13 are included in Volume 190 under the heading "Wetlands and Natural Resource Management" and 12 in Volume 191 under the heading "Wetlands: Functioning, Biodiversity Conservation, and Restoration".

With these books, we had the aim to summarize the most important recent scientific results in wetland science, their applications in wetland and water resource management and their implications for the development of global, national and regional policies in the perspective of the ever-progressing deterioration of natural wetlands and the major impacts that future climate change will have. We hope that the integrated content of the chapters on such a wide scope of different fields in wetland science will serve as a valuable source of information, both for professionals in environmental science and natural resource management and for students and young professionals seeking to familiarize themselves with these fields. We also hope that the interaction between scientists from different disciplines, resource managers and policy makers will be stimulated by the content of these publications.

We as editors have worked according to a strict time schedule and we want to thank the authors for their timeliness in producing inspiring manuscripts and the scientists who have contributed to the peer reviews of the chapters for their active and prompt participation, which has enabled us to complete our task more or less according to this schedule. We acknowledge the series editor of the Ecological Studies book series, Prof. Dr. Ulrich Sommer, for his invitation to produce these volumes as one of the outcomes of the INTECOL Conference. We also thank Dr. Andrea Schlitzberger of Springer for her advice and help. We would like to take the opportunity to thank all key people who made the conference into such a success. In particular we want to thank Prof. Dr. Eugene Turner and the other members of the INTECOL Wetlands Working group, as well as the

International and National Scientific Committees for their support. We are most indebted to the team that organized the conference, in particular the inner circle, Fred Knol, René Kwant, Nienke Pot and Miranda Motshagen. The members of the Landscape Ecology Group at Utrecht University are thanked for their enormous efforts during the conference.

These two volumes are the most tangible, durable result of the conference. It is our wish that they will find their way to wetland professionals and students worldwide and will contribute to the wise use and conservation of the still large wetland resources that remain on our planet.

Utrecht, June 2006

The Editors

Roland Bobbink,
Boudewijn Beltman,
Jos T.A. Verhoeven,
Dennis F. Whigham

These two volumes are major contributions from a well-run meeting inspired by the collegiality and good will of the hosts. This meeting sparked professionalism through the expression of the finer parts of Dutch culture and, indeed, of all cultures. The 7th International Wetland Conference, like the preceding meetings, was successful because people care about living systems – i.e., people, landscapes, science culture, political structures, birds, etc. – as they go about trying to make things a little better and a little sooner than when wetlands were first appreciated in their collective minds. The successes from the meetings, exemplified by these two volumes, is partly because they enhance the possibilities for clarity and develop a strong scientific enterprise amidst the interactions of people in neutral spaces and a sometimes strong gradient of personalities and cultures. We never quite know ahead of time what the results of the meetings will be, although it has always been wonderful to see them evolve to closure.

It is humbling to know how small things influence others, which is a lesson in being careful, thoughtful and open. These efforts and successes are an explicit recognition of the interdependency of our discipline interests, but also the fabric of human interactions through politics, science, economics, etc. This interdependency suggests that being involved in wetland science and management is a great way to improve the quality of the natural world, but also society. The world needs, whether it knows it or not, the expertise and clear thinking of experts of general and detailed understanding to contribute to the social good. These two volumes do exactly that. Kudos to the Editors!

R. Eugene Turner, Chair
On behalf of the INTECOL Wetland Working Group

Contents

1	**Wetland Functioning in Relation to Biodiversity Conservation and Restoration** R. Bobbink, D.F. Whigham, B. Beltman, and J.T.A. Verhoeven	1
1.1	Introduction	1
1.2	Functioning of Plants and Animals in Wetlands	2
1.3	Biodiversity Conservation and Wetlands	6
1.4	Ecological Restoration of Wetlands	8
1.5	Synthesis	11
References		12

Section I: Functioning of Plants and Animals in Wetlands

2	**Plant Survival in Wet Environments: Resilience and Escape Mediated by Shoot Systems** M.B. Jackson	15
2.1	Introduction	15
2.2	How Excess Water Threatens Plant Life	17
2.2.1	Excluding and Trapping Effects of Water	18
2.2.3	The Energy Crisis	19
2.3	Resilience	19
2.3.1	Oxygen Shortage	19
2.3.2	Shortage of Carbon Dioxide	22
2.4	Escape	26

2.4.1 Aerobic Shoot Extension (the Aerobic Escape) 26
2.4.2 Anaerobic Shoot Extension (the Anaerobic Escape) 29
2.5 Summary 30
References 32

3 Center Stage: The Crucial Role of Macrophytes in Regulating Trophic Interactions in Shallow Lake Wetlands 37
R.L. Burks, G. Mulderij, E. Gross, I. Jones, L. Jacobsen, E. Jeppesen, and E. van Donk

3.1 Introduction 37
3.2 Central Position of Aquatic Vegetation 38
3.2.1 Central Themes: Zooplankton Depend on Macrophytes as Habitats 39
3.2.2 Central Themes: Chemical Ecology Spans Trophic Levels . . 42
3.2.3 Central Themes: Impacts of Grazer–Epiphyton Interactions with Macrophytes 45
3.2.4 Central Themes: Prevalance of Fish Influence in Shallow Lakes 49
3.3 In the Wings: Research Areas Worthy of Attention 51
3.3.1 Predictability of Macrophyte Function in Trophic Interactions Across a Climatic Gradient 51
3.3.2 Relative Importance of Chemical Ecology Across Trophic Levels 52
3.3.3 Disproportional Impacts of Certain Invertebrates and Exotic Species 52
3.4 Returning to Center Stage: Macrophytes are Common Players in Trophic Interactions 53
References 53

4 Biological Invasions: Concepts to Understand and Predict a Global Threat 61
G. van der Velde, S. Rajagopal, M. Kuyper-Kollenaar, A. bij de Vaate, D.W. Thieltges, and H.J. MacIsaac

4.1 Introduction 61
4.2 What is a Biological Invasion? 62
4.3 Impacts of Biological Invasions 63
4.3.1 Ecological Impacts 63
4.3.2 Evolutionary Impacts 64
4.3.3 Economic Impacts 64

4.3.4 Human Health Impacts 65
4.3.5 Measuring Impacts 65
4.4 Examples of Biological Invasions 66
4.5 Understanding and Predicting Biological Invasions 68
4.5.1 Invading Species Approach 68
4.5.2 Invaded Ecosystem Approach 70
4.5.3 Relationship Between Invader and Invaded Ecosystem (Key-Lock Approach) 77
4.5.4 Invasion Processes Differentiated in Time 79
4.5.5 Comparative Historical Approach 83
4.6 Shadows on the Prospects of Prediction 84
4.7 Conclusion 85
References 85

Section II: Conservation and Management of Wetlands

5 Wetland Conservation and Management: Questions for Science and Society in Applying the Ecosystem Approach 93
E. Maltby

5.1 Introduction 93
5.2 Wetlands at the Interface 93
5.3 Recognising a New Paradigm in Ecosystem Management . . 97
5.4 The Ecosystem Approach 97
5.4.1 Principle 1: The Management of Land, Water and Living Resources is a Question of Societal Choice 99
5.4.2 Principle 3: Ecosystem Managers Should Consider the Effects of Their Activities on Adjacent and Other Ecosystems; and Principle 7: The Ecosystem Approach Should be Undertaken at the Appropriate Scale . 103
5.4.3 Principle 4: There is a Need to Understand the Ecosystem in an Economic Context 107
5.4.4 Principle 9: Management must Recognise that Change is Inevitable 109
5.4.5 Principle 10: The Ecosystem Approach Should Seek the Appropriate Balance Between Conservation and Use of Biological Diversity 111
5.5 Conclusion 113
References 114

6 **Wetlands in the Tidal Freshwater Zone** 117
A. Barendregt, D.F. Whigham, P. Meire, A.H. Baldwin, and S. Van Damme

6.1 Characteristics of Tidal Freshwater Wetlands 117
6.2 Human Activities 121
6.2.1 Historical Development 121
6.2.2 Water Quality Changes 122
6.3 Biological Variation Within the Freshwater Tidal Ecosystem 122
6.3.1 Vegetation Zonation *123*
6.3.2 The Vegetation of European Tidal Freshwater Wetlands . . . 124
6.3.3 The Vegetation of North American Tidal Freshwater Wetlands 126
6.3.4 Wildlife 131
6.3.5 Fish Species 131
6.3.6 Other Biota 132
6.4 Chemical and Physical Processes: the Wetland as a Filter . . 134
6.5 Restoration and Future Outlook 138
6.5.1 Europe *138*
6.5.2 United States 140
6.6 Conclusions 141
References 142

7 **Biodiversity in European Shallow Lakes: a Multilevel–Multifactorial Field Study** 149
L. De Meester, S. Declerck, J.H. Janse, J.J. Dagevos, R. Portielje, E. Lammens, E. Jeppesen, T. Lauridsen, K. Schwenk, K. Muylaert, K. Van der Gucht, W. Vyverman, G. Zwart, E. van Hannen, P.J.T.M. van Puijenbroek, J.M. Conde-Porcuna, P. Sánchez-Castillo, J. Vandekerkhove, and L. Brendonck

7.1 Introduction 149
7.2 Lake Selection 151
7.3 Sampling and Analysis 153
7.4 Lake Characteristics 154
7.5 Multidimensionality of System-Wide Biodiversity 157
7.6 Macrophytes and Nutrient Concentrations 158
7.7 Model and Expert Tools 159
7.7.1 Approach 159
7.7.2 PCLake 160

7.7.3 The Expert System BASIS 164
7.7.4 Combined Models; and PCLake and BASIS as Management Tools 164
7.8 Synthesis: Policy Implications of the Results 165
References 166

8 River Basin Management to Conserve Wetlands and Water Resources 169
J. Pittock, B. Lehner, and L. Lifeng

8.1 Introduction 169
8.2 Systematically Prioritising Wetland Conservation: Freshwater Ecoregion Conservation 172
8.2.1 Freshwater Ecoregions 172
8.2.2 Planning Conservation of Freshwater Ecoregions 172
8.2.3 Rapid Assessment of Watersheds and Landscapes in Data-Short River Basins 175
8.3 Using Treaties to Conserve Wetlands and River Basin 176
8.3.1 The Ramsar Convention on Wetlands 176
8.3.2 Convention on Biological Diversity 178
8.3.3 European Union's Water Framework Directive 178
8.4 Poverty Reduction Through Wetlands Conservation 180
8.4.1 Yangtze River and Dongting Lake 180
8.4.2 Xipanshanzhou Polder 181
8.4.3 Quinshan Polder and Lessons Learned 182
8.5 Conservation and Wise Use of Wetlands: a Regional Partnership Approach 182
8.5.1 Ramsar Convention and Regional Initiatives 183
8.5.2 MedWet – The Mediterranean Wetlands Initiative 184
8.5.3 Great Asian Mountains Wetlands 185
8.6 Target-Driven Wetland Conservation: Lessons from WWF's Program FY99–FY04 187
8.6.1 Target 1: Protect and Sustainably Manage 250 × 10[6] ha of Freshwater Ecosystems Worldwide 187
8.6.2 Target 2: Restore and Conserve Ecological Processes in More Than 50 River or Lake Basins 189
8.6.3 Target 3: Best Practices in Water management are Adopted in Key Water-Using Sectors 192
8.7 Conclusion 193
References 194

9 Aspects of Adaptive Management of Coastal Wetlands: Case Studies of Processes, Conservation, Restoration, Impacts and Assessment . . . 197
P.E.R. Dale, M.B. Dale, J. Anorov, J. Knight, M.C. Minno, B. Powell, R.C. Raynie, and J.M. Visser

9.1 Introduction . . . 197
9.2 Diverse Tools to Identify Processes and Long-Term Changes 200
9.2.1 Geomorphic Evolution and Vegetation History . . . 201
9.2.2 Human Modification of Carbrook Wetlands . . . *203*
9.2.3 Conclusion . . . 205
9.3 Managing for Conservation: Monitoring Ecological Changes in Coastal Wetlands in Northeast Florida, USA . . . 205
9.3.1 Approach . . . 206
9.3.2 Environmental Characteristics . . . 206
9.3.3 Conclusion . . . 209
9.4 Managing for Restoration: a Multi-Scale Adaptive Approach in Restoring Coastal Wetlands in Louisiana, USA . . . 209
9.4.1 Approach . . . 210
9.4.2 Lessons Learned . . . 210
9.4.3 Program Recommendations . . . 213
9.4.4 Conclusion . . . 214
9.5 Managing the Environment to Reduce Insect Pests: a Multivariate Approach to Assess Impacts of Disturbance on Saltmarsh Processes in Subtropical Australia . . . 214
9.5.1 Approach . . . 215
9.5.2 Conclusion . . . 216
9.6 Use of Remote Sensing to Monitor Hydrologic Processes in Mangrove Forests and to Integrate Across the Adaptive Management Framework . . . 217
9.6.1 Approach . . . 217
9.6.2 Conclusions . . . 219
References . . . 220

Section III: Wetland Restoration and Creation

10 Contrasting Approaches to the Restoration of Diverse Vegetation in Herbaceous Wetlands 225
A.M. Boers, C.B. Frieswyk, J.T.A. Verhoeven, and J.B. Zedler

10.1 Introduction 225
10.2 Restoration Contexts 228
10.2.1 Restoration Targets 230
10.3 Discouraging Undesirable Species 230
10.3.1 Undesirable Invaders 231
10.3.2 Controlling Invasives 233
10.3.3 Minimizing Eutrophication 233
10.3.4 Establishing Appropriate Hydrology 235
10.4 Encouraging Desirable Species 235
10.4.1 Site Modifications 236
10.4.2 Natural Recruitment 237
10.4.3 Sowing Seed 238
10.4.4 The Decision to Plant 240
10.4.5 Suitable Sources for Propagules 241
10.5 Emerging Principles 242
References 243

11 Fen Management and Research Perspectives: An Overview . 247
B. Middleton, A. Grootjans, K. Jensen, H. Olde Venterink, and K. Margóczi

11.1 Introduction 247
11.2 Hydrological Systems of Fens 249
11.2.1 Large and Small Hydrological Systems 250
11.2.2 Natural Fens Can Be Very Stable 251
11.2.3 Hydrochemical Processes Stabilizing the Biodiversity of Fens 253
11.3 Eutrophication in Fens 254
11.3.1 Change in Management 254
11.3.2 Change in Nutrient Budgets 254
11.3.3 Internal Eutrophication 256
11.4 Seed Bank and Seed Dispersal 258

11.4.1 Seed Banks 259
11.4.2 Seed Dispersal 260
11.5 Fen Restoration: An Example From Hungary 260
11.5.1 Introduction 260
11.5.2 Destruction and Restoration of a Fen System in Hungary . . 261
11.5.3 Monitoring and Evaluation of the Created Wetland 262
11.6 Concluding Remarks 263
References 265

12 Social Learning in Wetland Development 269
E. Van Slobbe, E. D. Morris, N. Röling, R. Torenbeek, K. Broker, and H. Heering

12.1 Introduction 269
12.2 Socio-Technical Characteristics of Wetlands 270
12.3 Different Perspectives on Planning 273
12.4 Social Learning as Part of a Mix of Governance Approaches 275
12.5 Social Learning in Wetland Development 277
12.6 Conclusions 281
References 282

13 Eco-Hydrological Functioning of the Biebrza Wetlands: Lessons for the Conservation and Restoration of Deteriorated Wetlands 285
M.J. Wassen, T. Okruszko, I. Kardel, J. Chormanski, D. Swiatek, W. Mioduszewski, W. Bleuten, E.P. Querner, M. El Kahloun, O. Batelaan, and P. Meire

13.1 Introduction 285
13.2 General Characteristics of the Biebrza Valley 287
13.2.1 Introduction 287
13.2.2 Geomorphology, Lithology, and Geo-Hydrology 287
13.2.3 Vegetation 289
13.2.4 Birds and Mammals 290
13.3 Hydrology of the Biebrza Valley 290
13.3.1 Surface Water: Hydrography and Hydrology 290
13.3.2 Groundwater 292
13.3.3 Flooding 292

13.3.4 Drainage 294
13.4 Relation Between Hydrology and Vegetation Zoning 297
13.5 Productivity and Nutrient Limitation of Marsh
and Fen Vegetation 301
13.6 Discussion and Conclusions 304
References 307

Subject Index 311

Contributors

JULIE ANOROV

Environmental Management Systems, Environmental Heritage and Risk Branch, Department of Defence, Brindabella Park, Canberra 2600, Australia, e-mail: julie.anorov@defence.gov.au

A.H. BALDWIN

University of Maryland, Department of Biological Resources Engineering, 1423 Animal Science Building, College Park, MD 20742, USA, e-mail: baldwin@umd.edu

A. BARENDREGT

Utrecht University, Department of Environmental Sciences, Copernicus Institute for Sustainable Development and Innovation,PO Box 80115, 3508 TC Utrecht, The Netherlands, e-mail: a.barendregt@geo.uu.nl

OKKE BATELAAN

Department of Hydrology and Hydraulic Engineering, Vrije Universiteit Brussel, Pleinlaan 2, 1050 Brussels, Belgium, e-mail: batelaan@vub.ac.be

BOUDEWIJN BELTMAN

Landscape Ecology, Institute of Environmental Biology, Utrecht University, PO Box 800.84, 3508 TB Utrecht, The Netherlands, e-mail: b.beltman@bio.uu.nl

ABRAHAM BIJ DE VAATE

Waterfauna Hydrobiological Consultancy, Oostrandpark 30, 8212 AP Lelystad, The Netherlands, e-mail: vaate@waterfauna.nl

Wladimir Bleuten

Department of Physical Geography, Faculty of Geosciences, Utrecht University, PO Box 80.115, 3508 TC Utrecht, The Netherlands, e-mail: w.bleuten@geo.uu.nl

Roland Bobbink

Landscape Ecology, Institute of Environmental Biology, Utrecht University, PO Box 800.84, 3508 TB Utrecht, The Netherlands, e-mail: r.bobbink@bio.uu.nl

Aaron M. Boers

Department of Botany, 430 Lincoln Drive, University of Wisconsin, Madison, WI 53706-1381, USA, e-mail: amboers@wisc.edu

Luc Brendonck

Laboratory of Aquatic Ecology, Katholieke Universiteit Leuven, Ch. De Beriotstraat 32, 3000 Leuven, Belgium, e-mail: Luc.Brendonck@bio.kuleuven.ac.be

Koen Broker

Arcadis NV, Postbus 673, 7300 AR Apeldoorn, The Netherlands, e-mail: k.broker@arcadis.nl

Romi L. Burks

Southwestern University, Department of Biology, Georgetown, TX 78626, e-mail: burksr@southwestern.edu

Jarek Chormanski

Department of Hydraulic Engineering and Environmental Recultivation, Faculty of Engineering and Environmental Science, Warsaw Agricultural University, Ul Nowoursynowska 166, 02-787 Warsaw, Poland, e-mail: J.Chormanski@levis.sggw.waw.pl

José-Maria Conde-Porcuna

Institute of Water Research, University of Granada, Ramon y Cajal 4, 18071 Granada, Spain ,e-mail: jmconde@goliat.ugr.es

Jeroen J. Dagevos

Netherlands Environmental Assessment Agency, National Institute of Public Health and the Environment (RIVM),Antonie van Leeuwenhoeklaan 9, 3720 BA Bilthoven, The Netherlands, e-mail: JJ.Dagevos@rivm.nl

MIKE B. DALE

Australian School of Environmental Studies, Griffith University, Nathan, Queensland 4111, Australia, e-mail: m.dale@griffith.edu.au

PAT E.R. DALE

Australian School of Environmental Studies, Griffith University, Nathan, Queensland 4111, Australia, e-amil: p.dale@griffith.edu.au

STEVEN DECLERCK

Laboratory of Aquatic Ecology, Katholieke Universiteit Leuven, Ch. De Beriotstraat 32, 3000 Leuven, Belgium, e-mail: Steven.Declerck@bio.kuleuven.ac.be

LUC DE MEESTER

Laboratory of Aquatic Ecology, Katholieke Universiteit Leuven, Ch. De Beriotstraat 32, 3000 Leuven, Belgium, e-mail: Luc.DeMeester@bio.kuleuven.ac.be

MOHSSINE EL KAHLOUN

Ecosystem Management Research Group, Department of Biology, University of Antwerp, Universiteitsplein 1, 2610 Wilrijk, Belgium, e-mail: mohssine.elkahloun@ua.ac.be

CHRISTIN B. FRIESWYK

Department of Botany, 430 Lincoln Drive, University of Wisconsin, Madison, WI 53706-1381, USA, e-mail: cbfrieswyk@wisc.edu

AB GROOTJANS

Community and Conservation Ecology group, University of Groningen, PO Box 14, 9750 AA Haren, The Netherlands, e-mail: a.p.grootjans@biol.rug.nl

ELISABETH GROSS

University of Konstanz, Limnological Institute, PO Box M659, 78457 Konstanz, Germany, e-mail: Elisabeth.Gross@uni-konstanz.de

HERO HEERING

Arcadis NV, Postbus 673, 7300 AR Apeldoorn, The Netherlands, e-mail: h.herring@arcadis.nl

Michael B. Jackson

School of Biological Sciences, University of Bristol, Woodland Road, Bristol BS18 1UG, UK, e-mail: mike.jackson@bristol.ac.uk

Lene Jacobsen

Danish Institute for Fisheries Research, Department of Inland Fisheries, Vejlsøvej 39, DK-8600 Silkeborg, Denmark, e-mail: lj@dfu.min.dk

Jan H. Janse

Netherlands Environmental Assessment Agency, National Institute of Public Health and the Environment (RIVM),Antonie van Leeuwenhoeklaan 9, 3720 BA Bilthoven, The Netherlands, e-mail: JH.Janse@rivm.nl

Kai Jensen

AG Populations- und Vegetationsökologie, Biozentrum Klein Flottbek, Nutzpflanzenbiologie und Angewandte Ökologie, Universität Hamburg, Ohnhorststrase 18, 22609 Hamburg, Germany, e-mail: kjensen@iangbot.uni-hamburg.de

Eerik Jeppesen

Department of Freshwater Ecology, The National Environmental Research Institute (NERI), Vejlsøvej 25, PO Box 314, 8600 Silkeborg, Denmark, e-mail: ej@dmu.dk

Iwan Jones

Centre for Ecology and Hydrology, Winfrith Technology Centre, Winfrith Newburgh, Dorchester, Dorset DT2 8ZD, UK, e-mail: iwaj@ceh.ac.uk

Ignacy Kardel

Department of Hydraulic Engineering and Environmental Recultivation, Faculty of Engineering and Environmental Science, Warsaw Agricultural University, Ul Nowoursynowska 166, 02-787 Warsaw, Poland, e-mail: I.Kardel@levis.sggw.waw.pl

Jon Knight

Department of Geographical Sciences, University of Queensland, St Lucia, Queensland 4067, Australia, e-mail: jon.knight@uq.edu.au

Mirjam Kuyper-Kollenaar

Department of Animal Ecology and Ecophysiology, Institute for Wetland and Water Research, Radboud University Nijmegen, Toernooiveld 1, 6525 ED Nijmegen, The Netherlands, e-mail: mjkuyper@home.nl

EDDY LAMMENS

Department of Ecology, Institute for Inland Water Management and Wastewater Treatment (RIZA), Maerlant 17, 8200 AA Lelystad, The Netherlands, e-mail:E.Lammens@RIZA.RWS.minvenw.nl

BERNHARD LEHNER

Conservation Science Program, WWF US, 1250 24th Street NW, Washington, DC 20037-1193, USA, e-mail: Bernhard.lehner@wwfus.org

LI LIFENG

Yangtze Programme, WWF China Programme Office, Room 1609, Wen Hua Gong, Beijing Working People's Culture Palace, Beijing 100006, China, e-mail: lfli@wwfchina.org

HUGH J. MACISAAC

Great Lakes Institute for Environmental Research, University of Windsor, Windsor, Ontario N9B 3P4, Canada, e-mail: hughm@uwindsor.ca

EDWARD MALTBY

Institute for Sustainable Water, Integrated Management and Ecosystem Research (SWIMMER), Nicholson Building, University of Liverpool, Liverpool L69 3GP, UK, e-mail:e.maltby@liv.ac.uk

KATALIN MARGÓCZI

Department of Ecology, University of Szeged, Hungary, e-mail: margoczi@bio.u-szeged.hu

PATRICK MEIRE

Ecosystem Management Research Group, Department of Biology, University of Antwerp, Universiteitsplein 1, 2610 Wilrijk, Belgium, e-mail: patrick.meire@ua.ac.be

BETH MIDDLETON

USGS National Wetlands Research Center, 700 Cajundome Boulevard, Lafayette, LA 70506, USA, e-mail: beth_middleton@usgs.gov

MARC C. MINNO

St Johns River Water Management District, PO Box 1429, Palatka, FL 32177, USA, e-mail: mminno@sjrwmd.com

Waldemar Mioduszewski

Water Resources Department, Institute for Land Reclamation and Grassland Farming, IMUZ Falenty, 05-090 Raszyn, Poland,
e-mail: w.mioduszewski@imuz.edu.pl

E. Dick Morris

The Centre for Complexity and Change, Faculty of Technology Systems Discipline, The Open University, Walton Hall MK7 6AA, UK,
e-mail: R.M.Morris@open.ac.uk

Gabi Mulderij

Department of Food Web Studies, Centre for Limnology, Netherlands Institute of Ecology, 3631 AC Nieuwersluis, The Netherlands,
e-mail: g.mulderij@nioo.knaw.nl

Koenraad Muylaert

Laboratory of Aquatic Ecology and Protistology, University of Gent,
K.L. Ledeganckstraat 35, 9000 Gent, Belgium,
e-mail: Koenraad.Muylaert@UGent.be

Tomasz Okruszko

Department of Hydraulic Engineering and Environmental Recultivation, Faculty of Engineering and Environmental Science, Warsaw Agricultural University, Ul Nowoursynowska 166, 02-787 Warsaw, Poland,
e-mail: T.Okruszko@levis.sggw.waw.pl

Jamie Pittock

Global Frehswater Programme, WWF International, 15/71 Constitution Ave, Campbell, ACT 2612, Australia, e-mail: jpittock@wwf.org.au

Rob Portielje

Department of Ecology, Institute for Inland Water Management and Wastewater Treatment (RIZA), Maerlant 17, 8200 AA Lelystad, The Netherlands,
e-mail: R.Portielje@RIZA.RWS.minvenw.nl

Bernie Powell

Land and Environmental Assessment, Natural Resource Sciences, Department of Natural Resources and Mines, 80 Meiers Road, Indooroopilly, Queensland 4068, Australia, e-mail: bernie.powell@dnr.qld.gov.au

Erik P. Querner

Alterra, Wageningen University and Research Centre, PO Box 47, 6700 AA Wageningen, The Netherlands, e-mail: Erik.Querner@wur.nl

Sanjeevi Rajagopal

Department of Animal Ecology and Ecophysiology, Institute for Wetland and Water Research, Radboud University Nijmegen, Toernooiveld 1, 6525 ED Nijmegen, The Netherlands, e-mail: S.Rajagopal@science.ru.nl

Richard C. Raynie

Louisiana Department of Natural Resources, Coastal Restoration Division, PO Box 44027, Baton Rouge, LA 70804-4027, USA, e-mail: richard.raynie@la.gov

Niels Röling

Communication & Innovation Studies Group, Department of Social Sciences, Wageningen University Research, Hollandseweg 1, 6706 KN Wageningen, The Netherlands, e-mail: N.roling@inter.nl.net

Pedro Sánchez-Castillo

Department of Plant Biology, University of Granada, 18071, Granada, Spain, e-mail: psanchez@ugr.es

Klaus Schwenk

Department of Ecology and Evolution, Zoological Institute, J.W. Goethe-University Frankfurt am Main, Siesmayerstrasse 70, 60054 Frankfurt am Main, Germany, e-mail: K.Schwenk@zoology.uni-frankfurt.de

Dorota Swiatek

Department of Hydraulic Engineering and Environmental Recultivation, Faculty of Engineering and Environmental Science, Warsaw Agricultural University, Ul Nowoursynowska 166, 02-787 Warsaw, Poland, e-mail: Dorotams@levis.sggw.waw.pl

David W. Thieltges

Alfred Wegener Institute for Polar and Marine Research, Wadden Sea Station Sylt, Hafenstraße 43, 25992 List, Germany, e-mail: dthieltges@awi-bremerhaven.de

Reinder Torenbeek

Arcadis NV, Postbus 673, 7300 AR Apeldoorn, The Netherlands,
e-mail: r.torenbeek@arcadis.nl

S. Van Damme

Department of Biology, Ecosystem Management Research Group, University of Antwerp, Universiteitsplein 1, BE-2610 Antwerpen, Belgium,
e-mail: stefan.vandamme@ua.ac.be

Jochen Vandekerkhove

Laboratory of Aquatic Ecology, Katholieke Universiteit Leuven,
Ch. De Beriotstraat 32, 3000 Leuven, Belgium, e-mail: Jochen.Vandekerkhove@bio.kuleuven.ac.be

Katleen Van der Gucht

Laboratory of Aquatic Ecology and Protistology, University of Gent,
K.L. Ledeganckstraat 35, 9000 Gent, Belgium, e-mail: Katleen.VanderGucht@UGent.be

Gerard van der Velde

Department of Animal Ecology and Ecophysiology, Institute for Wetland and Water Research, Radboud University Nijmegen, Toernooiveld 1,
6525 ED Nijmegen, The Netherlands, e-mail: G.vandervelde@science.ru.nl

Ellen Van Donk

Netherlands Institute of Ecology, Centre for Limnology, Department of Food Web Studies, 3631 AC Nieuwersluis, The Netherlands,
e-mail: e.vandonk@nioo.knaw.nl

Erik van Hannen

Netherlands Institute of Ecology (NIOO-KNAW), Centre for Limnology, Rijksstraatweg 6, 3631 AC Nieuwersluis, The Netherlands,
e-mail: e.vanhannen@antonius.net

Peter J.T.M. van Puijenbroek

Netherlands Environmental Assessment Agency, National Institute of Public Health and the Environment (RIVM),Antonie van Leeuwenhoeklaan 9,
3720 BA Bilthoven, The Netherlands, e-mail: peter.van.puijenbroek@rivm.nl

Erik Van Slobbe

Arcadis NV, Postbus 673, 7300 AR Apeldoorn, The Netherlands,
e-mail: e.j.j.slobbe@arcadis.nl

Harry Olde Venterink
Utrecht University, Utrecht, The Netherlands,
e-mail: harry.oldeventerink@ethz.ch

Jos T.A. Verhoeven
Landscape Ecology, Institute of Environmental Biology, Utrecht University, PO Box 800.84, 3508 TB Utrecht, The Netherlands,
e-mail: j.t.a.verhoeven@bio.uu.nl

Jenneke M. Visser
Ecology Institute, School of the Coast and Environment, Louisiana State University, Baton Rouge, LA 70803, USA, e-mail: comvss@lsu.edu

Wim Vyverman
Laboratory of Aquatic Ecology and Protistology, University of Gent, K.L. Ledeganckstraat 35, 9000 Gent, Belgium,
e-mail: Wim.Vyverman@Ugent.be

Martin J. Wassen
Copernicus Institute for Sustainable Development and Innovation, Environmental Sciences, Faculty of Geosciences, Utrecht University, PO Box 80.115, 3508 TC Utrecht, The Netherlands, e-mail: m.wassen@geo.uu.nl

D.F. Whigham
Smithsonian Environmental Research Center, Box 28, Edgewater, MD 21037, USA, e-mail: whighamd@si.edu

Joy B. Zedler
Department of Botany, 430 Lincoln Drive, University of Wisconsin, Madison, WI 53706-1381, USA, e-mail: jbzedler@wisc.edu

Gabriel Zwart
Netherlands Institute of Ecology (NIOO-KNAW), Centre for Limnology, Rijksstraatweg 6, 3631 AC Nieuwersluis, The Netherlands,
e-mail: g.zwart@nioo.knaw.nl

1 Wetland Functioning in Relation to Biodiversity Conservation and Restoration

ROLAND BOBBINK, DENNIS F. WHIGHAM, BOUDEWIJN BELTMAN, and JOS T.A. VERHOEVEN

1.1 Introduction

Wetland ecosystems are a natural resource of global significance. Historically, their high level of plant and animal (especially bird) diversity is perhaps the major reason why wetland protection has become a high priority worldwide, supported by international agreements such as the Ramsar Convention and the International Convention of Biological Diversity. More recently, a number of goods and services provided specifically by wetland ecosystems have been identified that may even outweigh biodiversity in terms of their importance for human welfare and sustainable natural resource management worldwide. Wetlands, as transitional zones between land and water, provide a natural protection against extreme floods and storm surges. They may also store freshwater to be used for drinking water preparation or for irrigation. Wetlands bordering streams, rivers and lakes have a water quality enhancement function that is increasingly recognized. Because riverine and lacustrine wetlands often provide a spawning habitat, their importance as a source of juvenile fish for adjacent aquatic ecosystems should not be underestimated. In addition to these local and regional benefits, wetlands as a global resource provide a net sink of carbon dioxide. The world's peatlands are the only type of terrestrial ecosystem with a long-term net carbon storage function. However, the large amounts of carbon that have accumulated historically in peatlands may be released as a result of degradation, such as drainage, excavation, or fertilization.

Wetlands do produce a striking variety of goods and services and it is no wonder that, more often than any other terrestrial ecosystem, they are used by environmental economists to illustrate ecosystem functions and their values to mankind. However, in spite of the high biodiversity and the high impor-

Ecological Studies, Vol. 191
R. Bobbink, B. Beltman, J.T.A. Verhoeven, and D.F. Whigham (Eds.)
Wetlands: Functioning, Biodiversity Conservation, and Restoration

tance of the goods and services of wetland ecosystems, their global status is poor and, in many situations, they are degrading at ever faster rates as a result of a wide variety of human impacts. Many wetlands, particularly river floodplains, deltas and estuaries, are especially impacted by human activities. Early civilizations have been successful particularly because of their utilization of wetland resources. Where agriculture has thrived, the natural fertility of the soils and transport has been favored by river channels and associated wetlands. In the industrial era, human impacts have become dramatically negative as a result of floodplain reclamation, poldering, construction of flood control structures, drainage for agriculture, excavation of peat for fuel and the modification and straightening of river channels in favor of navigation. Worldwide, more than 50 % of the wetland resource has been lost because of these reasons. In some densely populated regions in Europe, North America and East Asia, more than 80 % of the wetlands have been lost or severely degraded. This clearly stresses the urgent need to restore or (re)create wetlands ecosystems.

This volume presents an integrated account from several of the major symposia presented at the 7th INTECOL International Wetlands Conference (Utrecht, The Netherlands). In this introductory chapter, we will give an overview of recent advances in the comprehension of how plants and animal function in wetlands, the biodiversity conservation of wetland ecosystems and their ecological restoration. We will also identify remaining gaps in scientific knowledge and understanding that need to be addressed to optimize biodiversity conservation and restoration of wetlands across the globe.

1.2 Functioning of Plants and Animals in Wetlands

Plant species inhabiting wetlands possess physiological and morphological adaptations that support long-term survival of flooding, even sometimes with remarkably high plant productivity (e.g. Mitsch and Gosselink 2000; Larcher 2003). The most widespread physiological adaptation to tolerate short-term anoxia is the ability to switch to fermentation as the main pathway for extracting energy. This takes place in combination with energy conservation measures derived from a complex down-regulation of non-essential energy-consuming processes. Furthermore, wetland species may have developed mechanisms to tolerate the toxic end-products of fermentation, especially by diversification of end-products or production of less toxic compounds (malate). The most important strategy of wetland plants against anoxic conditions is, however, the presence of air spaces (aerenchyma) in roots and stems, allowing oxygen diffusion from the aerial plant parts to the roots. In non-wetland species root porosity is rather low (2–7 %), whereas wetland species have much higher root porosities (on average 20–50 %). This internal pathway may sustain internal aerobic conditions across the root, even reaching the root tips. Well aerated roots of many wetland species may even release oxygen to the surrounding soil, forming an oxidized rhizosphere. This is an important mechanism to detoxify harmful soluble reduced ions such as manganese and sulfide. Red-brown deposits (Fe^{3+} oxides) around roots of wetland species are a clear indication of this phenomenon. Another morphological adaptation to waterlogged conditions is the formation of adventitious roots just above the anoxic zone of the soil, thus functioning in a more or less aerobic situation. This is especially triggered by increased ethylene concentrations after inundation in both wetland and terrestrial species.

Escape-mediated plant survival in wet environments is described by Jackson (Chapter 2). It is shown that the shoots of several plant species possess a means of escaping submergence through prompt, gravity-directed upward extension growth. The growth is usually O_2-dependent and mostly entrapped ethylene is the triggering signal. This gas initiates and sustains an abnormally fast rate of shoot elongation underwater. A rapid decrease in the growth-inhibiting hormone abscisic acid, mediated by ethylene, is often a pre-requisite for the escape to start. Thus, there is an increased probability of shoots regaining contact with the atmosphere before reserves become exhausted. This mechanism may especially be of importance for survival in ecosystems with unpredictable periods of complete submergence of the vegetation. In this situation, the ability to have some underwater photosynthesis may be of importance for the survival of normally non-aquatic plant species. In addition, it has been found that the shoots of certain species that over-winter as dormant tubers, rhizomes or turions are stimulated to elongate very rapidly, even under the complete absence of O_2. In contrast, some species have evolved suppression of this fast shoot extension, resulting in an underwater quiescence that may confer an advantage for survival where the prevailing water depth is too great for escape.

The crucial role of macrophytes in regulating trophic interactions in shallow lakes is clearly demonstrated by Burks et al. (Chapter 3). They define shallow lakes as permanently flooded wetlands that may be surrounded by emergent vegetation (i.e. marshy habitat). Water depth is typically less than 3 m. Studies on the role of trophic interactions with submerged plants in shallow lakes are not novel, but the study of other growth forms of macrophytes (i.e. floating-leaved, emergent, freely-floating) is an expanding field of research. Four key issues can be identified in this context. Firstly, zooplankton (and planktivorous fish) depend on macrophytes as habitat refuge, although the role of emergent vegetation is still unclear. Second, biochemical interactions ("allelopathy") between macrophytes and competing primary producers can be of major importance for the structure of the plant community; and the deterrent chemical nature of some aquatic plants may even strongly influence the biological community. Furthermore, macrophytes form a substrate for epiphyton. In eutrophic to hypertrophic lakes, a negative relationship exists between macrophyte biomass and epiphyton growth. However, the density of epiphyton, in turn, depends on the amount of grazing macroinvertebrates (e.g. snails), but not on nutrients. These grazers may help maintain littoral communities by continuously removing unwanted algae or sediments. Finally, the interaction between macrophytes and fish can be of major importance for the functioning of shallow lakes, but the complexity of these interactions is large and, in many situations, not known. It is suggested that the presence of aquatic macrophytes is a driving force for interactions within the rest of the food web, and thus determines which trophic interactions play a role in shifts between alternative shallow lake states.

The structure and functioning of wetlands can be (strongly) influenced by vertebrate herbivory, as recently reviewed by Van den Wyngaert and Bobbink (2006). The most important large herbivores of wetland ecosystems are several species of rodents and waterfowl, though the latter are mostly restricted to wetlands where open water is also present. The effects of wetland herbivores, which use above-ground plant parts, are in general more or less comparable with those found in terrestrial ecosystems. Two typical differences are distinguished between the effects of vertebrate herbivory in wetlands compared with those in terrestrial systems.

First, herbivory of leaves and shoots does not severely restrict annual above-ground plant production in wetland ecosystems, as long as there is oxygen transport to the below-ground storage organs. Damage to the shoots of emergent macrophytes below water level by foraging herbivores is in most cases lethal and strongly reduces above-ground plant production. In this way, the structure of the plant community can be strongly affected despite the removal of only a small part of the biomass. Second, the consequences of grubbing and consumption of below-ground storage organs by natural herbivores in wetlands with their wet and soft soils can be obvious. Grubbing for below-ground storage organs can severely disturb the vegetation; and, when the regeneration time is long, grubbed vegetation is much more sensitive to increasing grazing pressure. Increased grubbing may lead to denudation rather than replacement of the plant species by a less palatable one (Fig. 1.1). The affected wetland ecosystems may evolve to a "low steady state" or, if grazing pressure is above carrying capacity, can become completely deteriorated. It is clear that the effects of herbivores can be of major importance to the structure and dynamics of various wetlands, although their nutrient cycling is considered to be detritus-based. In contrast, herbivory of aquatic macrophytes is, in general, of minor importance in shallow lake ecosystems, although large invertebrate grazers (especially crayfish) affect the macrophyte cover in a few cases (Chapter 3).

Interest in biological invasions has rapidly increased in recent decades and today they are a major concern in ecology and conservation. Particularly dramatic consequences of invasions have been observed on island ecosystems where endemic species suffered severely, but wetlands (marshes,

Fig. 1.1 Grubbing of Canadian subarctic salt marsh by lesser snow geese (*Anser caerulescens caerulescens*) has created bare soil, where an increase in salinity inhibits revegetation (photo kindly made available by Peter Kotanen)

lakes, rivers) and estuaries are also among the most affected systems. The nature and impacts of invasions in these ecosystems are described in detail by Van der Velde et al. in Chapter 4. Human-mediated dispersal of transport of species is nowadays clearly much higher than natural dispersal in historical periods. The number of introduced species is related to the number of introduction events and to the number of individuals per event. Freshwater, estuarine and coastal wetlands are amongst the most invaded systems worldwide, because of the numerous introduction vectors and activities that facilitate invasions in these environments. Impacts of invasions may occur at all levels of ecological organization and are especially severe when the introduced species function as an ecosystem engineer (e.g. Crooks 2002). In addition, Van der Velde et al. give an overview of approaches that are used to understand and predict biological invasions. They clearly show that the relational key–lock approach is most promising, because it integrates the importance of ecosystem characteristics for the success of the invader, who in turn must possess the "right" characteristics to invade a particular ecosystem. Recent studies on propagule pressure generally met with success in explaining vulnerability of ecosystems to invasion, but in reality it is very likely that more than one mechanism simultaneously affects invasion success. Therefore, predictions by the different models are still quite inaccurate and have an observational nature. It is concluded that more experimentation is needed to verify the predictions of theoretical models with respect to (wetland) invasions and that we should not forget to learn from all historical "experiments".

1.3 Biodiversity Conservation and Wetlands

Biodiversity of wetland and freshwater ecosystems is currently under high risk, with a very high proportion of species threatened with extinction (Millennium Ecosystem Assessment 2005). Thus, wetland management and conservation is a huge challenge in the near future. Pittock et al. (Chapter 8) examine large-scale mechanisms for wetlands conservation based on: ecoregion prioritization and vision setting, integrated river basin management, poverty reduction, multilateral treaties, regional collaboration between countries and target-driven work by a non-government organization. These methods show some promise but also highlight the complexity of this task and long-term investments required to establish sustainable conservation initiatives. Integration of the expertise of hydrologists and that of biologists is needed; and starting at the catchment scale is clearly better than starting at small "hotspots". This conclusion is also drawn by Maltby in Chapter 5, in his extensive description of the "ecosystem approach" for the conservation and management of wetlands. Both chapters also highlight the need to build partnerships and capacities to meet the social and economic needs of local communities to sustain wetlands conservation, and thus to come to a "wise use" of these systems. In general, the conservation of biodiversity is not the primary motivation for most governments or stakeholders in managing wetlands, especially where local people live in poverty, as shown for the Yangtze river polders in China by Pittock et al. and for the restoration of Mesopotamian marshes by Maltby. Futhermore, it is obvious that conservation of ecological processes at the river basin scale or catchment scale are needed to protect wetland biodiversity and thus large-scale planning and strategic interventions are needed at the national and regional scale. One of the greatest challenges in the near future is to reach out and engage the many sectors of our society that have to be part of these large-scale solutions and help them choose for wetland biodiversity conservation. Not only the adequate use of scientific knowledge and integrative management is necessary to reach these objectives, but also the concept of "social learning" can be used to reach these objectives in wetland development, as discussed by Van Slobbe et al. (Chapter 12) using wetland management projects in the Netherlands, Belgium and Sri Lanka.

Conservation of biodiversity is thus one of the main objectives for the management of wetland ecosystems. This implies an urgent search in conservation policy for the identification of simple ecosystem-wide indicators for biodiversity in these systems. De Meester et al. (Chapter 7) have studied simultaneously the relationships between biodiversity among organism groups at different trophic levels and several environmental variables in shallow lakes across Europe (from Denmark to Spain). The patterns of association observed for biodiversity strongly indicate that biodiversity tends to be rather unrelated between different organism groups and trophic levels, although clear-water lakes are generally more species-rich for several groups (macrophytes, zooplankton, macro-invertebrates, amphibians, birds) than turbid lakes. This implies that the search for one simple index reflecting overall ecosystem diversity has little relevance. Rather, useful indicators may only be found for one or a restricted subset of organism groups that tend to be associated with a similar gradient. To assess richness at the ecosystem level, several of these indicators have to be combined and weighed according to the final aim of the assessment. Therefore, lake managers first have to define priorities, including, for instance, groups deserving special attention, the importance of rarity, etc. Second, given the low association in diversity between groups, future biodiversity studies would gain from the inclusion of taxonomic or functional groups that are often ignored in wetland science, such as microbial communities and periphyton. Finally, the obvious multidimensionality of diversity also has important implications for the definition of management objectives: management measures that may increase the diversity in one organism group may be ineffective or even counter-productive for diversity in other groups. Probably no management technique exists that leads to an enhancement of diversity at all trophic levels and for all taxonomic groups. One implication is that a good biodiversity strategy should involve variation in the management of shallow lakes in a region, so as to increase regional diversity.

Barendregt et al. (Chapter 6) describe the biodiversity and ecological functioning of tidal freshwater wetlands in both Europe and the United States, a relatively little-studied system but under high anthropogenic pressure in the past as well as presently. Tidal freshwater wetlands occur in the upper part of estuaries and can have tides of up to several meters in amplitude twice a day. They occur at the interface between the brackish zone in the estuary and the river; and where brackish and fresh water mix there is an area of maximum suspended matter. The tidal freshwater zone within the estuary plays an important role in overall patterns of nutrient cycling for the whole estuary and the pattern appears to differ in the brackish and saline sections. Although tidal freshwater wetlands do not include many endemic or restricted species, they are characterized by high species and habitat diversity. There is distinct zonation in flora and fauna species responding to the relationship between surface elevation and tidal amplitude. The tidal freshwater wetlands in Europe and North America also have a common history of being highly influenced by human activities, resulting in altered hydrology, losses in wetland area and high levels of sediment and nutrient input. Recently, management activities have been initiated on both sides of the Atlantic to maintain or restore its characteristic biodiversity and other free ecological services to man.

In many parts of the world, coastal wetlands are under pressure from increasing human populations and from predicted sea level rise. These wetlands are dynamic ecosystems with a characteristic species composition and include the freshwater–intertidal interface. Because of this, they are complex systems and at present limited knowledge is available for the processes operating in these coastal wetlands, making appropriate management a real challenge. Dale et al. (Chapter 9) demonstrate that "adaptive management strategies" are needed for the conservation and management of these coastal wetlands. They provide case studies from Australia and the United States that show the role science can (and has to) play in informing the adaptive management process. Several common themes emerge from these studies with respect to wetland conservation management. Hydrology appears to be a driving variable in all cases and may be impacted by various forms of development. The projects highlight the importance of monitoring water table levels and water quality (including salinity). Because of the complexity of coastal wetlands, all projects have involved interdisciplinary teams, bringing together skills from a wide range of areas including pedology, geomorphology, palynology, hydrology, plant science, entomology, remote sensing and, for all of them, aspects of management. They conclude that the sustainable management of coastal wetlands relies on knowledge of ecosystem processes, so that the rates and direction of ecosystem change can be assessed, whether due to ongoing environmental changes or direct human impacts. The adaptive management approach is strongly recommended, to improve the use of both existing and new knowledge of ecosystem processes in order to inform wetland management actions.

1.4 Ecological Restoration of Wetlands

Human activities have long caused the loss of large areas of wetlands because they become degraded as a result of changing their structure or function. Overviews of their status and the threats to wetlands in a global perspective are provided by Brinson and Malvarez (2002) for temperate freshwater wetlands, by Moore (2002) for bogs and by Junk (2002) for tropical wetlands. It is obvious that, besides reduction in the rate of wetland loss, restoration of degraded wetlands or creation of new wetland habitats is urgently needed to improve wetland condition and area.

Contrasting approaches to the ecological restoration of diverse herbaceous wetlands are evaluated by Boers et al. (Chapter 10). These species-rich herbaceous wetlands can shift to species-poor wetlands by alternations in water supply and/or the nutrient regime in response to human activities within the watershed. Both external eutrophication (e.g. phosphorus loading by surface water or atmospheric nitrogen deposition) and internal eutrophication by inputs of water with a changed chemical composition (e.g. higher alkalinity or increased sulfate concentration) can drastically impact the nutrient status of a wetland system, leading to dominance of aggressive species and loss of biodiversity. In addition, invasive non-native species may establish and outcompete native species. By comparison of restoration efforts in degraded herbaceous wetlands in Winconsin (USA) and the Netherlands, the authors have come to some general principles of restoration. Many of the constraints on restoration are abiotic. Hydrologic conditions, water chemistry and nutrient status are primary examples, e.g. a decline in groundwater levels at a restoration site can limit the ability to restore vegetation that depends on a consistent subterranean water supply. Furthermore, changes in the biota can limit a site's potential to be restored, e.g. the seed bank found at a site will degrade over time. For example, former wetlands that have been converted to agriculture for many years will lack a rich seed bank, constraining the development of a typical wetland vegetation even under appropriate abiotic conditions. In addition, the dispersal of wetland species can be low, especially in modern fragmented landscapes.

As shown for conservation management (see also Section 1.3), managers become better equipped to prioritize sites for restoration as the science of restoration advances and tools become available to deal with abiotic and biotic constraints. When degradation is not reversible within a reasonable time-scale, another site might have a higher priority for restoring diversity. In landscapes with many former wetlands and limited resources for restoration, it is strategic to develop management priorities first, then prioritize the sites where each goal can best be achieved. It may not be possible to recover exactly what was lost; in such cases, recovery of a general wetland vegetation type and recovery of ecosystem functions might be more realistic goals. Project goals can be adjusted to suit the site's potential to be restored. Projects can be designed to allow learning as restoration proceeds. To date, our understanding of the underlying processes that influence ecosystem development is poor; and specific outcomes are not highly predictable. If restoration projects were implemented as experiments, alternative approaches could be tested and results monitored; then, approaches that best achieve the goals could be used in future projects (Chapter 10).

Ecohydrological knowledge is of increasing importance in wetland management and restoration. It includes the study of the origin, flow and quality of groundwater and surface water and their implications for wetland functioning. It also takes in account the functional interrelations between hydrology and biota at the catchment scale. Undisturbed reference wetland ecosystems are sometimes needed to reveal environmental conditions necessary for the restoration of these wetlands in highly stressed or changed landscapes, where these conditions disappeared before quantification. Wassen et al. (Chapter 13) summarize the ecohydrological research in the Polish Bierbrza valley as an undisturbed reference system for comparable floodplain and fen systems elsewhere in Europe. They conclude that the information from reference areas may be useful to demonstrate general potentials for recovery or rehabilitation, but has to be treated with care to avoid too high expectations by wetland managers. In addition, experimentation is needed in many situations to verify the correlative data from most reference studies.

Fens are peat-forming mires that are fed by ground water or surface water and are found in both North America and Europe. In general, the vegetation of fens is rich in vascular plant species and bryophytes. Nowadays, many fens in the United States or Europe are lost through habitat destruction or are heavily degraded in structure or function. A prerequisite for successful restoration of degraded fens is the creation of suitable habitat conditions (nutrient status, hydrology), as shown in Chapter 11 by Middleton et al. There will be no adequate fen restoration without proper assessment of the hydrological functioning of the system. It is important that stagnation of surface water in the summer is prevented in restoration projects dealing with desiccated fens, in particular when the groundwater or surface water is rich in sulfate. The best option, which is not always available, is to stimulate the discharge of unpolluted (sulfate- and nitrate-poor) groundwater in such a way that this groundwater can flow through the fen system, leaving iron behind to bind phosphates in the top soil. Restoration measures are also needed to counterbalance, for instance, the increased atmospheric nitrogen deposition in northwestern European fens (Bobbink et al. 1998). Annual cutting of the vegetation, with removal of the hay can be necessary in this nitrogen-enriched situation, but sometimes has induced a shift to phosphorus limitation, which also may change biodiversity. Recently, it became obvious that the impact of reduced forms of nitrogen (ammonium and ammonia) on fen bryophytes is much more severe than the effects of oxidized nitrogen (Paulissen et al. 2004). In addition, Middleton et al. conclude that both soil seed persistence and seed dispersal capacity is rather underestimated, although the data in Chapter 10 are partly in contrast to this conclusion. Both types of seed sources can be influenced by restoration projects. For successful restoration, both temporal (time since alteration of the habitat) and spatial (degree of fragmentation/isolation) aspects have clearly to be taken into account.

Large areas on low-lying peaty soils in large parts of Europe and North America can no longer be maintained as intensely used agricultural areas. This offers good opportunities for the (re)creation of wetlands. However, in most cases the development of eutrophic wetlands is likely, as found in the Hungarian wetland restoration project described in Middleton et al. (Chapter

11). The reflooding of former agricultural areas usually results in very nutrient-rich soil conditions and in a persistent, highly-productive marsh vegetation, suitable to maintain waterfowl, but remarkable poor in other plant and animal species. Recreation of the peat-forming function is still unlikely in this situation, although it is a highly wanted target by managers in respect to carbon sequestration in natural systems. Full recovery of all fen qualities and functions is certainly not to be expected in cases where restoration of the hydrological system is not possible. Furthermore, the phosphorus status of the sediment or soil is highly increased. Setting clear targets, which can meet the opportunities that are being presented, combined with a realistic strategy for future possibilities may prevent much disappointment. A well explained modest result sometimes does more good for the public support than a scientifically sound project that lacks community or administrative support.

1.5 Synthesis

As shown by the examples in Section 1.3, the biodiversity conservation and management of wetlands requires clear insight into the ecosystem processes that maintain wetland functions. As we gain more knowledge about wetland functions, we also increase our understanding of the complex interactions between wetland biota and human activities. The results of the Utrecht symposia that are summarized in this book are yet another step in the directon of understanding how wetlands function. The syntheses also clearly demonstrate that effective management, restoration and conservation of wetland biodiversity will require the development of management tools that can be applied at a variety of scales. In some instances, ecological principles need to be integrated into assessment and management tools that can effectively result in the conservation of wetlands that have not been degraded to the point that they no longer provide high-quality ecological functions. Similarly, challenges continue to exist that require the applications of more effective tools to restored and create wetland as global attempts are made to reverse the long-standing trend of wetland loss. Now that the citizens of the world are increasingly asking for effect-driven management of wetland resources, the lessons learned and summarized in these chapters will hopefully move us closer to more successful application of knowledge about wetland ecology in the pursuit of a world in which human needs and natural wetland functions and biodiversity can coexist.

References

Bobbink R, Hornung M, Roelofs, JGM (1998) The effects of air-borne nitrogen pollutants on species diversity in natural and semi-natural vegetation. J Ecol 86:717–738

Brinson MM, Malvarez AI (2002) Temperate freshwater wetlands: types, status and threats. Environ Conserv 29:115–133

Crooks JA (2002) Characterizing ecosystem-level consequences of biological invasions: the role of ecosystems engineers. Oikos 97:153–166

Junk WJ (2002) Long-term environmental trends and the future of tropical wetlands. Environ Conserv 29:414–435

Larcher W (2003) Physiological plant ecology, 4th edn. Springer, Berlin, Heidelberg, New York

Millennium Ecosystem Assessment (2005) Ecosystems and humane well-being: synthesis. Island Press, Washington, D.C.

Mitsch WJ, Gosselink JG (2000) Wetlands, 3rd edn. Wiley, New York

Moore PD (2002) The future of temperate bogs. Environ Conserv 29:3–20

Paulissen MPCP, Van der Ven PJM, Dees AJ, Bobbink R (2004). Differential effects of nitrate and ammonium on three fen bryophyte species in relation to pollutant nitrogen input. New Phytol 164:551–458

Van den Wyngaert IJJ, Bobbink R (2006) The influence of vertebrate herbivory on ecological dynamics in wetland ecosystems. In: Maltby E (ed) The wetlands handbook, section 2. Blackwell Scientific, London (in press)

Section I:

Functioning of Plants and Animals in Wetlands

2 Plant Survival in Wet Environments: Resilience and Escape Mediated by Shoot Systems

MICHAEL B. JACKSON

2.1 Introduction

Species inhabiting marshland, river banks and similar places possess metabolic and developmental features that support long-term survival of flooding, sometimes in combination with remarkably fast rates of dry matter accumulation (Maltby 1991).This success in wet environments can be attributed to a combination of several physiological, metabolic and structural characteristics. Their blending and relative prominence vary widely between species and help explain species distribution across regions experiencing different levels of flooding severity (van Eck et al. 2004), including tree species in the Amazon basin (Parolin et al. 2004). The mechanism conferring tolerance can depend on the time of year, an effect that is unavoidably linked to stage of plant development (Crawford 2003).

The present article concentrates on shoot systems. Although roots do contribute to the survival of flooding, it seems that, in the last resort, it is the shoots (including rhizomes and related perennating organs) that count, because only they are a source of renewable energy. Furthermore, roots themselves seldom regenerate shoots, while shoot tissue can often regenerate replacement root systems adventitiously (Fig. 2.1). Nevertheless, roots of well-adapted species do display features that aid survival of both shallow and deep flooding and these are an absolute requirement for success during flooding. This aspect was reviewed recently (Jackson 2004). Of particular importance is the possession, by roots, of extensive aerenchyma and intercellular spaces (Jackson and Armstrong 1999). These can be present even at the root apex (Aschi-Smiti et al. 2003) as part of an interconnected and largely constitutive feature that links to aerenchyma in the shoots and thus helps to sustain the entire plant (Fig. 2.2). This arrangement creates internal pathways for aeration that substitute for the radial gas exchange that an external water cover suppresses. Aerenchyma formation is spatially regulated (Matsukura et al. 2000)

Ecological Studies, Vol. 191
R. Bobbink, B. Beltman, J.T.A. Verhoeven, and D.F. Whigham (Eds.)
Wetlands: Functioning, Biodiversity Conservation, and Restoration

Fig. 2.1 Prominent stem hypertrophy and prolific adventitious rooting of *Sesbania rostrata* growing in water approximately 2 m deep in a crop of deep-water rice at The Prachinburi Rice Research Center, Thailand. Swollen and highly aerenchymatous and spongy short roots of *Lugwigia adscendens* are also visible floating on the water surface

and is the outcome of selective cell death (lysigeny), particular patterns of cell division and cell separation (schizogeny), or differential cell expansion (expansigeny; Seago et al. 2005). Its presence underpins the adaptations discussed in the present article. The aeration via aerenchyma is achieved mainly by diffusion. However, in some cases, this leads to internal pressurization that may induce mass flow of gas. This can be effective over longer distances than can diffusion alone (Grosse et al. 1996).

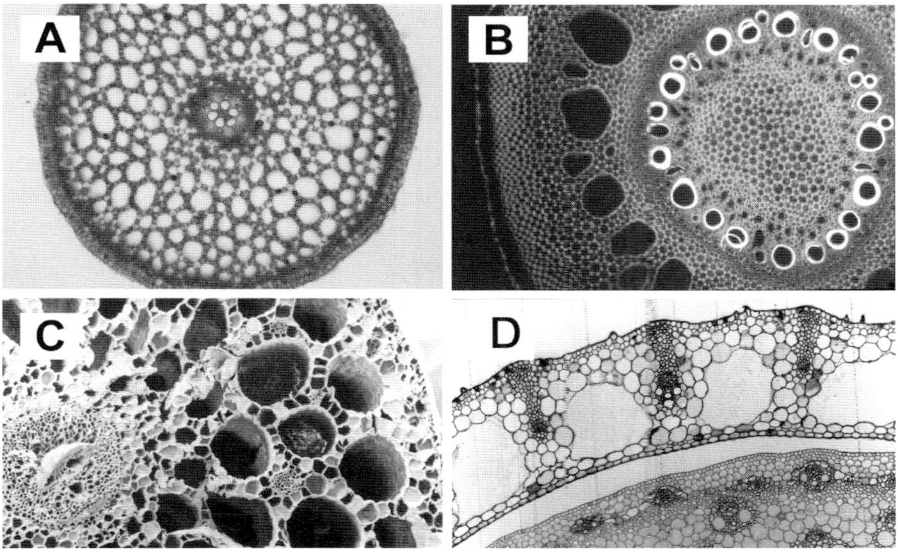

Fig. 2.2 Examples of the constitutive aerenchyma of root, rhizome, stem and leaf sheath in wetland species. **A** Root of *Acorus calamus*, free-hand section stained with toluidine blue, **B** rhizome of the water chestnut (*Trapa natans*), stained in toluidine blue and viewed under UV light, **C** stem of pond weed (*Potamogeton pectinatus*), **D** leaf sheath of common reed (*Phragmites australis*). **A** Courtesy of Ales Soukup (Charles University, Prague, Czech Republic), **B** courtesy of J. Seago (State University of New York, Oswego, USA), **C** from Summers and Jackson (1994), **D** courtesy of J. Armstrong (University of Hull, UK)

2.2 How Excess Water Threatens Plant Life

Plants suffer from inundation because of certain physical characteristics of liquid water that impose a rate of gas diffusion in and out of the plant that is approximately 10 000 times slower than in air. This isolates inundated tissues from adequate external supplies of O_2 or CO_2 and traps gaseous products of metabolism within the plant. The effect is exacerbated by the development of substantial unstirred boundary layers, especially in still water (Armstrong 1979). In addition, water can have a more overt physical impact through wave and hydraulic forces that can strongly limit the extent of colonization. Such effects have been modelled mathematically (Schutten et al. 2004).

2.2.1 Excluding and Trapping Effects of Water

Oxygen

In addition to slowing O_2 diffusion towards the plant, water lowers the concentration of O_2 around the plant 33 times compared to air on a per unit volume basis. This is because O_2 is not very soluble (8.31 mol m^{-3} in air but only 0.25 mol m^{-3} in water at 25 °C; Armstrong 1979). Thus, roots and other underground organs are especially vulnerable to external oxygen shortage. They are also vulnerable to toxins, including hydrogen sulfide (H_2S; Armstrong et al. 1996), generated by anaerobic soil bacteria that utilize inorganic ions as alternative electron acceptors to O_2 in energy generation (Laanbroek, 1990).

Carbon Dioxide

Although CO_2 is over 25 times more soluble than O_2, underwater photosynthesis is severely CO_2-limited by the low aqueous diffusivity of the gas, the small absolute concentrations involved and the relatively low affinity for CO_2 of the CO_2-capturing enzyme ribulose bisphosphate carboxylase– oxygenase (Rubisco). Thus, for underwater photosynthesis to rival the rate in air, water needs to be bubbled with 3–20% (v/v) CO_2 (Setter et al. 1989), which is many times greater than the small concentration of normal air (air contains approx. 360 ppm CO_2). Photosynthesis also suffers underwater because of light exclusion (Adkins et al. 1990) caused by reflection and absorption. The latter increases logarithmically with depth (Madsen 1993).

Entrapping Effects of Water

As well as taking in O_2 and CO_2 from their surroundings, plants also generate these gases. Any photosynthetically generated O_2 that is trapped within the plant can help offset any shortage of externally sourced O_2. Furthermore, any entrapped respiratory CO_2 can be fixed by photosynthesis and can also influence growth and metabolic processes, even in non-photosynthetic organs. Ethylene gas, generated from methionine, is also trapped in submerged tissues and quickly attains physiologically active (ppm level) concentrations (Musgrave et al. 1972; Voesenek et al. 1993). Entrapped ethylene activates the principal escape mechanism by which well-adapted species recognize the submerged condition and respond positively to it (see Section 2.4.1).

2.2.3 The Energy Crisis

When shoots are underwater, a shortfall in available energy is brought about, in part, by the suppression of photosynthesis. This suppresses the synthesis of respirable sugars and O_2, both of which support respiration and thus the ATP production needed for cell maintenance. If, in addition, exclusion of externally sourced O_2 renders tissues anaerobic (especially at night and in the roots), an overall energy shortfall is inevitable, if only because of the small yield of ATP obtained from anaerobic respiration (Gibbs and Greenway 2003; Jackson and Ricard 2003). The energy crisis injures or kills cells in several ways, including by loss of membrane selectivity and integrity, loss of essential proteins and other macromolecules, or loss of control of damaging reactive oxygen species when oxygen re-enters previously anaerobic cells. Once mitochondria degenerate, there is no future, energetically speaking, for the cell (Vartapetian 1991).

Thus, flooding and submergence inflict blows to three of the most fundamental live-sustaining processes in plants, namely aerobic respiration, photosynthesis and cellular maintenance. This damages even well-adapted species such as rice (Jackson and Ram 2003) but begs the question of how the thriving wetland flora effectively overcomes these problems So far, no life-saving metabolic route to tolerance has been identified that is unique to wetland plants. Instead, each species possesses a compilation of structural and biochemical characteristics that, together, endow a plant with increased ability to survive flooding. These can be constitutive or response-activated features that increase resilience to stresses such as energy shortage, or minimize exposure to flooding by promoting physical escape. Examples of these attributes constitute the remainder of this article.

2.3 Resilience

2.3.1 Oxygen Shortage

Rapid metabolic adaptability to oxygen deprivation is evident even in non-wetland plants. As O_2 becomes less available, considerable re-adjustments to the pattern of gene expression and protein synthesis are being revealed by new technologies, such as two dimensional MALDI-DE-TOF mass spectrometry (Chang et al. 2000), multi-gene expression analysis (Dolferus et al. 2003) and molecular transformations involving over expression, antisensing (Ismond et al. 2003) or transposon tagging (Baxter-Burrell et al. 2003). The work (mostly with maize root tips and *Arabidopsis thaliana*) reveals that, in the run-down to full anoxia, partial oxygen shortage trains tissue to tolerate later anoxia for longer (Saglio et al. 1988). How cells initially sense the partial oxygen shortage that triggers this biochemical acclimation remains unknown. Clearly, the O_2-sensing mechanism has a lower affinity for O_2 than cytochrome oxidase that chemically reduces O_2 in mitochondria. Accordingly, binding to haemoglobin has been ruled out as the sensing step, on the grounds that its affinity for O_2 is excessive (K_d is approx. 3 nM). But, haemoglobin remains important for anoxia tolerance in other ways not yet fully understood (Dordas et al. 2003).

The initial training, under partial oxygen shortage, is achieved by means of molecular and metabolic adjustments (summarized in Fig. 2.3). These start to take effect within 2 h and are predicated on changes in gene expression detectable after only 30 min (Klok et al. 2002) and mediated by changes in internal cytosolic calcium, pH, active oxygen species and protein phosphorylations. These adjustment involve upregulation of genes (e.g., myb-related genes) coding for proteinaceous *trans*-acting factors (transcription factors) involved in regulating key genes involved in sustaining an anaerobic existence (Hoeren et al. 1998). This leads to genes being expressed that ensure the gly-

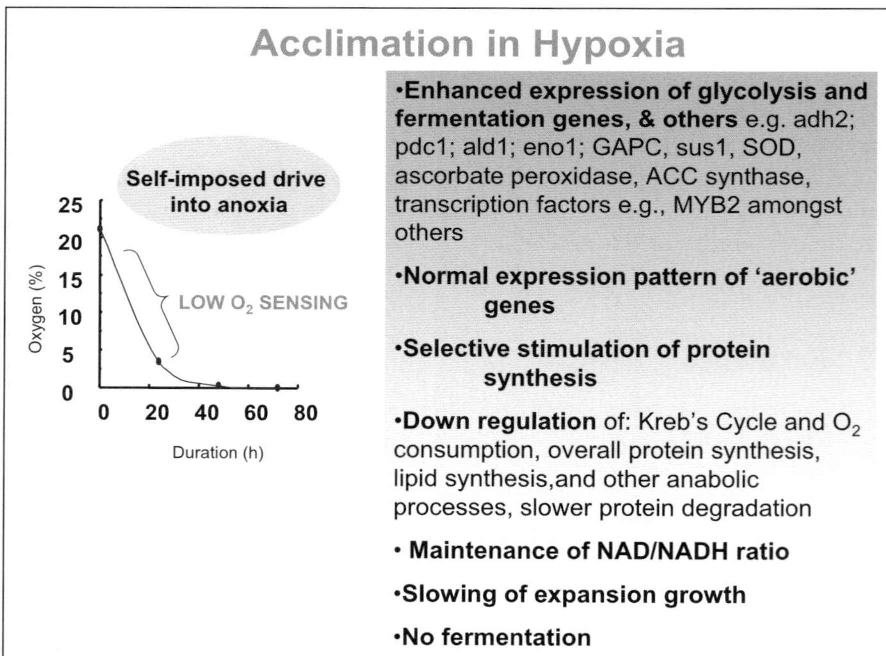

Fig. 2.3 Summary of some metabolic and molecular responses to early partial oxygen shortage caused by flooding. These changes are thought to increase resilience to subsequent anoxic conditions that would develop if the submerged tissue continued to consume oxygen faster than oxygen entered from external sources or via photosynthesis

colysis and alcoholic fermentation pathways still function, should anoxia set in later. For example, in the wetland species *Acorus calamus*, transcript levels for the glycolysis enzyme fructose-1,6-bisphosphate aldolase (ALD) and the fermentation enzyme alcohol dehydrogenase (ADH) are especially high during winter submergence when internal O_2 is likely to be absent (Bucher et al. 1996). In the short term, a fast rate of fermentation during anoxia may be more effective than a slower one, exemplified by light-grown rice seedlings (Mustroph and Albrecht 2003), but more subtle accommodation of the energy crisis is probably more typical. This phenomenon is not well-understood. It involves changes during hypoxia that ensure some fermentation during later anoxia, while suppressing sugar consumption and lessening ATP demand through suppressing pathways not essential for maintaining cell integrity. Suppression can begin before all O_2 is exhausted and thus may delay the onset of anoxia. It can also set in before the onset of fermentation or any increase in the NADH:NAD$^+$ ratio (Geigenberger et al. 2000). Examples of energy-saving metabolic adjustments include decreasing overall protein synthesis and minimizing potentially damaging cytoplasmic acidification. The latter effect would also decrease the level of reactionary ATP-driven outward proton pumping needed to maintain the normal cytoplasmic pH of approximately 7.5. Slower use of ATP would also help avoid cytoplasmic acidification, since ATP hydrolysis itself liberates acidifying protons. Other possible ATP-conserving measures include a shift from invertase-based conversion of sucrose to hexose phosphates to one based on the hypoxically inducible enzyme sucrose synthase (Bologa et al. 2003). This pathway requires half the ATP of the more usual invertase-based route. Similarly, the ATP-consuming step in glycolysis can be accomplished without expenditure of ATP if the phosphorylating enzyme {pyrophosphate:fructose 6-phosphate 1-phosphotransferase [PFK(PP_i)]} and inorganic diphosphate are available (Mohanty et al. 1993). Along similar lines, starch degradation mediated by starch phosphorylase instead of amylases can save ATP thereby increasing the ATP yield 50 %, assuming onward processing by glycolysis and fermentation. This seems to be the dominant pathway in anaerobic rhizomes of the common club-rush *Schoenoplectus lacustris* (Steinmann and Braendle 1984). These and other processes of metabolic adaptation were reviewed recently (Geigenberger 2003; Greenway and Gibbs 2003) and are summarized in Fig. 2.3. Overall, such changes ease the imbalance between ATP consumption and production, thereby retaining the cellular integrity that is a prerequisite for recovery when oxygen re-enters, once flooding subsides.

The importance of down-regulating activities that consume ATP and scarce respirable resources has been highlighted in some whole plant studies. In *Ranunculus repens*, the 'turlough' form naturally tolerates several months of total submergence each autumn and winter (Lynn and Waldren 2003) in association with a suppression of the energy-requiring acceleration of petiole elongation that submergence normally induces in this species. Similarly, in rice, the normal acceleration of leaf extension by young rice plants when they are submerged is suppressed in a submergence-tolerant cultivar FR13A (Jackson et al. 1987). In rice, the genes involved have been fine mapped mainly to one locus on Chromosome 9 (Toojinda et al. 2003). This finding holds promise for the future development of rice cultivars with superior submergence tolerance.

The need to cope with oxidative stresses when O_2 re-enters after a period of anoxia adds a further dimension to understanding tolerance, especially for perennial aquatics re-emerging after their anoxic winter hibernation. In rhizomes of anoxia-tolerant *Iris pseudocorus,* post-anoxic peroxidative damage to unsaturated fatty acids of lipid membranes is much less than in anoxia-sensitive *Iris germanica* (Monk et al. 1989). The damaging factors are reactive oxygen species (ROS) such as superoxide. Their formation is suppressed more strongly in tolerant species; and this supression can involve hypoxia-inducible enzymes such as superoxide dismutase (SOD). SOD accumulates more in tolerant *I. pseudocorus* than in a less tolerant *I. germanica* (Monk et al. 1989). Ethane production gives an indirect measure of lipid peroxidation by ROS. Accordingly, post-submergence ethane emission is less from submergence tolerant rice FR13A than from a less tolerant type (Santosa 2002), suggesting a stronger control of oxygen free radicals. For rice, it has been postulated that acetaldehyde production from ethanol, mediated by the enzyme catalase, can help deplete re-aerated tissues of free radicals (Boamfa et al. 2005); The subject of post-anoxic oxidative damage is comprehensively reviewed by Blokhina et al. (2003).

It is important to recognize that O_2 has key roles in addition to acting as an acceptor of reductant in mitochondrial respiration or as a source of damaging free radicals on re-aeration. One such role is in the unsaturation (elimination of hydrogens to form double bonds) of fatty acids that are essential components of membrane lipids. No such desaturation can take place in anaerobic cells because key oxygenation reactions are impossible. Thus, long-term survival without O_2 may depend on lowering of lipid turnover rates to conserve the limited pool of desaturated fatty acids (Heinzi and Brändle 1993).

2.3.2 Shortage of Carbon Dioxide

The energy-saving features described above that are thought to enhance resilience to oxygen-deficient conditions also protect against energy shortage imposed by slow or reduced photosynthesis caused by impeded CO_2 entry and utilization. In addition, several constitutive response strategies minimize the deleterious impact on photosynthesis of inadequate delivery of external CO_2. It can be argued that these constitute an escape rather than enhanced resilience. However, it is convenient to consider them here in terms of resilience because no actual physical escape from the floodwater is involved.

Internal Carbon Dioxide-Concentrating Mechanisms

Some submersed aquatics have evolved C4 acid-mediated photosynthesis to concentrate CO_2 at fixation sites. Several aquatic plants adopt crassulacean acid metabolism (CAM). This is unexpected since CAM is most commonly found in species frequenting dry places. CAM allows aquatic species such as *Isoetes*, *Crassula* and the plantain *Littorella uniflora* to fix CO_2 at night as well by more conventional means in the day. In CAM plants, dark fixation is achieved by means of PEP carboxylase activity that generates malic acid from phosphoenolpyruvate and CO_2 once the latter has crossed the plasma membrane. This malate is concentrated in the vacuoles of chloroplast-containing cells and is then decarboxylated to CO_2 the next day and re-fixed by a conventional light-driven Calvin cycle. The CO_2-concentrating effect of this mechanism confers a further advantage by suppressing the oxidase activity of Rubisco, thereby suppressing photorespiration (Madsen 1993). In contrast, a small number of aquatic species appear to adopt a modified Kranz-less C4 type of daytime fixation of CO_2 that involves cytoplasmic fixation of CO_2 into malate by PEP carboxylase. In *Hydrilla verticillata*, malate then diffuses into chloroplasts of the same cells where NADP malic enzyme decarboxylates malate back to CO_2. Here, ample Rubisco refixes it via the Calvin cycle. This mechanism would also fix respiratory CO_2 with improved efficiency since the high CO_2:O_2 ratio would inevitably suppress photorespiration. It also appears to be inducible by low CO_2 availability, when a particular isoform of PEP carboxylase is expressed. In *H. verticillata*, photosynthesis reverts to the conventional C3 pathway when CO_2 is plentiful (Bowes et al. 2002).

Another mechanism of enhancing carbon harvesting involves substituting bicarbonate (HCO_3^-) for CO_2. HCO_3^- can dissolve in water to higher concentrations than CO_2 (e.g., 2 mol m^{-3}), even at neutral pHs but more so when floodwater is alkaline. In these conditions, HCO_3^- is a potentially rich source of external inorganic carbon for the select number of species able to utilize it effectively, e.g., *Myriophyllum spicatum* (Bowes 1987). The mechanism that allows HCO_3^- to be used in photosynthesis is not entirely clear. In *Potamogeton lucens* and *Elodea canadensis*, ATPase-mediated proton extrusion acidifies the apoplast and boundary layer of the abaxial leaf surface, thus favouring conversion of HCO_3^- to CO_2 at the leaf surface, where it more readily enters photosynthesizing cells. Internal electrical neutrality is maintained by a balancing proton uptake by cells of the adaxial leaf surface (Staal et al. 1989). The presumed competitive advantage that HCO_3^- usage confers may be offset to some extent by higher energy costs of this more complex photosynthetic machinery. Accordingly, the number of species that adopt HCO_3^- as a carbon source is small, even amongst habitually submersed species. In ways not understood, low light and impeded CO_2 availability can increase the affinity of photosynthesis for HCO_3^-, as shown for *E. canadensis* (Sand-Jensen 1987). Such an effect may thus constitute an adaptive response mechanism.

Sediment Carbon Dioxide

A few species can supplement foliar intake of dissolved CO_2 with uptake from the sediment, e.g., *Littorella uniflora*, *Lobelia dortmanna* and species of the primitive vascular genus *Isoetes* (Richardson et al. 1984). In such plants, CO_2-harvesting via the roots and internal diffusive transport to photosynthesizing leaves is enhanced by high root:shoot ratios, short shoots and leaves and prominent, longitudinally connected, gas-filled lacunae that permeate both roots and shoots and are lined by cells rich in chloroplasts, lateral losses of CO_2 being minimized by a thick outer cuticle. *Isoetes* species are especially successful in soft water lakes where HCO_3^- concentrations are below 200 μM. This concentration is too low for competing bicarbonate users, thus leaving the way open for efficient CO_2-utilizers such as members of the genus *Isoetes* (Brouwer et al. 2002).

Leaf Heterophylly

Aquatic species commonly develop morphologically distinct leaves when inundated (Sculthorpe 1967). Some amphibious species such as *Ranunculus flabelaris* are also capable of this. These 'water' leaves may favour underwater carbon fixation as well as minimizing potentially damaging drag from flowing currents, although this latter aspect requires more experimental testing. Typically, water leaves are highly dissected, with a larger perimeter:leaf area ratio than terrestrial leaves, and are thin. Sometimes, these leaves are only three to four cells thick and have remarkably small intercellular spaces. Typical water leaves also possess thin epidermal cells, have very little cuticle and often fail to form a palisade layer (e.g., *R. flabelaris*; Young et al. 1987). In several species of aquatic *Callitriche* species, lath-shaped leaves rather than a dissected outline typify water leaves, while the leaves of floating rosettes are ovate (Fig. 2.4; Jones 1955). This change in leaf form is also susceptible to other environmental factors, such as photoperiod, temperature and the red/far-red ratio of the irradiance, suggesting the existence of several upstream signal transduction pathways that converge to induce the same developmental outcome. Such pathways include hormone action since applying gibberellin promotes lath-like water leaves in non-submerged *Callitriche platycarpa* (Musgrave et al. 1972), while abscisic acid favours floating-type ovate leaves in submerged *Potamogeton nodus* (Anderson 1978). A similar picture has been shown in amphibious *R. flabelaris* (Young et al. 1987). These pharmacological experiments are complimented by results showing that submergence can enhance naturally occurring gibberellin levels and rapidly depress abscisic acid concentrations (Hoffmann-Benning and Kende 1992; Benschop et al. 2005). Although ethylene is known to increase in submerged

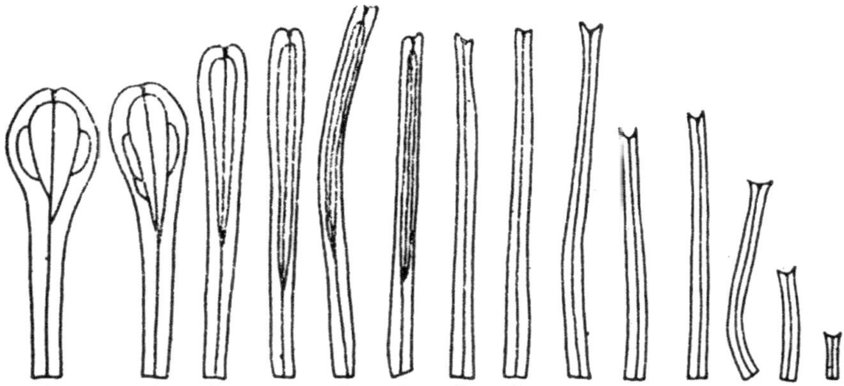

Fig. 2.4 Leaf heterophylly induced in *Callitriche intermedia* by four weeks or 12 weeks of total submergence in a flowing river. *Leaves 1–14* indicate oldest to youngest leaves with *leaf 1* being an original 'floating' leaf at the start of four weeks submergence. *Leaf 14* is after 12 weeks underwater (Jones 1955)

tissues, there is no evidence that this hormone influences heterophylly (Musgrave et al. 1972).

Photosynthetic Responses of Amphibious Plants

Although total submergence usually leads to loss of respirable substrate (Singh et al. 2001) because of low CO_2 assimilation, residual slow rates of underwater photosynthesis can still enhance survival (Vervuren et al. 2003). Accordingly, species with a greater ability to sustain underwater photosynthesis survive submergence better than others. For example, after one week of submergence, underwater photosynthesis by *Arrhenatherum elatius* at realistically low dissolved CO_2 levels (0.11 mol m^{-3}) was inhibited by over 60 %, while in more tolerant *Phalaris arundinacea* L. the loss was only 12 % (Vervuren et al. 1999). Similarly, leaves of *Rumex crispus* formed underwater have improved rates of carbon fixation compared with those originating in air, as a consequence of faster uptake of dissolved CO_2 linked to a thinner cuticle barrier and leaf lamina (Vervuren et al. 1999). Such features not only allow underwater photosynthesis to utilize lower external CO_2 concentrations than unadapted leaves, but also enable them to harvest external O_2 more effectively (Mommer et al. 2004). These adaptive changes in morphology by amphibious species are reminiscent of the more structurally distinct water leaves described in the preceding section. More work is now needed on the relationships between submergence tolerance improved photosynthesis, changed external leaf morphology, improved internal oxygenation and enhanced carbohydrate supply.

2.4 Escape

The shoots of many aquatic and amphibious species elongate faster when submerged (Sculthorpe 1967). Resultant renewed contact with the air constitutes the escape and permits a return to more the normal gas exchange processes and to full illumination. The whole plant, including roots and rhizomes, then benefits indirectly from the improved aeration via extensive interconnected aerenchyma (the Schnorkel effect; Kordan, 1974) and the renewed supply of respirable substrates. For submersed aquatic species, escape leads to a return to slower shoot extension rates (Jackson 1982). However, in some amphibians, strong emergent growth is sustained even after contact with the air is regained (Voesenek et al. 1993).

2.4.1 Aerobic Shoot Extension (the Aerobic Escape)

Submergence-induced acceleration of elongation can begin within 20–30 min of inundation and normally involves the stem or petiole although, in rice, the coleoptile, leaf base and mesocotyl also respond in this way (Jackson and Pearce 1991). The response usually requires oxygen and the initiating signal is most often entrapped ethylene gas, as originally shown in *Callitriche platycarpa* and *Ranunculus sceleratus* (Musgrave et al. 1972). The entrapped ethylene promotes longitudinal cell expansion and sometimes cell division, in concert with gibberellin (Musgrave et al. 1972) or auxin (Cookson and Osborne 1978). Species that sense submergence via ethylene and respond with an elongation-based aerobic escape includes monocots, dicots, a liverwort and a fern. Thus the system appears to have evolved separately several times, reflecting the dependability of increased ethylene as a submergence signal. This dependability is predicated on the high physiological activity of ethylene at sub-ppm concentrations and an absence of metabolic pathways for its rapid breakdown. This is in marked contrast to CO_2 and O_2. Nevertheless, depending on species, decreased O_2 concentrations and increased CO_2 can play a subsidiary role. Low O_2 and high CO_2 can each favour underwater elongation by an independent promoting action on elongation (Raskin and Kende 1983), or enhancing responsiveness to ethylene (Voesenek et al. 1997), or promoting (Raskin and Kende 1984), or maintaining effective rates of ethylene biosynthesis underwater (Banga et al. 1996). In at least two species (*Potamogeton pectinatus*, *Rumex acetosa*), ethylene is not the signal that triggers faster underwater shoot extension (Jackson et al.

1997; Benschop 2004). In these cases, O_2 or CO_2 may be mainly responsible, but the explanation remains unclear. It is notable that, in the rice-mimic *Echinochloa oryzoides*, vigorous upward elongation of the coleoptile remains largely unaffected by anoxia or increases in CO_2 or ethylene that submergence can impose (Jackson and Pearce 1991). In this plant, the escape mechanism is therefore constitutively enabled in the sense that upward growth remains undiminished by the stress.

Intensive study of submergence stimulated elongation in the stem of deepwater rice and the petioles of *R. palustris* revealed many details concerning the actions and interactions that mediate between the initial sensing of ethylene and resulting faster extension growth. These were reviewed by the principal authors involved (Kende et al. 1998; Voesenek et al. 2003a) and aspects are summarized in Fig. 2.5. The rice response differs from that of *R. palustris* in that cell division is activated by ethylene. In addition, starch breakdown is also induced. Both species are similar in that the faster elongation is linked to a very rapid decrease in the growth-inhibiting hormone abscisic acid that precedes the growth response. This decrease in abscisic acid is induced by ethylene. Both species also display a slow rise in endogenous gibberellin and an upregulation of a gene coding for a non-enzymic protein (expansin) thought to loosen cell walls. The outcome of cell wall loosening is an enhancement of tugor-driven cell expansion according to the Lockhart equation (see Eq. 2.1).

$$\text{Rate of cell wall expansion} = \phi(\rho - \gamma) \qquad (2.1)$$

Here, ϕ is a wall yield coefficient (i.e., the wall extensibility that extensin can modify), ρ is cell turgor and γ is the minimum turgor needed to initiated wall expansion.

Recent work with *R. palustris* showed that increases in cell wall acidity are required for expansin-mediated cell wall loosening (Vreeburg 2004), as first reported for *Nymphoides peltata* (Malone and Ridge 1983). To add to the complexity, there is probably an input by the hormone auxin, since endogenous concentrations rise within 4 h of submergence in association with faster petiole extension (Voesenek et al. 2003a). Taken together, these results suggest the operation of interacting pathways of signal transduction, comprising independent and inter-dependent components that, in addition to ethylene, can involve abscisic acid, gibberellins (especially at later stages of the response), auxin, expansins and apoplastic acidification (Fig. 2.5).

For the accelerated underwater elongation to constitute an effective escape, it must be upwardly directed. *R. palustris* leaves are normally horizontally orientated. Futile sideways extension underwater is avoided by a short period of ethylene-induced hyponastic re-adjustment to a more upright angle, brought about by faster elongation of abaxial cells at the petiole base. Once a 40–50 degree angle above the horizontal has been attained, longitudinal, ethylene-driven elongation of the entire petiole then becomes enabled

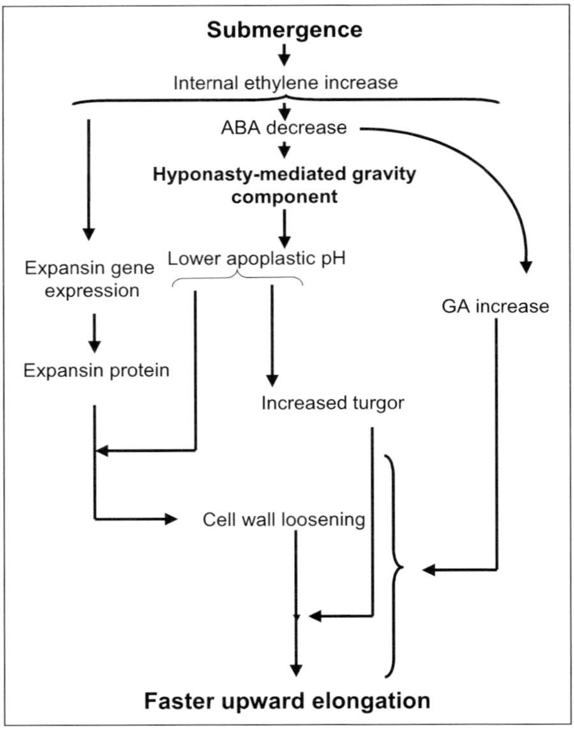

Fig. 2.5 Diagrammatic summary of the response network of events and factors linking perception of ethylene as a signal for submergence and the subsequent stimulation of upward extension that constitutes the aerobic escape. The network is based primarily on work with *Rumex palustris* (Benschop 2004; Cox et al. 2004; Vreeburg 2004)

(Cox et al. 2003). A key step is the rapid decline in ABA concentrations that de-represses ethylene action on hyponasty. There are also inputs from gibberellin and auxin. The latter is re-distributed under gravity to the abaxial motor cells driving hyponastic growth (Cox et al. 2004).

Once contact with the water surface is regained, elongation usually slows in response to a lowering of internal ethylene by ventilation to the air (e.g., *C. platycarpa*; Jackson 1982). However, in *R. palustris*, post-emergence elongation continues at a fast rate. This too appears to be ethylene-promoted. The extra gas is generated from a surge of the precursor of ethylene (ACC) transported from the roots in xylem sap of the renewed transpiration flow. There is also post-submergence ethylene biosynthesis in the leaves resulting from increased expression of an ACC synthase gene, an ACC oxidase gene and increases in ACC synthase and ACC oxidase enzyme activity (Voesenek et al. 2003b). The outcome is a prolonged above-water escape from total submergence.

2.4.2 Anaerobic Shoot Extension (the Anaerobic Escape)

The most extreme deprivation that submergence can impose is the complete absence of both O_2 and light. Section 2.2 suggests that, in these circumstances, long-term survival is mostly a question of balancing a small amount of available energy and respirable substrates with much curtailed energy consumption centred on preserving cell integrity. However, this strategy is inappropriate for wetland perennial species at the time they emerge from anaerobic surroundings each spring to re-start their annual life cycle. Instead, shoots from perennating organs of several species are capable of elongating in the absence of O_2, albeit more slowly than in air, to effect their release (e.g., *Schoenoplectus lacustris*). Unexpectedly perhaps, there are also examples of an acceleration of shoot extension in response to anoxia, exemplified by the internodes of overwintered tubers of the aquatic monocot *Potamogeton pectinatus* (Summers and Jackson 1994). Fuelled by ample starch reserves and sucrose synthesis, anaerobic stimulation of internode extension of several centimetres can be sustained by *P. pectinatus* in the dark over at least 14 days in a gravity-directed upward direction (Fig. 2.6) that is mediated by auxin and ABA but not by ethylene or gibberellin (Summers and Jackson 1996). Part of the promoting mechanism involves apoplastic acidification (presumably this is an aid to cell wall loosening) while almost totally avoiding the cytoplasmic

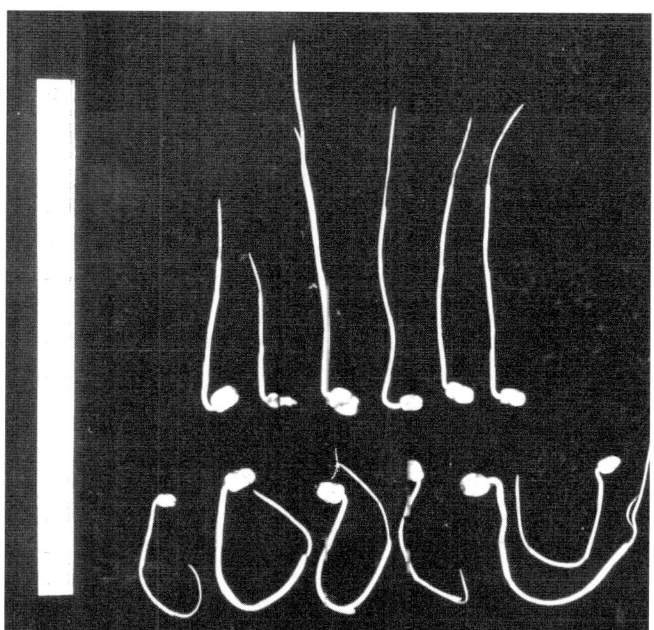

Fig. 2.6 Anaerobically promoted stem extension by the shoot of over-wintering tubers of *Potamogeton pectinatus* (three days growth). The *lower set* of tubers indicates that the ability to respond to gravity is retained in the absence of O_2, ensuring that the direction of elongation is upward and thus towards a potential source of oxygen, thereby achieving the anaerobic escape. *Bar* 15 cm. Photograph by J.E. Summers

acidification (Summers et al. 2000) that threatens cells of less well-adapted species. Energy to support these vigorous activities is generated by strong alcoholic fermentation. However, the fast rate of fermentation remains insufficient to form more than a small proportion of the ATP yield from aerobic respiration. How vigorous stem extension is achieved in the face of such a large energy shortfall remains unclear. Some energy is saved by a halving of the rate of leaf extension and of cell division (Summers and Jackson 1994). A strong down-regulation of protein synthesis measured by incorporation of ^{35}[S] methionine that generates a largely anoxia-specific set of proteins will also conserve energy (Dixon et al. 2006). Many of these findings were substantiated and extended by studies of *Potamogeton distinctus* and *Sagittaria pygmaea*, although the contribution of apoplastic acidity and cell division may be less than in *P. pectinatus*. In particular, it was shown that fast-growing anaerobic shoots of one or other of these species maintain a high energy status when measured in terms of energy charge (ATP + 0.5ADP/ATP+ ADP+ AMP). The anaerobic escape appears to depend on calcium uptake and is linked to new kinds of proteins synthesized only in anoxia. These include sucrose synthase (Tamura et al. 2001; Sato et al. 2002; Harada and Isahizawa 2003). Despite the seemingly irreconcilable combination of a fast rate of elongation and a shortfall in energy generation via glycolysis, no unusual or specific metabolic pathways of energy generation have been detected. Thus, the mechanism by which these remarkable species square the energetic circle has still not been fully explained.

2.5 Summary

Although water is chemically benign, the tendency of liquid water to exclude key environmental inputs essential for energy generation (CO_2, O_2, light) imposes major constraints on plant productivity and survival. Nevertheless, the evolution of metabolic and morphological features has endowed a sizeable minority of plant species with the ability to survive, reproduce and often grow vigorously in flooded conditions fatal to the majority. There are many different kinds of flooding stress, extending from short- or long-term waterlogging of the soil in winter, to flash-flooding in summer, prolonged winter submergence or permanent flooding of variable depth such as that experienced by riverside- and lakeside-dwelling plants. Each situation imposes its own particular combination of subsidiary stresses which evolution has overcome to varying degrees, leading to select species or ecotypes that survive and even thrive when flooded. The most widespread adaptation to deprivations imposed by flooding is an ability to tolerate short-term anoxia by switching to fermentation as the main pathway for extracting energy from respirable substrates. This takes place in combination with energy conservation measures derived from a complex down-regulation of unessential energy-consuming processes. These changes are achieved through altered patterns of gene expression and post-translational modification brought about by responses to the partial O_2 shortage that precedes anoxia and for which the sensing mechanism remains uncertain. Such metabolic adaptations are found, to some extent, in dryland species in addition to wetland plants and are well-developed in both shoots and roots. More particular to wetland species are adaptations peculiar to shoots. Shoot systems are of overarching importance in these species since, as photosynthetic organs, they are sources of both O_2 and respirable substrates and possess superior powers of regeneration. Even very slow rates of photosynthesis underwater can prolong survival. Accordingly, many specialist aquatic and amphibious species possess metabolic or morphological adaptations to promote underwater photosynthesis by raising the CO_2 concentration in chloroplast-containing cells. Mechanisms for achieving this include CO_2-harvesting from the rooting substrate and hormone-mediated formation of morphologically distinct water leaves. These present maximal absorbing area per unit mass while minimizing diffusive resistances to CO_2 ingress. In addition to features that enhance resilience to O_2 and CO_2 shortage, the shoots of many well-adapted species possess means of escaping inundation through prompt, gravity-directed upward extension growth that increases the probability of shoots regaining contact with the aerial atmosphere before reserves become exhausted. The growth is usually O_2-dependent and most but not all relevant species use entrapped ethylene as the triggering signal. This gas initiates and sustains an abnormally fast rate of shoot elongation underwater sometimes in a complex interaction with low O_2 and high CO_2 and with other hormones, notably gibberellin and auxin. A rapid decrease in the growth-inhibiting hormone abscisic acid, mediated by ethylene, is often a pre-requisite for the escape to commence. More recently, it was found that the shoots of certain species that over-winter as dormant tubers, rhizomes or turions are stimulated to elongate very rapidly by the complete absence of O_2. This remarkable achievement appears to be based on an ability to harness considerable reserves of starch and to synthesize sucrose to fuel fast rates of glycolysis following transport from storage tissue. This underpins, energetically, a hormone-sensitive gravity-responsive escape, while dampening potentially damaging cytoplasmic acidosis and other usually lethal consequences of anoxia. Ethylene is not involved in the anaerobic escape. Both the aerobic and anaerobic escape strategies are rendered effective beneficiaries of the whole plant because of the existence of a highly porous and interconnected aerenchyma that can pass on the benefits of improved gas exchange at the re-surfaced shoot apex to the remaining and still-submerged parts below. In complete contrast, some ecotypes of escaping species have evolved to suppress the fast rates of shoot elongation normal for the species when submerged. This results in an underwater quiescence that may confer an advantage where the prevailing depth of water is too large for a dependable escape.

References

Adkins SW, Shiraishi T, McComb JA(1990) Submergence tolerance of rice – a new glasshouse method for the experimental submergence of plants. Physiol Plant 80:642–646

Anderson LWJ (1978) Abscisic acid induces formation of floating leaves in the heterophyllous aquatic angiosperm *Potamogeton nodus*. Science 201:1135–1138

Armstrong J, Afreen Zobayed F, Armstrong W (1996) Phragmites die-back: sulphide- and acetic acid-induced bud and root death, lignifications, and blockages within aeration and vascular systems. New Phytol 134:601–614

Armstrong W (1979) Aeration in higher plants. Adv Bot Res 7:225–332

Aschi-Smiti S, Chaibi W, Brouquisse R, Ricard B, Saglio P (2003) Assessment of enzyme induction and aerenchyma formation as mechanisms for flooding tolerance in *Trifolium subterraneum* 'Park'. Ann Bot 91:195–204

Banga M, Slaa EJ, Blom C, Voesenek LACJ (1996) Ethylene biosynthesis and accumulation under drained and submerged conditions (a comparative study of two *Rumex* species). Plant Physiol 112:229–237

Baxter-Burrell A, Cang R, Springer P, Bailey-Serres J (2003) Gene and enhancer trap transposable elements reveal oxygen deprivation-regulated genes and their complex patterns of expression in arabidopsis. Ann Bot 91:129–141

Benschop J (2004) The role of abscisic acid in ethylene-induced elongation. PhD thesis, Utrecht University, Utrecht

Benschop JJ. Jackson MB, Guhl K, Vreeburg RAM, Croker SJ, Peeters AJM, Voesenek LACJ (2005). Contrasting interactions between ethylene and abscisic acid in *Rumex* species differing in submergence tolerance. Plant J 44:756–768

Blokhina OB, Virolainen E, Fagerstedt KV (2003) Antioxidants, oxidative damage and oxygen deprivation stress: a review. Ann Bot 91:179–194

Boamfa EI, Veres AH, Ram PC, Jackson MB, Reuss J, Harren FJM (2005) Detailed kinetics of ethanol and acetaldehyde release suggest a role for acetaldehyde production in tolerance of rice seedlings to micro-aerobic conditions. Ann Bot 96:727–736

Bologa KL, Fernie AR, Leisse A, Loureiro ME, Geigenberger P (2003) A bypass of sucrose synthase leads to low internal oxygen and impaired metabolic performance in growing potato tubers. Plant Physiol 132:2058–2072

Bowes G (1987) Aquatic plant photosynthesis: strategies that enhance carbon gain. In: Crawford RMM (ed) Plant life in aquatic and amphibious habitats. Blackwell, Oxford, pp 79–98

Bowes G, Rao SK, Estavillo GM, Reiskind JB (2002) C-4 mechanisms in aquatic angiosperms: comparisons with terrestrial C-4 systems. Funct Plant Biol 29:379–392

Brouwer E, Bobbink R, Roelofs JGM (2002) Restoration of aquatic macrophyte vegetation in acidified and eutrophied softwater lakes: an overview. Aquat Bot 73:405–431

Bucher M, Brandle R, Kuhlemeier C (1996) Glycolytic gene expression in amphibious *Acorus calamus* L. under natural conditions. Plant Soil 178:75–82

Chang WWP, Huang L, Shen M, Webster C, Burlingame AL, Roberts JKM (2000) Patterns of protein synthesis and tolerance of anoxia in root tips of maize seedlings acclimated to a low-oxygen environment, and identification of proteins by mass spectrometry. Plant Physiol 122:295–318

Cookson C, Osborne DJ (1978) The stimulation of cell extension by ethylene and auxin in aquatic plants. Planta 144:39–47

Cox MCH, Millenaar FF, Jong van Berkel YEM de, Peeters AJM, Voesenek LACJ (2003) Plant movement. submergence-induced petiole elongation in *Rumex palustris* depends on hyponastic growth. Plant Physiol 132:282–291

Cox MCH, Benschop JJ, Vreeburg RAM, Wagemaker CAM, Moritz T, Peeters AJM, Voesenek LACJ (2004) The roles of ethylene, auxin, abscisic acid, and gibberellin in the hyponastic growth of submerged *Rumex palustris* petioles. Plant Physiol 136:2948–2960

Crawford RMM (2003) Seasonal differences in plant responses to flooding and anoxia. Can J Bot 81:1224–1246

Dixon MH, Hill SA, Jackson MB, Ratcliffe RG, Sweetlove LJ (2006) Physiological and metabolic adaptations of *Potamogeton pectinatus* L. tubers support rapid elongation of stem tissue in the absence of oxygen. Plant Cell Physiol 47:128–140

Dolferus R, Klok EJ, Delessert C, Wilson S, Ismond KP, Good AG, Peacock WJ, Dennis ES (2003) Enhancing the anaerobic response. Ann Bot 91:111–117

Dordas C, Rivola J, Hill RD (2003) Plant haemoglobins, nitric oxide and hypoxic stress. Ann Bot 91:173–178

Eck WHJM van, Steeg HM van de, Blom CWPM, Kroon H de (2004) Is tolerance to summer flooding correlated with distribution patterns in river floodplains? A comparative study of 20 terrestrial grassland species. Oikos 107:393–405

Geigenberger P (2003) Response of plant metabolism to too little oxygen. Curr Opin Plant Biol 6:247–256

Geigenberger P, Fernie AR, Gibon Y, Christ M, Stitt M (2000) Metabolic activity decreases as an adaptive response to low internal oxygen in growing potato tubers. Biol Chem 381:723–740

Gibbs J, Greenway H (2003) Mechanisms of anoxia tolerance in plants. I. Growth and survival. Funct Plant Biol 30:1–47

Greenway H, Gibbs J (2003) Mechanisms of anoxia tolerance in plants. II. Energy requirements for maintenance and energy distribution to essential processes. Funct Plant Biol 30:999–1036

Grosse W, Armstrong J, Armstrong W (1996) A history of pressurised gas-flow studies in plants. Aquat Bot 54:87–100

Harada T, Ishizawa K (2003) Starch degradation and sucrose metabolism during anaerobic growth of pondweed (*Potamogeton distinctus* A. Benn.) turions. Plant Soil 253:125–135

Heinzi T, Brändle R (1993) Long term survival of rhizomatous species under oxygen deprivation. In: Jackson MB, Black CR (eds) Interacting stresses on plants in a changing climate. Springer, Berlin Heidelberg New York, pp 305–314

Hoeren FU, Dolferus R, Wu Y, Peacock WJ, Dennis ES (1998) Evidence for a role of AtMYB2 in the induction of the *Arabidopsis* alcohol dehydrogenase (ADH1) gene by low oxygen. Genetics 149:479–490

Hoffmann-Benning S, Kende H (1992) On the role of abscisic acid and gibberellin in the regulation of growth in rice. Plant Physiol 99:1156–1161

Ismond KP, Dolferus R, De Pauw M, Dennis ES, Good AG (2003) Enhanced low oxygen survival in arabidopsis through increased metabolic flux in the fermentative pathway. Plant Physiol 132:1292–1302

Jackson MB (1982) Ethylene as a growth promoting hormone under flooded conditions. In: Wareing PF(ed) Plant growth substances. Academic, London, pp 291–301

Jackson MB (2004) The impact of flooding stress on plants and crops. PlantStress: A website addressing plant environmental stress issues in agriculture, plant physiology, breeding, genetics and biotechnology. Available at http://www.plantstress.com/Articles/waterlogging_i/waterlog_i.htm

Jackson MB, Armstrong W (1999) Formation of aerenchyma and the process of plant ventilation in relation to soil flooding and submergence. Plant Biol 1:274–287

Jackson MB, Pearce DME (1991) Hormones and morphological adaptation to aeration in rice. In: Jackson MB, Davies DD, Lambers H (eds) Plant life under oxygen deprivation: ecology, physiology and biochemistry. SPB Academic, The Hague, pp 47–67

Jackson MB, Ram PC (2003) Physiological and molecular basis of susceptibility and tolerance of rice plants to complete submergence. Ann Bot 91:227–241

Jackson MB, Ricard B (2003) Physiology, biochemistry and molecular biology of plant root systems subjected to flooding of the soil. In: Kroon H, Visser EJW (eds) Root ecology. Springer, Berlin Heidelberg New York, pp 193–211

Jackson MB, Waters I, Setter TL, Greenway H (1987) Injury to rice plants caused by complete submergence; a contribution by ethylene. J Exp Bot 38:1826–1838

Jackson MB, Summers JE, Voesenek LACJ (1997) *Potamogeton pecinatus*: a vascular plant that makes no ethylene. In: Kanellis AK, Chang C, Kende H, Grierson D (eds) Biology and biotechnology of the plant hormone ethylene. Kluwer, Dordrecht, pp 229–237

Jones HG (1955) Heterophylly in some species of *Callitriche*, with especial reference to *Callitriche intermedia*. Ann Bot 19:225–245

Kende H, Knaap E van der, Cho H-T (1998) Deepwater rice: a model plant to study stem elongation. Plant Physiol 118:1105–1110

Klok EJ., Wilson IW, Wilson, D, Chapman SC, Ewing RM, Somerville SC, Peacock WJ, Dolferus R, Dennis ES(2002) Expression profile analysis of the low-oxygen response in *Arabidopsis* root cultures. Plant Cell 14:2481–2494

Kordan HA (1974) Patterns of shoot and root growth in rice seedlings germinating under water. J Appl Ecol 11:685–690

Laanbroek HJ (1990) Bacterial cycling of minerals that affect plant growth in waterlogged soils: a review. Aquat Bot 38:109–125

Lynn DE, Waldren S (2003) Survival of *Ranunculus repens* L. (creeping buttercup) in an amphibious habitat. Ann Bot 91:75–84

Madsen TV (1993) Inorganic carbon assimilation and growth of aquatic macrophytes. In: Jackson MB, Black CR (eds) Interacting stresses on plants in a changing climate. Springer, Berlin Heidelberg New York, pp 267–285

Malone M, Ridge I (1983) Ethylene-induced growth and proton excretion in the aquatic plant *Nymphoides peltata*. Planta 157:71–73

Maltby E (1991) Wetlands – their status and role in the biosphere. In: Jackson MB, Davies DD, Lambers H (eds) Plant life under oxygen deprivation: ecology, physiology and biochemistry. SPB Academic, The Hague, pp 3–21

Matsukura C, Kawai M, Toyofuku K, Barrero RA, Uchimiya H, Yamaguchi J (2000) Transverse vein differentiation associated with gas space formation – fate of the middle cell layer in leaf sheath development of rice. Ann Bot 85:19–27

Mohanty B, Wilson PM, Rees T (1993) Effects of anoxia on growth and carbohydrate metabolism in suspension cultures of soybean and rice. Phytochemistry 34:75–82

Mommer L, Pederson O, Visser EJW (2004) Acclimation of a terrestrial plant to submergence facilitates gas exchange under water. Plant Cell Environ 27:1281–1287

Monk LS, Fagerstedt KV, Crawford RMM (1989) Oxygen toxicity and superoxide dismutase as an antioxidant in physiological stress. Physiol Plant 76:456–459

Musgrave A, Jackson MB, Ling E (1972) *Callitriche* stem elongation is controlled by ethylene and gibberellin. Nat New Biol 238:93–96

Mustroph A, Albrecht G (2003) Tolerance of crop plants to oxygen deficiency stress: fermentative activity and photosynthetic capacity of entire seedlings under hypoxia and anoxia. Physiol Plant 117:508–520

Parolin P, De Simone O, Hasse K, Waldhoff D, Rottenberger S, Khun U, Kesselmeierk J, Kleiss B, Schmidt W, Piedade MTF, Junk WJ (2004) Central amazonian floodplain forests: tree adaptations in a pulsing system. Bot Rev 70:357–380

Raskin I, Kende H (1983) Regulation of growth in rice seedlings. J Plant Growth Regul 2:193–203

Raskin I, Kende H (1984) Regulation of growth in stem sections of deep-water rice. Planta 160:66–72

Richardson K, Griffiths H, Reed ML, Griffiths NM (1984) Inorganic carbon assimilation in the isoetids, *Isoetes lacustris* L. and *Lobelia Dortmanna* L. Oecologia 61:115–121

Saglio PH, Drew MC, Pradet A (1988) Metabolic acclimation to anoxia induced by low (2–4 kPa) partial pressure oxygen pre- treatment (hypoxia) in root tips of *Zea mays*. Plant Physiol 86:61–67

Sand-Jensen K (1987) Environmental control of bicarbonate use among freshwater and marine macrophytes. In: Crawford RMM (ed) Plant life in aquatic and amphibious habitats. Blackwell, Oxford, pp 99–112

Santosa E (2002) Oxidative stress and pathogenic attack in plants, studied by laser based photoacoustic trace gas detection. PhD thesis, University of Nijmegen, Nijmegen

Sato T, Harada T, Ishizawa K (2002) Stimulation of glycolysis in anaerobic elongation of pondweed (*Potamogeton distinctus*) turions. J Exp Bot 53:1847–1856

Schutten J, Dainty J, Davy AJ (2004) Wave-induced hydraulic forces on submerged aquatic plants in shallow lakes. Ann Bot 93:333–341

Sculthorpe CD (1967) The biology of aquatic vascular plants. Arnold, London

Seago JL Jr, Marsh LC, Stevens KJ, Soukup A, Votrubova O, Enstone DE (2005) A re-examination of the root cortex in wetland flowering plants with respect to aerenchyma. Ann Bot 96:565–579

Setter TL, Waters I, Wallace I, Bhekasut P, Greenway H (1989) Submergence of rice I. Growth and photosynthetic response to CO_2 enrichment of floodwater. Aust J Plant Physiol 16:251–263

Singh HP, Singh BB, Ram PC (2001) Submergence tolerance of rainfed lowland rice: search for physiological marker traits. J Plant Physiol 158:883–889

Staal M, Elenga JT, Prins HBA (1989) ^{14}C fixation by leaves and leaf cell protoplasts of the submerged aquatic angiosperm *Potamogeton lucens*: carbon dioxide or bicarbonate? Plant Physiol 90:1035–1040

Steinmann F, Braendle R (1984) Carbohydrate and protein metabolism in the rhizomes of the bulrush (*Schoenoplectus lacustris* L.) in relation to natural development of the whole plant. Aquat Bot 19:53–63

Summers JE, Jackson MB (1994) Anaerobic conditions strongly promote extension by stems of overwintering tubers of *Potamogeton pectinatus* L. J Exp Bot 45:1309–1318

Summers JE, Jackson MB (1996) Anaerobic promotion of stem extension in *Potamogeton pectinatus*. Roles for carbon dioxide, acidification and hormones. Physiol Plant 96:615–622

Summers JE, Ratcliffe RG, Jackson MB (2000) Anoxia tolerance in the aquatic monocot *Potamogeton pectinatus*: absence of oxygen stimulates elongation in association with an unusually large Pasteur effect. J Exp Bot 51:1413–1422

Tamura S, Kuramochi H, Ishizawa K (2001) Involvement of calcium ion in the stimulated shoot elongation of arrowhead tubers under anaerobic conditions. Plant Cell Physiol 42:717–722

Toojinda T, Siangliw M, Tragoonrung S, Vanavichit A (2003) Molecular genetics of submergence tolerance in rice: QTL analysis of key traits. Ann Bot 91:243–253

Vartapetian BB (1991) Flood-sensitive plants under primary and secondary anoxia: ultrastructural and metabolic responses. In: Jackson MB, Davies DD, Lambers H (eds) Plant life under oxygen deprivation; physiology, biochemistry and molecular biology. SPB Academic, The Hague, pp 201–216

Vervuren PJA, Beurskens SMJH, Bom CWPM (1999) Light acclimation, CO_2 response and long-term capacity of underwater photosynthesis in three terrestrial plant species. Plant Cell Environ 22:959–968

Vervuren PJA, Blom CWPM, Kroon H de (2003) Extreme flooding events on the Rhine and the survival and distribution or riparian plant species. J Ecol 91:135–146

Voesenek L, Banga M, Thier RH, Mudde CM, Harren F, Barendse G, Blom CWPM (1993) Submergence-induced ethylene synthesis, entrapment, and growth in two plant species with contrasting flooding resistances. Plant Physiol 103:783–791

Voesenek L, Vriezen WH, Smekens M, Huitink F, Bogemann GM, Blom C (1997) Ethylene sensitivity and response sensor expression in petioles of *Rumex* species at low O_2 and high CO_2 concentrations. Plant Physiol 114:1501–1509

Voesenek LACJ, Benschop JJ, Bou J, Cox MCH, Groenveld HW, Millenaar FF, Vreeburg RAM, Peeters AJM (2003a) Interactions between plant hormones regulate submergence-induced shoot elongation in the flooding-tolerant dicot *Rumex palustris*. Ann Bot 91:205–211

Voesenek LACJ, Jackson MB, Toebes AHW, Huibers W, Vriezen WH, Colmer TD (2003b) De-submergence-induced ethylene production in *Rumex palustris*: regulation and ecophysiological significance. Plant J 33:341–352

Vreeburg RAM (2004) Expansins in submergence-induced petiole elongation of *Rumex palustris*: kinetics and regulation. PhD thesis, Utrecht University, Utrecht.

Young JP, Dengler NG, Horton RF (1987) Heterophylly in *Ranunculus flabellaris*: the effect of abscisic acid on leaf anatomy. Ann Bot 60:117–125

3 Center Stage: The Crucial Role of Macrophytes in Regulating Trophic Interactions in Shallow Lake Wetlands

ROMI L. BURKS, GABI MULDERIJ, ELISABETH GROSS, IWAN JONES, LENE JACOBSEN, ERIK JEPPESEN, and ELLEN VAN DONK

3.1 Introduction

Hydrophilic, or water-loving, macrophytes characterize wetland ecosystems, indicating prerequisite conditions of hydric soils and sufficient hydrology. The presence of such macrophytes is a key descriptor in multiple wetland definitions (Lewis 2001a) and macrophytes may be further used to actually describe particular types of wetlands, such as cattail marshes. Macrophytes contribute significant biomass to wetland systems and represent a critical component of wetland biogeochemistry as primary producers and drivers of organic matter cycling within aquatic systems. In this chapter, we argue that macrophytes occupy the center of trophic interactions in shallow lakes, influencing outcomes through structural, behavioral and chemical interactions. We define shallow lakes as permanently flooded wetlands that often contain submerged or floating macrophytes and that may be surrounded by emergent vegetation (i.e. marshy habitat). Shallow remains a relative term in limnology circles, but typically is less than 3 m average depth, such that macrophytes can fill a substantial portion of the water column and stratification is neither predictable nor long-term. Such systems may be termed lakes, ponds or wetlands, depending on their size and the ecological context. Macrophytes may regulate trophic interactions in ephemeral systems without permanent inundation.

Ecological Studies, Vol. 191
R. Bobbink, B. Beltman, J.T.A. Verhoeven, and D.F. Whigham (Eds.)
Wetlands: Functioning, Biodiversity Conservation, and Restoration

3.2 Central Position of Aquatic Vegetation

Terrestrial ecologists historically have focused on vegetation. However, early aquatic studies also noted the dominance of plants in aquatic systems. For example, in *The lake as a microcosm*, Forbes (1925) derided areas 'clogged with weeds', but then elaborated on how these weeds (*Chara, Potamogeton, Ceratophyllum*) swarmed with life, including fishes, invertebrates, zooplankton and algae. Lindeman (1942) included pondweeds as an important component of lake food webs and Teal (1962) went a step further and quantified the energy flow between trophic levels in a salt marsh. Thus, historical precedent exists for focusing attention on macrophytes as part of the aquatic community. First to emphasize ecosystem processes, Carpenter and Lodge (1986) summarized the impacts of submerged macrophytes on abiotic and biotic components of aquatic systems. With a similar aim, Jeppesen et al. (1997a) published a comprehensive book documenting the structuring role of submerged macrophytes in lakes. We extend their discussion of trophic interactions and also consider other macrophyte architectures (i.e. floating-leaved, emergent, freely-floating).

Reinforced by multiple mechanisms, macrophytes promote water clarity, help stabilize a clear-water state (Jeppesen 1998; Scheffer 1998) and enhance biodiversity (Declerck et al. 2005). Several of those feedbacks depend on structural or chemical effects that, in turn, influence behaviors of multiple trophic levels. To evaluate our progress in understanding the role of macrophytes in wetland systems since Jeppesen et al. (1997a), we performed an extensive literature search (1997–2004) regarding aquatic macrophytes. Using four search engines [Web of Science, Science Direct (Elsevier), BioOne, Kluwer-Online] our inquiry discovered 126 papers that involved trophic interactions with macrophytes.

The vast majority (74 %) of recent studies focused on behavioral interactions with macrophytes, with an additional 13 % exploring this within a chemical context. The remaining 13 % focused on single structural–chemical interactions, largely between macrophytes and phytoplankton. Macrophytes, fish and zooplankton composed the most common suite of organisms studied together (12 %), although 35 % of the studies involved three groups and 56 % only examined a single interaction. Thirteen studies (10 %) examined more complex food webs in shallow wetland systems; however, only four of these studies (3 %) examined five trophic levels. Surprisingly, similar numbers of single-interaction studies involved both fish and invertebrates, dominated by studies including fish (44 %). Only 10 % of the studies included an additional primary producer (periphyton or phytoplankton) besides macrophytes. These complex trophic connections that occur with macrophytes dramatically alter the carbon flow through the system (Fig. 3.1). With this context, we used the literature survey to identify four key themes regarding the regulation of trophic interactions by macrophytes.

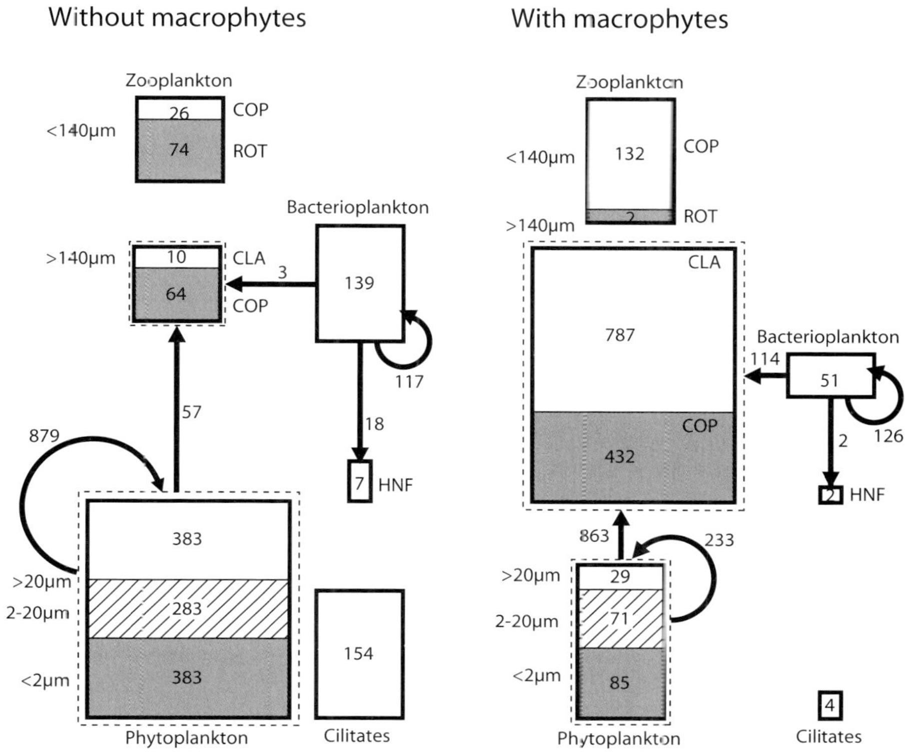

Fig. 3.1 Illustration of the dramatic alteration of energy flow in the presence of macrophytes (originally published by Jeppesen et al. 2002). The figure illustrates the cascading trophic interactions within enclosures placed in the littoral zone of a shallow Danish lake (Stigsholm). The data illustrate carbon flow between trophic levels in enclosures with and without submerged macrophytes for the first diel periods where day and night sampling occurred. *Boxes* represent the biomass (µg C l^{-1}) of different trophic components and the *arrows* measure flux per day. *COP* Copepods, largely cyclopoids, *ROT* rotifers, *CLA* cladocerans, *HNF* heterotrophic nanoflagellates. *Broken box outlines* indicate that phytoplankton production and zooplankton grazing are community estimates for all groups within the boxes. Note that zooplankton grazing by far exceeded phytoplankton production in the macrophyte state, suggesting extremely high grazing on phytoplankton and likely benthic facilitation of the zooplankton. The latter may, in turn, ensure a steady high abundance of zooplankton and thus high grazing on phytoplankton when zooplankton migrate to the open water at night

3.2.1 Central Themes: Zooplankton Depend on Macrophytes as Habitats

While many historical studies noted the abundance of macroinvertebrates among macrophytes (e.g. for damselflies, see Lombardo 1997; for snails, see Brönmark and Vermaat 1997), fewer studies recognized the role of macrophytes as providing refuge for zooplankton. Limnologists generally restrict their discussion of refuges for zooplankton to the hypolimnetic refuge provided by deep lakes for large-bodied zooplankton undergoing diel vertical migration (i.e. *Daphnia*). However, littoral-associated cladocerans (i.e. *Alona, Chydorus, Simocephalus, Ceriodaphnia*) commonly occur among macrophytes in large lacustrine wetlands (Hann and Zrum 1997) and small glacial kettleholes (Gaiser and Lang 1998). Recent studies from shallow lake and wetland systems demonstrated that macrophyte presence can bolster pelagic zooplankton, as large-bodied species such as *Daphnia* migrate horizontally into macrophytes to seek refuge from predators (for a review, see Burks et al. 2002).

Zooplankton dependence on macrophyte beds acts as a positive feedback mechanism that helps maintain clear-water conditions in lakes, especially eutrophic systems with high macrophyte biomass (Jeppesen et al. 1999; although see Blindow et al. 2000). Multiple scenarios exist, however, for the relationship between zooplankton size (at the individual or population scale) and macrophyte density. One possibility is a non-linear relationship based on the assumptions that increased prey size leads to increased vulnerability (Brooks and Dodson 1965) and that increased structural complexity makes foraging more difficult for some predators in a non-linear way (Jeppesen et al.1997a; Burks et al. 2001a). This refuge scenario suggests that some threshold of macrophytes is required, which is corroborated by a large mesocosm experiment that showed smaller rotifers and cyclopoid copepods comprised the zooplankton in plant-free controls while large-bodied cladocerans dominated enclosures with plants, when plant density exceeded a certain threshold (Fig. 3.1; Jeppesen et al. 1997a, 2002).

With horizontal migration, the extent of entry into macrophytes by zooplankton varies, depending on the costs of inhabiting macrophytes. These costs include competition with other zooplankton, food scarcity, chemical inhibition, or adverse abiotic conditions (Burks et al. 2002). Such costs might be responsible for the horizontal patchiness that occurred in Rybak and Weglenska's (2003) study, where sparse beds showed more limnetic species of Cladocera while dense beds held more littoral species. Intense predation pressure from macrophyte-associated invertebrates (Burks et al. 2001b) also may factor into the non-linear relationship between macrophyte density and zooplankton size and contribute to patterns commonly observed in the field. Invertebrate density across vertical and horizontal dimensions varies among macrophytes. The substantial surface area of submerged macrophytes likely supports higher abundances of macroinvertebrates than floating or emergent plants (Cattaneo et al. 1998). Marklund et al. (2001) found higher densities of invertebrate grazers in the layers closer to the sediment than near the surface of *Chara* beds. Their results also showed higher densities at the macrophytes' edge versus the innermost parts for some species. The latter corroborates earlier results by Lauridsen et al. (1996). Thus, for zooplankton, macrophyte beds may form a horizontal transition zone in predation, from invertebrate predation inshore to fish predation offshore (Tolonen et al. 2001).

The use of macrophytes as a refuge by zooplankton is also likely to decline with increasing turbidity. While several studies document the use of submerged and floating plants as a refuge, the ability of emergent plants to provide a refuge for zooplankton remains questionable. Nurminen and Horppila (2002) investigated the importance of emergent macrophytes as refugia for littoral cladocerans in lakes of differing trophic status. No tendency toward horizontal migration into emergent macrophytes occurred in a clearwater system. In the turbid system, Nurminen and Horppila (2002) found reverse migration by large-bodied filter feeders (*Limnosida frontosa*, *Sida crystallina*), possibly in response to high densities of small fishes among the vegetation. This agrees well with a recent study of changes along a chlorophyll *a* (turbidity) gradient in the proportion of fish near the shore versus the open water. Numerous Danish lakes showed an increasing share of the most abundant fish species in the littoral as turbidity increased (Jeppesen et al. 2006). In contrast, the anti-predator behavior of juvenile fish, such as using macrophytes for refuge, may occur less frequently in turbid water, where hunting by sight-dependent predator fish is less successful (Abrahams and Kattenfeld 1997). Under clear-water conditions, juvenile pike (*Esox lucius*) spent more time foraging among complex structures compared to an even distribution between complexity and open water under turbid conditions (Skov et al. 2002). In a survey of 33 interconnected shallow ponds, Cottenie et al. (2001) found that zooplankton community structure depended strongly on predation intensity and macrophyte cover and that the clear-water state coincided with macrophytes, piscivorous fish and large *Daphnia* species.

Besides the influence of turbidity, the ability of macrophytes to serve as a refuge for zooplankton also declines when macrophytes play host to other predators, suggesting the avoidance of macrophytes by zooplankton. If one considers its root network, water hyacinth (*Eichhornia crassipes*) may possess a comparable surface area to submerged plants. However, *Eichhornia* does not support a similar high diversity of invertebrates, including zooplankton (Meerhoff et al. 2003). Lower colonization of macrophytes than predicted may be due to chemical deterrence (Burks et al. 2000, 2001a), differences in water chemistry from the floating plant canopy or high fish predation (Meerhoff et al. 2003 and unpublished results). However, the few studies on free-floating plants make such generalizations difficult and more research is needed to understand if this architectural type differs in function from submerged or emergent macrophytes. Despite any potential differences, high fish predation can occur outside all types of macrophyte architectures. Romare et al. (2003), for example, found that roach (*Rutilus rutilus*) and perch (*Perca fluviatilis*) congregated outside dense emergent vegetation in a small Danish lake, thus likely eliminating the advantage of horizontal migration for cladocerans. This negative relationship may exist with the carnivorous macrophyte bladderwort, *Utricularia* (Guisande 2000; Englund and Harms 2003), depending on zooplankton size. Furthermore, larval dragonflies also pose a serious predation threat to zooplankton undergoing diel horizontal migration, particularly at low macrophyte densities (Burks et al. 2001b). Alternatively, littoral fishes may reinforce the refuge effect for zooplankton if they prefer macroinvertebrates (Sagrario and Balseiro 2003). Overall, macrophytes act as effective refuges for zooplankton or planktivorous fishes when the diversity or density is sufficient enough to depress the predation rates of planktivores (Priyadarshana et al. 2001) or piscivores, respectively (e.g. largemouth bass, *Micropterus salmoides*; Valley and Bremigan 2002). Further experiments and field studies that investigate patterns of zooplankton size, macrophyte density and composition will help resolve the ambiguity associated with these multiple scenarios and lend further insight into how carbon cycling occurs across multiple trophic interactions (Fig. 3.1).

3.2.2 Central Themes: Chemical Ecology Spans Trophic Levels

Biochemical interactions between (submerged) macrophytes and competing primary producers, termed 'allelopathy' (*sensu* Molisch 1937), often appear in the literature, but at the same time ignite much controversy. Allelopathy is unlikely to act as a determining factor of a clear versus turbid state, but it certainly influences the interactions between the main opponents. Multiple laboratory studies document inhibitory effects of extracts, exudates or intact macrophytes on algae (summarized e.g. in Gross 2003a; for a recent example from freely-floating plants, see Mulderij et al. 2005a). Most studies show differential sensitivity among algal groups, with cyanobacteria often the most sensitive and chlorophytes and diatoms less susceptible (e.g. Gross et al. 1996; but see Mulderij et al. 2005b). Macroalgae, such as charophytes, also employ allelopathic chemicals in their interactions with other primary producers (Van Donk and van de Bund 2002; Mulderij et al. 2003). Relative to other macrophyte growth forms, the likelihood of allelopathic interactions increases in dense stands of submerged macrophytes where more stagnant water prevails. Based on our literature survey, we provide a ranked list of macrophyte species that we think best represent allelopathic species (Table 3.1).

Knowledge of chemical ecology in freshwater systems lags behind marine and terrestrial systems (Burks and Lodge 2002), but we see substantial progress. Our literature survey indicated that 25 % of recent studies included an aspect of chemical influence, suggesting that more investigations of this nature are underway. The practicality of both realistic laboratory and in situ experiments will increase as we identify more compounds. For now, we noticed particularly the variety of ways in which chemical cues or allelopathic compounds altered trophic interactions. The impacts often depended on species identity and commonly involved macroinvertebrates.

Table 3.1 Ranking allelopathic potential of aquatic macrophytes (based on evidence presented in studies cited by Wium-Andersen 1987; Gopal and Goel 1993; Gross 2003a). Evidence was considered high when different authors reported the same or similar observations and/or when proof for the release of active compounds occurred. The asterisk (*) indicates that evidence was not as strong as for "high" species. Note that not all *Chara* species showed indications of allelopathic properties

Allelopathic potential	Species	Common name
High	*Myriophyllum spicatum*	Eurasian water milfoil
↓	*Chara globularis*, *Ch. aspera*, *Ch. contraria*	Stonewort or charophyte
↓	*Ceratophyllum demersum*	Coontail
↓	*Stratiotes aloides*	Water soldier
Medium/high*	*Elodea* spp	Waterweed, water pest
↓	*Eleocharis acicularis*	Spike rush
↓	*Myriophyllum brasiliense*	Parrotfeather, Brasilian milfoil
↓	*Myriophyllum verticillatum*	Whorl-leaf watermilfoil
Medium	*Berula erecta* (*Sium erectum*)	Water parsnip
↓	*Eichhornia crassipes*	Water hyacinth
↓	*Egeria densa*	Common waterweed
↓	*Najas marina* ssp. *intermedia*	Spiny naiad
↓	*Nuphar lutea*	Yellow water lily
Low	*Potamogeton* spp	Pondweeds

Some species showed higher tendencies to be allelopathic than others (Table 3.1). For example, Körner and Nicklisch (2002) demonstrated that *Myriophyllum spicatum* and *Ceratophyllum demersum* both inhibited photosystem II activity of phytoplankton, but they found no effect by *Potamogeton pectinatus*. Based on the absence of plant fragments in fish guts, Nurminen et al. (2003) also implied selective avoidance of the first two species (*M. spicatum*, *C. demersum*) by rudd (*Scardinius erythrophthalmus*). Macrophyte species may influence both behavioral and chemical interactions. For example, water incubated with *Stratiotes* (i.e. water soldier) demonstrated a highly significant inhibitory effect on the green alga *Scenedesmus obliquus* (Mulderij et al. 2005a), cyanobacterium *Microcystis* and eustigmatophyte *Nannochloropsis* (Mulderij et al. 2005b). Furthermore, the presence of *Stratiotes aloides*, especially young plants, increased colony formation by *S. obliquus*. Sinking rates of the algae also increased, leading to improved water clarity and light conditions for the macrophyte (Mulderij et al. 2005a). In the reverse direction, cyanobacterial compounds can have detrimental impacts on submerged vegetation. For example, Microcystin-LR shows allelopathic properties against *M. spicatum* and *C. demersum* (Pflugmacher 2002).

The deterrent chemical nature of some macrophytes may strongly regulate the biological community. One study suggests that aquatic plants repellent to grass carp could serve as areas of amphibian rehabilitation (Murphy et al. 2002). Endangered odonate larvae (*Aeshna viridis*) also preferred *Stratiotes aloides*, as it also served as a refuge from perch predation (Rantala et al. 2004). Other work connects herbivores and predators through the chemical nature of the macrophyte. For example, *Hyallela azteca* (Amphipoda) readily grazes on roots of emergent macrophytes (*Berula erecta*, water parsnip), even though they may be chemically defended (Rowell and Blinn 2003). Consequently, these amphipods exhibited lower susceptibility to fish predation. Although common in the marine literature, this study provided the first experimental evidence of reduced predation on a freshwater herbivore that consumed a chemically defended plant (Rowell and Blinn 2003).

Of all the aquatic macrophytes, one species (*Myriophyllum spicatum*, Eurasian watermilfoil; EWM) clearly stood out as the most influential chemically (Table 3.1) because of multiple interactions that occur with herbivores and competing phototrophs. On the whole-lake scale, macroinvertebrate biomass on dominant plant species in a lake may decline as the percentage of EWM cover increases (Cheruvelil et al. 2001). Furthermore, Lindén and Lehtiniemi (2005) found that chemicals excreted by dense EWM resulted in high mortality (73–89 %) of mysids (*Neomysis integer*, *Praunus felxuosus*) and lowered the feeding efficiencies of three-spined sticklebacks (*Gasterosteus aculeatus*). At a smaller scale, *Acentria* (Lepidoptera) exhibited higher feeding rates on *Myriophyllum* than on *Elodea* (Gross et al. 2001). In choice experiments, *Acentria* larvae clearly preferred EWM. By selectively removing apical tissue with the highest concentrations of allelopathically active compounds (Choi et al. 2002), herbivory by *Acentria* might weaken the allelopathic strength of this species, indirectly altering competitive interactions and the plant's susceptibility to bacterial infection. Overall, the magnitude of herbivory on EWM may shape competitive interactions between submerged macrophytes, change invertebrate biomass and perhaps alter ecosystem processes, such as primary production (Gross 2003a).

Remarkably, many invasive species, such as *M. spicatum* in Northern America or *Elodea* spp in Europe (Gross et al. 1996; Erhard and Gross 2005), act as allelopathically active species (Table 3.1). Significant differences in allelopathic activity between native and invasive specimens remain uncertain for aquatic systems, although Bais et al. (2003) documented such an occurrence with invasive, spotted knapweed (*Centaurea maculosa*). In addition, the extent to which allelopathic interactions add significantly to successful invasion also merits further attention. The concentration of active compounds in EWM may vary strongly, depending on environmental conditions (Gross 2003b) and may vary between native and invaded ranges of plants. Therefore, the role of macrophytes as chemical signalers under field conditions strongly depends on the macrophyte species present, the density or spatial heterogeneity of their patches, the velocity of breakdown of active compounds and possibly the nutrient state of both donor (macrophyte) and target organisms (e.g. phytoplankton).

The magnitude of the deterrent effect of macrophytes on trophic interactions also may differ among species and in the presence of other predatory cues. For example, *Daphnia* actively avoided submerged macrophytes in the absence of fish, but then overcame this repellency and sought refuge when fish arrived (Burks et al. 2001a), despite the possible costs of reduced zooplankton growth (Burks et al. 2000). In the absence of fish, Van de Meutter et al. (2004a) found an additional, although weak, avoidance of macrophytes that housed *Ischnura elegans* (10 % less within vegetation). Unfortunately, the predictability of macrophyte avoidance by *Daphnia* varies, perhaps depending on the clone, the presence of conspecific cues or the presence of other predators. In a similar study, Van de Meutter et al. (2004b) found no horizontal migration response toward the open water in the presence of *Ischnura elegans* alone. However, when *Ischnura* co-occurred with *Chaoborus*, daphnids showed an increased avoidance of plants. Van de Meutter et al. (2004b) attributed this to an increased risk of predation in the vegetation. If strong enough, chemical cues from littoral predators may contribute to the already repellent nature of macrophytes and override the tendency for zooplankton to seek refuge from fish, suggesting a hierarchal nature of chemical cues.

3.2.3 Central Themes: Impacts of Grazer–Epiphyton Interactions with Macrophytes

Although the term periphyton encompasses the whole community of organisms that colonize surfaces, we use epiphyton to stress the competitive interactions between algae and macrophytes. Macrophytes can serve as a habitat, a refuge from predators, or can act as a substrate which potential food sources can colonize (Fig. 3.2). Most epiphytes belong to the classes Bacilliophyta (diatoms) and Chlorophyta (green algae), but nuisance, grazing-resistant forms, such as filamentous blue-greens (Cyanobacteria) and *Cladophora,* can occur when the community is grazed by invertebrates which are selective in their feeding (Jones et al. 2000). Snail densities on macrophytes depend in part on plant palatability, periphyton density and algal quality and growth form (Brönmark and Vermaat 1997; Jones et al. 2000; Lombardo and Cooke 2002). Besides snails, other insects (including caddisflies, mayflies, beetles, corixids), along with several crustaceans (e.g. amphipods, cladocerans, ostracods) consume ample amounts of epiphyton (Jones et al. 1997).

We often recognize the importance of top-down control of phytoplankton by zooplankton grazing and its central role in maintaining the clear-water state (Jeppesen et al. 1997b). However, the distinction between the origin and maintenance of alternative equilibria is often not made; and it may be that

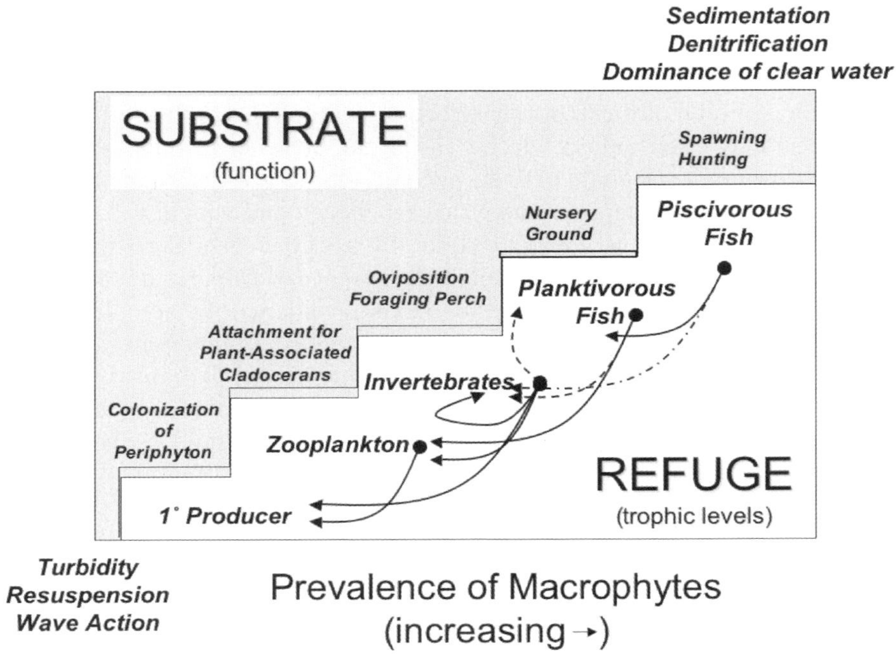

Fig. 3.2 Multi-faceted dual roles of macrophytes as both refuge and substrate across trophic levels. The *steps* suggest that the refuge potential of macrophytes increases with increasing macrophyte density after a threshold is established (i.e. the plateaus). *Arrows* show the predator–prey relationships that necessitate the need for macrophyte refuge. *Solid lines* illustrate the most predicted prey choice (*origin* with predator, *arrow* points to prey). *Dotted lines* illustrate prey choices that could occur with ontogenetic diet shifts. Four arrows originate from invertebrates because of the versatility of functional groups (i.e. predators, grazers, filterers). Increasing refuge provided by macrophytes leads to sedimentation and the dominance of clear-water, while the absence of macrophytes increases turbidity and promotes wave action and resuspension

phytoplankton plays a less important role in precipitating the shift from a turbid to a clear-water state. Macrophytes lead to increased sedimentation, through reduced water movement, which results in an influx of particles to the sediment. Consequently, there is less phytoplankton available to zooplankton in macrophytes, thus increasing the importance of epiphyton. Traditional approaches centered on the response of epiphyton to increased nutrient availability (Phillips et al. 1978), but recent work provides insight into how higher trophic levels affect the interaction between periphyton and plants, and accordingly shallow lake function.

Shading by periphyton shortened the clear-water phase associated with optimum growth conditions for submerged plants (e.g. *P. pectinatus*) in Lake Müggelsee (Germany; Roberts et al. 2003). Nutrients alone, however, do not appear to determine whether epiphyton can reach sufficient densities to negatively impact macrophytes. Experimental work suggests that other mechanisms are important for the loss of macrophytes (Jones et al. 2000; Williams et al. 2002). These experiments, previous work (Brönmark and Vermaat 1997) and a survey of 17 macrophyte-dominated lakes in the United Kingdom (Jones and Sayer 2003) have led to a proposal that the strength of a trophic cascade in the littoral zone determines the fate of macrophytes (Fig. 3.3). We agree with a recent review (Vadeboncoeur et al. 2002) that argued that benthic organisms, including epi- or periphyton and grazers, form integral links in lake food webs and deserve more emphasis in studies of trophic interaction.

In eutrophic to hypertrophic lakes, macrophyte biomass exhibits a negative relationship with epiphyton growth, with the density of epiphyton being, in turn, dependent on the density of grazing macroinvertebrates, but not nutrients (Fig. 3.3; Jones and Sayer 2003). A 20-fold increase in nutrient loading had no effect on the abundance of epiphyton growing on the surface of *Elodea nuttallii* in artificial ponds, whereas epiphyton density negatively correlated with increased grazers (Jones et al. 2002). In experimental ponds constructed to simulate reservoir wetlands, Balci and Kennedy (2003) also found that macroinvertebrate abundance correlated well with epiphyton biomass, although epiphyton biomass varied between exotic EWM and native water stargrass, *Heteranthera dubia* (Balci and Kennedy 2003). Fish biomass also served as a reliable predictor for the density of invertebrates (Jones and Sayer 2003) and accordingly epiphyton. Williams et al. (2002) showed that epiphyton chlorophyll *a* in experimental mesocosms increased significantly with increasing fish biomass (0–700 kg ha^{-1}). Interestingly, Jones and Waldron (2003) found that the proportion of periphyton in the diets of zooplankton increased with fish density, possibly as a result of increased horizontal migration. For these reasons, we highlight the strong potential for epiphyton–grazer interactions with macrophytes, mediated through interactions with fish, to facilitate shifts between turbid and clear states (see Section 3.2.4).

Colonization by epiphytes generally places macrophytes at a substantial disadvantage, due to increased competition for light, nutrients and carbon (Jones et al. 2000; Williams et al. 2002). By continuously grooming macrophyte hosts of unwanted algae and sediments, grazers may help maintain littoral communities (James et al. 2000). The periphyton removal rate generally increases allometrically with grazer biomass (Jones et al. 1997). Most grazing studies focus on snails (Gastropoda, see Brönmark and Vermaat 1997) because they are among the largest and most visible grazers on aquatic plants and can remove substantial biomass. While it may be advantageous for macrophytes to host snails, Jones et al. (2000) suggest that it is unlikely that macrophytes alter the community composition of epiphytes to increase their attractiveness to snails. However, see Brönmark's (1985) study that showed how macrophyte exudates (dissolved organic matter excreted from *Cerato-*

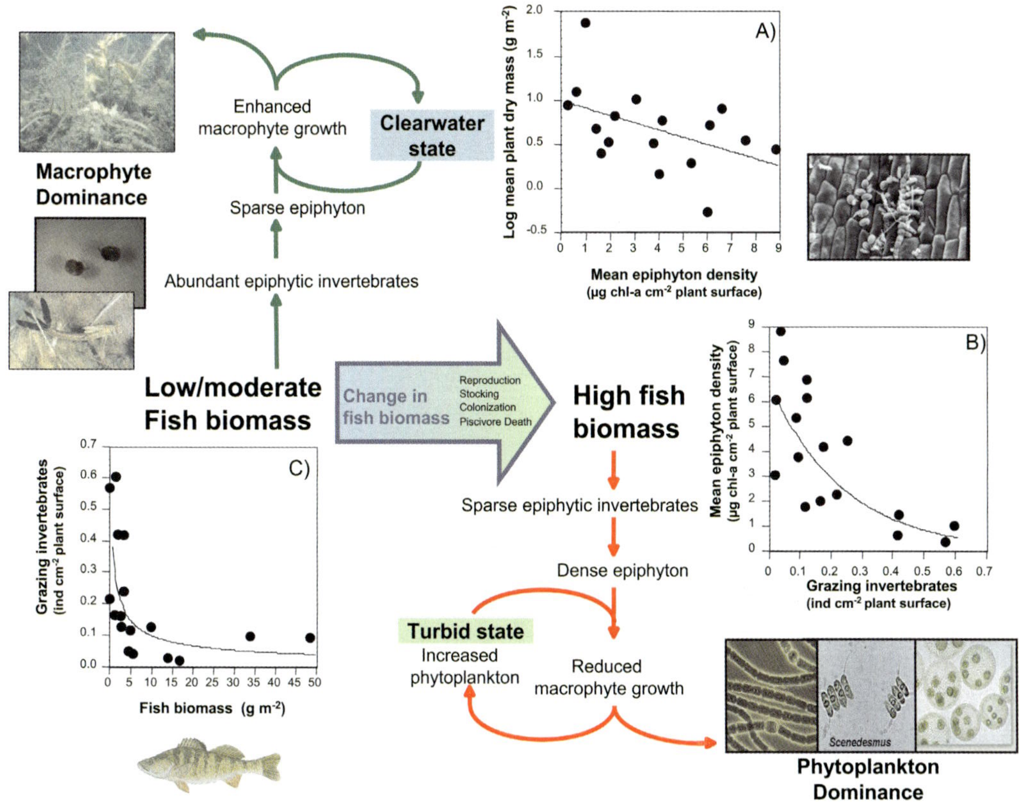

Fig. 3.3 Illustration of a littoral trophic cascade that could mediate the shift from turbid to clear conditions in shallow lakes (graphs originally published by Jones and Sayer 2003). In 17 British shallow lakes, plant biomass was negatively correlated with mean periphyton density (*graph A*: r^2=0.240, P=0.046). Periphyton density was then strongly negatively related to the density of grazing invertebrates (*graph B*: r^2=0.714, P<0.0001). Last, grazing invertebrate density declined significantly with increasing fish biomass (*graph C*: r^2=0.395, P=0.007). Overall, this suggests that a high abundance of fish would eliminate epiphyton grazers, allowing epiphyton to accumulate on plant surfaces. Consequently, macrophyte loss would occur due to light limitation and the absence of macrophytes would allow and a turbid state to dominate (*pictures* illustrate different types of phytoplankton blooms). *Images* show representative trophic levels. Note that the biomass of plankto-benthivorous fish increases with increasing nutrient levels. Thus, the scenario to the right is the most typical in eutrophic lakes unless high eutrophication leads to fish kills

phyllum demersum) attracted periphyton grazers (i.e. the snail *Lymnaea peregra*).

Direct consumption of macrophytes also alters trophic interactions. Some large invertebrates, including the invasive channeled applesnail *Pomacea canaliculata* (Lach et al. 2000) or crayfish (Lodge et al. 1994), may consume or destroy macrophytes copiously, decreasing the habitat available for colonization. For example, *Procambarus clarkii* reduced the macrophyte cover in a mesotrophic, Spanish shallow lake from 97 % to below 10 % in less than three years (Rodríguez et al. 2003). However, species-specific impacts occur, e.g. Hessen et al. (2004) demonstrated that the crayfish *Astacus astacus* failed to control rapid growth of *Elodea*. In addition, macrophytes differ in their palatability to snails and other invertebrates (Elger and Willby 2003), therefore changing plant species composition through selective herbivory (Elger et al. 2002).

The large size and omnivorous nature of crayfish differentiate them from other macroinvertebrates, underlining their importance in littoral habitats (Lodge et al. 1994). With a demonstrated preference for herbivory over carnivory and for native versus exotic macrophytes (i.e. *Eichhornia*), *P. clarkii* could be responsible for the extirpation of native aquatic macrophytes in Lake Naivasha, Kenya (Smart et al. 2002). In small Michigan ponds, vascular macrophytes failed to establish in the presence of crayfish, blue-green algae came to dominate and two epiphyton herbivores (i.e. tadpoles, snails) did not co-exist with crayfish (Dorn and Wojdak 2004). Beyond just consuming macrophytes, Dorn and Wojdak's (2004) study showed that crayfish preyed heavily on fish eggs, reducing fish recruitment (Fig. 3.2) and indirectly enhancing zooplankton biomass. Crayfish also influenced the life history traits and habitat choice of snails. Increased abundance led to the aggregation of snails on food-poor macrophytes and reduced snail growth (Lewis 2001b). Crayfish can structure the abundance and size distribution of thin-shelled snails through size-selective predation, although crayfish effects on snail size distribution may be less pronounced in complex habitats, such as macrophyte beds (Nyström and Pérez 1998). Collectively, these results indicate that crayfish can have dramatic direct and indirect impacts on littoral pond communities via feeding links with multiple trophic levels and non-trophic activities.

3.2.4 Central Themes: Prevalance of Fish Influence in Shallow Lakes

Tight coupling makes it challenging to discuss the role of macrophytes in shallow lakes without routinely acknowledging the prevalence of fishes in the littoral. Multiple interactions highlighted in our literature survey involved fish. Particular littoral fish communities exist in both temperate (Weaver et al. 1997; Lewin et al. 2004) and tropical systems (Vono and Barbosa 2001; Jeppesen et al. 2005). Such fishes spend their time predominately within macrophyte beds, although the proportion of that time may change with ontogenetic shifts (Persson and Crowder 1997). For example, juvenile pike (*Esox lucius*) inhabit littoral vegetation, but as they grow larger, pike associate with open water along the edges of macrophyte beds (Bry 1996) or less complex structure(s) of emergent macrophytes (i.e. *Typha*; Skov and Berg 1999).

Apart from ontogenetic habitat shift, the time spent in the littoral by fish also varies on a diel and a seasonal scale (Romare et al. 2003). For instance, many juvenile littoral fish perform diel horizontal migrations into the open water at night (e.g. Jacobsen and Perrow 1998; Shoup et al. 2003). Limited knowledge exists regarding winter behavior in shallow lake fish, but the littoral structure might act as a winter refuge, possibly resulting in a lower ratio of zooplankton to phytoplankton compared to summer (Jeppesen et al. 2004). This suggests a lower potential for zooplankton grazing. Habitat use of macrophytes by lower taxa strongly depends on the presence and species identity of littoral fishes. As suggested earlier, the water clarity of the system also influences interactions between macrophytes and fish. For example, Jacobsen et al. (2004) showed roach hide under water lilies during daytime in a clear-water lake inhabited by piscivorous bird species. Therefore, floating-leaved macrophytes might provide an effective cover for fish if the predation threat comes from above.

Macrophytes also serve as substrate for spawning (Fig. 3.2; Winfield 2004). Pike prefer a short dense structure, even flooded grasslands, for spawning. As for several other species, eggs require a structure on which to adhere. Some fish even use vegetation for nest-building during spawning, especially large sticklebacks (*Gasterosteus aculeatus*; Kraak et al. 2000). Littoral fish density and identity, however, fail to tell the whole story. The structural complexity of some macrophytes negatively impact the foraging rates of some fishes (Warfe and Barmuta 2004). Predation risk and competition can influence the distribution of fish species and age or size structure. For example, small bluegills (*Lepomis*) spent more time in artificial macrophytes during day than night when predators were present (Shoup et al. 2003), while large bluegill (>200 mm), in contrast used open water, emergent vegetation and submerged vegetation similarly during the day (Paukert and Willis 2002). No single factor controls habitat use of macrophytes by fishes, although macrophyte density, predation risk and prey availability may be the most influential.

Macrophytes also affect the trophic structure and interactions by enhancing fish kills under ice in lakes with long-lasting ice-cover. Canadian studies have shown higher oxygen depletion under ice in shallow lakes with abundant macrophyte biomass (Meding and Jackson 2001). This may, in turn, affect the entire trophic dynamic and water clarity quite substantially in such lakes (Bayley and Prather 2003). Grazing by fishes also may alter the biomass and species composition of submerged macrophytes (Van Donk and Otte 1996). Furthermore, some fishes may not intentionally consume macrophytes, but instead inflict damage at a critical developmental stage when they forage for macroinvertebrates (Körner and Dugdale 2003). Despite these known impacts, the complexity of interactions associated with macrophytes and fish seems limitless and warrants further study.

3.3 In the Wings: Research Areas Worthy of Attention

Our synthesis of the role of macrophytes in regulating trophic interactions suggests the need for more information to answer several key questions.

3.3.1 Predictability of Macrophyte Function in Trophic Interactions Across a Climatic Gradient

Most studies of macrophytes have been confined to North temperate lakes (Jeppesen et al. 2003), while little information exists from warm temperate to tropical lakes (Jeppesen et al. 2005). We might expect that higher temperature and favorable growth conditions during winter promote the growth of macrophytes (Rooney and Kalff 2000; Liboriussen et al. 2005). We must also consider that floating plants and floating-leaved plants are more prominent in the tropics and subtropics, but these architectures have not been as well studied as submerged plants. In dry areas, a lower water table also might promote macrophyte growth (Blindow et al. 1993). However, some macrophyte species, such as *Elodea,* may be broadly resilient to small increases associated with warming (Mckee et al. 2002). Recent enclosure studies along a climatic gradient in Europe suggested a less positive role of macrophytes on water clarity in warmer versus colder temperate lakes (Moss et al. 2004). Bachmann et al. (2002) draw similar conclusions from a cross-comparison of data from Florida lakes. One explanation may be that increased omnivory and a shift to small fish size in warm lakes also increase the use of macrophytes as a habitat for fish, making the plants a less useful refuge for zooplankton and macroinvertebrate grazers (Fig. 3.3; Blanco et al. 2003; Meerhoff et al. 2003; Jeppesen et al. 2005). In the tropics, the life cycles of some fish species are completely coupled to plants (Sazima and Zamprogno 1985; Agonstinho et al. 2003). More comparative studies of submerged, free-floating, floating-leaved and emergent plant communities are needed to understand exactly how architecture influences the dual roles of macrophytes (Fig. 3.2). We advise more studies to consider this temporal scale and to explore how macrophyte age and the ontogeny of other organisms (fish, zooplankton, invertebrates) affect the use of macrophytes as substrates, habitats, or refuges. We especially encourage more studies along climatic gradients to gain needed insight into how climate affects the role of macrophytes in lakes. Such studies are also needed to provide knowledge of how global warming may affect plants and their role in lake ecosystems.

3.3.2 Relative Importance of Chemical Ecology Across Trophic Levels

As others often propose, we also stress the need for collaboration between chemists and ecologists to identify allelopathic compounds of aquatic macrophytes. Such identification would produce incredible potential for experiments that provide definitive answers to testable hypotheses. Experimental evidence for in situ allelopathic activity on epiphyte density is challenging because of natural heterogeneity due to wave action, grazers and macrophyte structure. Thus, no general proof clearly exists for lower epiphyte density on allelopathically active macrophytes (Wium-Andersen 1987; Gross et al. 2003a), but different classes of compounds may differentially affect primary producers. For example, after investigating the allelopathic properties of *Ceratophyllum demersum* and *Najas marina*, Gross et al. (2003b) suggested that water-soluble allelochemicals may inhibit phytoplankton, whereas lipophilic allelochemicals may act through direct cell–cell contact, e.g. against epiphytes. Although difficult, we argue for more field studies that examine allelopathy. For example, we could benefit from more detailed knowledge on how changes in dissolved organic matter (i.e. concentration, composition, or source) or turbidity alter interactions inside and outside macrophyte beds.

3.3.3 Disproportional Impacts of Certain Invertebrates and Exotic Species

Fish undoubtedly drive interactions in pelagic communities (Jeppesen 1998). However, in littoral communities, macroinvertebrates, acting as grazers or predators, may be nearly as important in regulating trophic interactions among macrophytes. We recommend that studies continue to focus on invertebrates, especially large individuals such as snails and crayfish. Our literature review also indicated that epiphyton (in combination with grazers) and benthic–pelagic interactions may play a significant role in pushing systems toward a clear or turbid state. We recommend more field experiments that explore this possibility and its relation to lake/wetland management. Last, we encourage wetland scientists and ecologists to examine how macrophytes may facilitate the introduction of exotic species and the implications of macrophytes themselves as exotic species. The introduction of exotic species has the potential to alter primary production and nutrient cycling as well as native biodiversity. Furthermore, climate warming may further facilitate the invasiveness of exotic species.

3.4 Returning to Center Stage: Macrophytes are Common Players in Trophic Interactions

Any wetland textbook will enumerate the many ways in which macrophytes alter biogeochemical cycles. Recognition of the importance of macrophytes in regulating ecosystem processes is not novel (Carpenter and Lodge 1986). The take-home message of our chapter, however, focuses on examining macrophytes at the level of the community and emphasizing the multi-faceted interactions that result. Macrophytes compete with other primary producers, serve as "prey" for herbivores, function as habitat or substrate for multiple trophic levels and may serve as vectors for exotic species. Our chapter broadens the discussion from Jeppesen et al. (1997a) by speculating about how floating-leaved, free-floating and emergent macrophytes may structure interactions beyond what is known for submerged macrophytes. We also discuss how regulation of such interactions could differ between temperate and tropical systems. Regardless of locality, we strongly suggest that the presence of aquatic macrophytes drives interactions within the rest of the food web and determines which trophic interactions play a role in influencing shifts between alternative states.

Acknowledgments We thank Dr. Jos Verhoeven and an anonymous reviewer for guidance on this chapter. M. Bouche, A.M. Poulsen and K. Clovis assisted with our literature search and compilation. A Cullen Faculty Development grant from Southwestern University made R.B.'s participation in the INTECOL symposium possible. E.J. was supported by the research project "Consequences of weather and climate changes for marine and freshwater ecosystems" (CONWOY, 2052-01-0034), funded by the Danish Scientific Research Council and the EU Eurolimpacs project (GOCE-CT-2003-505540).

References

Abrahams MV, Kattenfeld MG (1997) The role of turbidity as a constraint on predator–prey interactions in aquatic environments. Behav Ecol Sociobiol 40:169–174

Agostinho AA, Gomes LC, Ferreira HJ Jr (2003) Relacoes entre macrófitas acquáticas e fauna de peixes. In: Thomas SM, Bini LM (eds) Ecologia e majejo de macrófitas aquáticas. EDUEM, pp 261–279

Bachmann RW, Horsburgh CA, Hoyer MV, Mataraza LK, Canfield DE Jr (2002) Relations between trophic state indicators and plant biomass in Florida lakes. Hydrobiologia 470:219–234

Bais HP, Vepachedu R, Gilroy S, Callway RM, Vivanco JM (2003) Allelopathy and exotic plant invasion: from molecules and genes to species interactions. Science 301:1377–1380

Balci P, Kennedy JH (2003) Comparison of chironomids and other macroinvertebrates associated with *Myriophyllum spicatum* and *Heteranthera dubia*. J Freshwater Ecol 18:235–247

Bayley SE, Prather CM (2003) Do wetlands exhibit alternate stable states? Submersed aquatic vegetation and chlorophyll in western boreal shallow lakes. Limnol Oceanogr 48:2335–2345

Blanco R, Romo S, Villena MJ, Martinez S (2003) Fish communities and food web interaction in some shallow Mediterranean lakes. Hydrobiologia 506/509:473–480

Blindow I, Andersson G, Hageby A, Johansson S (1993) Long-term pattern of alternative stable states in two shallow eutrophic lakes. Freshwater Biol 30:159–167

Blindow I, Hargeby A, Wagner BMA, Andersson G (2000) How important in the crustacean plankton for the maintenance of water clarity in shallow lakes with abundant submerged vegetation? Freshwater Biol 44:185–197

Brooks JL, Dodson SI (1965) Predation, body size and composition of plankton. Science 150:28–35

Brönmark C (1985) Interactions between epiphytes, macrophytes and herbivores: an experimental approach. Oikos 45:26–30

Brönmark C, Vermaat J (1997) Complex fish–snail–epiphyton interactions and their effects on submerged freshwater macrophytes. In: Jeppesen E, Søndergaard M, Søndergaard M, Christoffersen K (eds) The structuring role of submerged macrophytes in lakes (Ecological Studies 131) Springer, Berlin Heidelberg New York, pp 47–68

Bry C (1996) Role of vegetation in the life cycle of pike. In: Craig JF (ed) Pike: biology and exploitation. Chapman and Hall, London, pp 45–67

Burks RL, Lodge DM (2002) Cued in: advances and opportunities in freshwater chemical ecology. J Chem Ecol 28:1901–1917

Burks RL, Jeppesen E, Lodge DM (2000) Chemicals from macrophytes and fishes suppress *Daphnia* growth and alter life history traits. Oikos 88:139–147

Burks RL, Jeppesen E, Lodge DM (2001a) Littoral zone structures as *Daphnia* refugia against fish predators. Limnol Oceanogr 46:230–237

Burks RL, Jeppesen E, Lodge DM (2001b) Pelagic prey and benthic predators: impact of odonate predation on *Daphnia*. J N Am Benthol Soc 20:615–628

Burks RL, Lodge DM, Jeppesen E, Lauridsen TL (2002) Diel horizontal migration of zooplankton: costs and benefits of inhabiting the littoral. Freshwater Biol 47:343–365

Carpenter SR, Lodge DM (1986) Effects of submersed macrophytes on ecosystem processes. Aquat Bot 26:341–370

Cattaneo A, Galanti G, Gentinetta S, Romo S (1998) Epiphytic algae and macroinvertebrates on submerged and floating-leaved macrophytes in an Italian lake. Freshwater Biol 39:725–740

Cheruvelil KS, Soranno PA, Madsen JD (2001) Epiphytic macroinvertebrates along a gradient of Eurasian watermilfoil cover. J Aquat Plant Manage 39:67–72

Choi C, Bareiss C, Walenciak O, Gross EM (2002) Impact of polyphenols on growth of the aquatic herbivore *Acentria ephemerella*. J Chem Ecol 28:2245–2256

Cottenie K, Nuytten J, Michels E, De Meester L (2001) Zooplankton community structure and environmental conditions in a set of interconnected ponds. Hydrobiologia 442: 339–350

Declerck S, Vandekerkhove J, Johansson L, Muylaert K, Conde-Porcuna JM, Van der Gucht K, Pérez-Martínez C, Lauridsen T, Schwenk K, Zwart G, Rommens W, López-Ramos J, Jeppesen E, Vyverman W, Brendonck L, De Meester L (2005) Multi-group biodiversity in shallow lakes along gradients of phosphorus and water plant cover. Ecology 86:1905–1915

Dorn NJ, Wojdak JM (2004) The role of omnivorous crayfish in littoral communities. Oecologia 140:150–159

Elger A, Willby NJ (2003) Leaf dry matter content as an integrative expression of plant palatability: the case of freshwater macrophytes. Funct Ecol 17:58–65
Elger A, Barrat-Segretain MH, Amoros C (2002) Plant palatability and disturbance level in aquatic habitats: an experimental approach using the snail *Lymnaea stagnalis* (L.). Freshwater Biol 47:931–940
Englund G, Harms S (2003) Effects of light and microcrustacean prey on growth and investment in *Utricularia vulgaris*. Freshwater Biol 48:786–794
Erhard D, Gross EM (2005) Do environmental factors influence the composition of potential allelochemicals in the freshwater submersed macrophyte *Elodea nuttallii* (Hydrocharitaceae)? Verh Int Ver Limnol 29:287–291
Forbes SA (1925) The lake as a microcosm (reprint). INHS Bull 15:537–550
Gaiser EE, Lang KL (1998) Distribution of cladoceran zooplankton among prairie pothole wetlands in northwest Iowa. Lake Reserv Manage 14:37–51
Gopal B, Goel U (1993) Competition and allelopathy in aquatic plant communities. Bot Rev 59:155–210
Gross EM (2003a) Allelopathy of aquatic autotrophs. Crit Rev Plant Sci 22:313–339
Gross EM (2003b) Differential response of tellimagrandin II and total bioactive hydrolysable tannins in an aquatic angiosperm to changes in light and nitrogen. Oikos 103:497–504
Gross EM, Meyer H, Schilling G (1996) Release and ecological impact of algicidal hydrolysable polyphenols in *Myriophyllum spicatum*. Phytochemistry 41:133–138
Gross EM, Johnson RL, Hairston NG (2001) Experimental evidence for changes in submersed macrophyte species composition caused by the herbivore *Acentria ephemerella* (Lepidoptera). Oecologia 127:105–114
Gross EM, Feldbaum C, Graf A (2003a) Epiphyte biomass and elemental composition on submersed macrophytes in shallow eutrophic lakes. Hydrobiologia 506/509:559–565
Gross EM, Erhard D, Ivanyi E (2003b) Allelopathic activity of *Ceratophyllum demersum* L. and *Najas marina* ssp. *intermedia* (Wolfgang) Casper. Hydrobiologia 506/509:583–589
Guisande C (2000) Effects of zooplankton and conductivity on tropical *Utricularia foliosa* investment in carnivory. Aquat Ecol 34:137–142
Hann BJ, Zrum L (1997) Littoral microcrustaceans (Cladocera, Copepoda) in a prairie coastal wetland: seasonal abundance and community structure. Hydrobiologia 357:37–52
Hessen DO, Skurdal J, Braathen JE (2004) Plant exclusion of a herbivore; crayfish population decline caused by an invading waterweed. Biol Invasions 6:133–140
Jacobsen L, Perrow MR (1998) Predation risk from piscivorous fish influencing the diel use of macrophytes by planktivorous fish in experimental ponds. Ecol Freshwater Fish 7:78–86
Jacobsen L, Berg S, Jepsen N, Skov C (2004) Does roach behaviour differ between shallow lakes of different environmental state? J Fish Biol 65:135–147
James MR, Hawes I, Weatherhead M (2000) Removal of settled sediments and periphyton from macrophytes by grazing invertebrates in the littoral zone of a large oligotrophic lake. Freshwater Biol 44:311–326
Jeppesen E (1998) The ecology of shallow lakes – biological interactions in the pelagial. DSc dissertation (NERI Report 24), Ministry of Environment and Energy, National Environmental Research Institute
Jeppesen E, Søndergaard M, Søndergaard M, Christoffersen K (eds) (1997a) The structuring role of submerged macrophytes in lakes. (Ecological Studies 131) Springer, Berlin Heidelberg New York
Jeppesen E, Lauridsen TL, Kairesalo T, Perrow MR (1997b) Impact of submerged macrophytes on fish-zooplankton interactions in lakes. In: Jeppesen E, Søndergaard M,

Søndergaard M, Christoffersen K (eds) The structuring role of submerged macrophytes in lakes (Ecolological Studies 131) Springer, Berlin Heidelberg New York, pp 91–114

Jeppesen E, Jensen JP, Søndergaard M, Lauridsen T (1999) Trophic dynamics in turbid and clearwater lakes with special emphasis on the role of zooplankton for water clarity. Hydrobiologia 408/409:217–231

Jeppesen E, Søndergaard M, Søndergaard M, Christoffersen K, Theil-Nielsen J, Jurgens K (2002) Cascading trophic interactions in the littoral zone: an enclosure experiment in shallow Lake Stigsholm, Denmark. Arch Hydrobiol 153:533–555

Jeppesen E, Jensen JP, Jensen C, Faafeng B, Hessen DO, Søndergaard M, Lauridsen T, Brettum P, Christoffersen K (2003) The impact of nutrient state and lake depth on top-down control in the pelagic zone of lakes: a study of 466 lakes from the temperate zone to the arctic. Ecosystems 6:313–325

Jeppesen E, Jensen JP, Søndergaard M, Fenger-Grøn M, Bramm ME, Sandby K, Møller PH, Rasmussen HU (2004) Impact of fish predation on cladoceran body weight distribution and zooplankton grazing in lakes during winter. Freshwater Biol 49:432–447

Jeppesen E, Søndergaard M, Mazzeo N, Meerhoff M, Branco CC, Huszar V, Scasso F (2005) Lake restoration and biomanipulation in temperate lakes: relevance for subtropical and tropical lakes. In: Reddy V (ed) Tropical eutrophic lakes: their restoration and management. pp 331–359

Jeppesen E, Pekcan-Hekim Z, Lauridsen TL, Søndergaard M, Jensen JP (2006) Habitat distribution of fish in late summer: changes along a nutrient gradient in Danish lakes. Ecol Freshwater Fish (in press)

Jones JI, Sayer CD (2003) Does the fish–invertebrate–periphyton cascade precipitate plant loss in shallow lakes? Ecology 84:2155–2167

Jones JI, Waldron S (2003) Combined stable isotope and gut contents analysis of food webs in plant dominated, shallow lakes. Freshwater Biol 48:1396–1407

Jones JI, Moss B, Young JO (1997) Interactions between periphyton, nonmolluscan invertebrates and fish in standing freshwaters. In: Jeppesen E, Søndergaard M, Søndergaard M, Christoffersen K (eds) The structuring role of submerged macrophytes in lakes (Ecological Studies 131) Springer, Berlin Heidelberg New York, pp 69–90

Jones JI, Moss B, Eaton JW, Young JO (2000) Do submerged aquatic plants influence periphyton community composition for the benefit of invertebrate mutualists? Freshwater Biol 43:591–604

Jones JI, Young JO, Eaton JW, Moss B (2002) The influence of nutrient loading, dissolved inorganic carbon and higher trophic levels on the interaction between submerged plants and periphyton. J Ecol 90:12–24

Körner S, Dugdale T (2003) Is roach herbivory preventing re-colonization of submerged macrophytes in a shallow lake? Hydrobiologia 506/509:497–501

Körner S, Nicklisch A (2002) Allelopathic growth inhibition of selected phytoplankton species by submerged macrophytes. J Phycol 38:862–871

Kraak SBM, Bakker TCM, Hocevar S (2000) Stickleback males, especially large and red ones, are more likely to nest concealed in macrophytes. Behaviour 137:907–919

Lach L, Britton DK, Rundell RJ, Cowie RH (2000) Food preference and reproductive plasticity in an invasive freshwater snail. Biol Invasions 2:279–288

Lauridsen TL, Pedersen LJ, Jeppesen E, Søndergaard M (1996) The importance of macrophyte bed size for cladoceran composition and horizontal migration in a shallow lake. J Plank Res 18:2283–2294

Lewin WC, Okun N, Mehner T (2004) Determinant of the distribution of juvenile fish in the littoral area of a shallow lake. Freshwater Biol 49:410–424

Lewis WM (2001a) Wetlands explained: wetland science, policy and politics in America. Oxford University Press, New York

Lewis DB (2001b) Trade-offs between growth and survival: responses of freshwater snails to predacious crayfish. Ecology 82:758–765

Liboriussen L, Landkildehus F, Meerhoff M, Bramm ME, Søndergaard Mo, Christoffersen K, Richardson K, Søndergaard Ma, Lauridsen T, Jeppesen E (2005) Global warming: design of a flow-through shallow lake mesocosm climate experiment. Limnol Oceanogr Methods 3:1–9

Lindeman RL (1942) The trophic–dynamic aspect of ecology. Ecology 23:399–418

Lindén E, Lehtiniemi M (2005) The lethal and sublethal effects of the aquatic macrophyte *Myriophyllum spicatum* on Baltic littoral planktivores. Limnol Oceanogr 50:405–411

Lodge DM, Kershner MW, Aloi JE, Covich AP (1994) Effects of an omnivorous crayfish (*Orconectes rusticus*) on a freshwater littoral food web. Ecology 75:1265–1281

Lombardo P (1997) Predation by *Enallagma* nymphs (Odonata, Zygoptera) under different conditions of spatial heterogeneity. Hydrobiologia 356:1–9

Lombardo P, Cooke GD (2002) Consumption and preference of selected food types by two freshwater gastropod species. Arch Hydrobiol 155:667–685

Marklund O, Blindow I, Hargeby A (2001) Distribution and diel migration of macroinvertebrates within dense submerged vegetation. Freshwater Biol 46:913–924

Mckee D, Hatton K, Eaton J, Atkinson D, Atherton A, Harvey I, Moss B (2002) Effects of simulated climate warming on macrophytes in freshwater microcosm communities. Aquat Bot 74:71–83

Meding ME, Jackson LJ (2001) Biological implications of empirical models of winter oxygen depletion. Can J Fish Aquat Sci 58:1727–1736

Meerhoff M, Mazzeo N, Moss B, Rodríguez-Gallego L (2003) The structuring role of free-floating versus submerged plants in a subtropical shallow lake. Aquat Ecol 37:377–391

Molisch H (1937) Der Einfluss einer Pflanze au die Anderd-Allelopathie. Fischer, Jena

Moss B, Stephen D, Balayla DM, Bécares E, Collings SE, Fernández-Aláez C, Fernández-Aláez M, Ferriol C, García P, Gomá J, Gyllström M, Hansson L-A, Hietala J, Kairesalo T, Miracle MR, Romo S, Rueda J, Russell V, Ståhl-Delbanco A, Svensson M, Vakkilainen K, Valentín M, Van de Bund WJ, Van Donk E, Vicente E, Villena MJ (2004) Continental-scale patterns of nutrient and fish effects on shallow lakes: synthesis of a pan-European mesocosm experiment. Freshwater Biol 49:1633–1649

Mulderij G, Van Donk E, Roelofs JGM (2003) Differential sensitivity of green algae to allelopathic substances from *Chara*. Hydrobiologia 491:261–271

Mulderij G, Mooij WM, Van Donk E (2005a) Allelopathic growth inhibition and colony formation of the green alga *Scenedesmus obliquus* by the aquatic macrophyte *Stratiotes aloides*. Aquat Ecol 39:11–21

Mulderij G, Mooij WM, Smolders AJP, Van Donk E (2005b) Allelopathic inhibition of phytoplankton by exudates from *Stratiotes aloides*. Aquat Bot 82:284–296

Murphy JE, Beckmen KB, Johnson JK, Cope RB, Lawmaster T, Beasley VR (2002) Toxic and feeding deterrent effects of native aquatic macrophytes on exotic grass carp (*Ctenopharyngodon idella*). Ecotoxicology 11:243–254

Nurminen L, Horppila J (2002) A diurnal study on the distribution of filter feeding zooplankton: effect of emergent macrophytes pH and lake trophy. Aquat Sci 64:198–206

Nurminen L, Horppila J, Lappalainen J, Malinen T (2003) Implications of rudd (*Scardinius erythropthalmus*) herbivory on submerged macrophytes in a shallow eutrophic lake. Hydrobiologia 506/509:511–518

Nyström P, Pérez JR (1998) Fish predation on the common pond snail (*Lymnaea stagnalis*): the effect of habitat complexity and snail size on foraging efficiency. Hydrobiologia 368:201–208

Paukert CP, Willis DW (2002) Seasonal and diel habitat selection by bluegills in a shallow natural lake. Trans Am Fish Soc 131:1131–1139

Persson L, Crowder LB (1997) Fish-habitat interactions mediated via ontogenetic niche shifts. In: Jeppesen E, Søndergaard M, Søndergaard M, Christoffersen K (eds) The structuring role of submerged macrophytes in lakes. (Ecolological Studies 131) Springer, Berlin Heidelberg New York, pp 3–23

Pflugmacher S (2002) Possible allelopathic effects of cyanotoxins, with reference to microcystin-LR, in aquatic ecosystems. Environ Toxicol 17:407–413

Phillips GL, Eminson D, Moss B (1978) A mechanism to account for macrophyte decline in progressively eutrophicated freshwaters. Aquat Bot 4:103–126

Priyadarshana T, Aeaeda T, Manatunge J (2001) Foraging behaviour of planktivorous fish in artificial vegetation: the effects on swimming and feeding. Hydrobiologia 442:231–239

Rantala MJ, Ilmonen J, Koskimäki J, Suhonen J, Tynkkynen K (2004) The macrophyte, *Stratiotes aloides,* protects larvae of dragonfly *Aeshna viridis* against fish predation. Aquat Ecol 38:77–82

Roberts E, Kroker J, Körner S, Nicklisch A (2003) The role of periphyton during the re-colonization of shallow lake with submerged macrophytes. Hydrobiologia 506/509: 525–530

Rodríguez CF, Bécares E, Fernández-Aláez M (2003) Shift from clear to turbid phase in Lake Chozas (NW Spain) due to the introduction of American red swamp crayfish (*Procambarus clarkii*). Hydrobiologia 506/509:421–426

Romare P, Berg S, Lauridsen T, Jeppesen E (2003) Spatial and temporal distribution of fish and zooplankton in a shallow lake. Freshwater Biol 48:1353–1362

Rooney N, Kalff J (2000) Inter-annual variation in submerged macrophyte community biomass and distribution: the influence of temperature and lake morphometry. Aquat Bot 68:321–335

Rowell K, Blinn DW (2003) Herbivory on a chemically defended plant as a predation deterrent in *Hyalella azteca*. Freshwater Biol 48:247–254

Rybak JI, Weglenska T (2003) Temporal and spatial changes in the horizontal distribution of planktonic crustacea between vegetated littoral zone and the zone of open water. Pol J Ecol 51:205–218

Sagrario MDG, Balseiro E (2003) Indirect enhancement of large zooplankton by consumption of predacious macroinvertebrates by littoral fish. Arch Hydrobiol 158:551–574

Sazima I, Zamprogno C (1985) Use of water hyacinth as shelter, foraging pace, and transport by young piranhas, *Seeasalmus spilopleura*. Environ Biol Fish 12:237–240

Scheffer M (1998) Community dynamics of shallow lakes. Chapman and Hall, London

Shoup DE, Carlson RE, Heath RT (2003) Effects of predations risk and foraging return on the diel use of vegetated habitat by two size-classes of bluegills. Trans Am Fish Soc 132:590–597

Skov C, Berg S (1999) Utilization of natural and artificial habitats by YOY pike in a biomanipulated lake. Hydrobiologia 408/409:115–122

Skov C, Berg S, Jacobsen L, Jepsen N (2002) Habitat use and foraging success of 0+ pike (*Esox lucius* L.) in experimental ponds related to prey fish, water transparency and light intensity. Ecol Freshwater Fish 11:65–73

Smart AC, Harper DM, Malaisse F, Schmitz S, Coley S, Gouder de Beauregard A-C (2002) Feeding of the exotic Louisiana red swamp crayfish, *Procambarus clarkii* (Crustacea, Decapoda), in an African tropical lake: Lake Naivasha, Kenya. Hydrobiologia 488:129–142

Teal JM (1962) Energy flow in the salt marsh ecosystem of Georgia. Ecology 43:614–624

Tolonen KT, Hämäläinen H, Holopainen IJ, Karjalainen J (2001) Influences of habitat type and environmental variables on littoral macroinvertebrate communities in a large lake system. Arch Hydrobiol 152:39–67

Vadeboncoeur Y, Vander Zanden MJ, Lodge DM (2002) Putting the lake back together: reintegrating benthic pathways into food web models. Bioscience 52:44–55

Valley RD, Bremigan MT (2002) Effects of macrophyte bed architecture on largemouth bass foraging: implications of exotic macrophyte invasions. Trans Am Fish Soc 131:234–244

Van de Meutter F, Stoks R, De Meester L (2004a) Behavioral linkage of pelagic prey and littoral predators: microhabitat selection by *Daphnia* induced by damselfly larvae. Oikos 107:265–272

Van de Meutter F, Stoks R, De Meester L (2004b) Spatial avoidance of littoral and pelagic invertebrate predators by *Daphnia*. Oecologia, DOI: 10.1007/s00442-004-1738-5

Van Donk E, Bund WJ van de (2002) Impact of submerged macrophytes including charophytes on phyto- and zooplankton communities: allelopathy versus other mechanisms. Aquat Bot 72:261–274

Van Donk E, Otte A (1996) Effects of grazing by fish and waterfowl on the biomass and species composition of submerged macrophytes. Hydrobiologia 340:285–290

Vono V, Barbosa FAR (2001) Habitats and littoral zone fish community structure of two natural lakes in southeast Brazil. Environ Biol Fish 61:371–379

Warfe DM, Barmuta LA (2004) Habitat structural complexity mediates the foraging success of multiple predator species. Oecologia 141:171–178

Weaver MF, Magnuson JJ, Clayton MK (1997) Distribution of littoral fishes in structurally complex macrophytes. Can J Fish Aquat Sci 54:2277–2289

Williams AE, Moss B, Eaton J (2002) Fish induced macrophyte loss in shallow lake: top-down and bottom-up processes in mesocosm experiments. Freshwater Biol 47:2216–2232

Winfield I (2004) Fish in the littoral zone: ecology, threats and management. Limnologica 34:124–131

Wium-Andersen S (1987) Allelopathy among aquatic plants. Arch Hydrobiol Beih Erg Limnol 27:167–172

4 Biological Invasions: Concepts to Understand and Predict a Global Threat

Gerard van der Velde, Sanjeevi Rajagopal, Mirjam Kuyper-Kollenaar, Abraham bij de Vaate, David W. Thieltges, and Hugh J. MacIsaac

4.1 Introduction

Charles Elton was the modern founder of the science of biological invasions. He wrote that 'biological invasions are so frequent nowadays in every continent and island, and even in the oceans, that we need to understand what is causing them and try to arrive at some general viewpoint about the whole business' (Elton 1958). He tried to predict the outcome of global invasion processes and assumed that invasions would result in homogenization of regional floras and faunas. The prediction of homogenization was formulated earlier by Lyell (1832) who, in contrast to Elton (1958), did not consider the resulting human-caused extinctions to be a cause of concern because, in his opinion, this was a natural process (Wilkinson 2004). Interest in biological invasions has rapidly increased in recent decades and today biological invasions are a major concern in ecology and conservation. Particularly dramatic consequences of invasions have been reported from island ecosystems where endemic species suffered severely, but wetlands (marshes, lakes, rivers) and estuaries are also among the most affected systems (Moyle 1996; Williamson 1996; Ruiz et al. 1997). On the background of accelerating invasion rates, science has become increasingly interested in understanding the underlying mechanisms of biological invasions to predict invasion processes and impacts. Following a brief overview on the nature and impacts of invasions, we review different concepts regarding determinants of invasion success. We also highlight promising research areas to cope with this major threat to biodiversity in wetlands.

Ecological Studies, Vol. 191
R. Bobbink, B. Beltman, J.T.A. Verhoeven, and D.F. Whigham (Eds.)
Wetlands: Functioning, Biodiversity Conservation, and Restoration

4.2 What is a Biological Invasion?

Biological invasions are natural phenomena and simply refer to range expansions of species into new areas. This can happen, for example, by drifting dispersal stages that eventually reach new areas (e.g. Censky et al. 1998) or when biogeographic barriers disappearing over geological times, as with the Isthmus of Panama, thereby allowing the mixing of formerly separated biota (Vermeij 1991). However, what concerned Elton and today's scientists is the human-mediated disappearance of biogeographic barriers through global transport and trade. For example, human-mediated dispersal exceeds natural dispersal to the Great Lakes by nearly 50 000-fold (Hebert and Cristescu 2002). Invasions result in the presence of an increasing number of species in areas where they have not been recorded in historical times.

There is a multitude of terms used in the scientific literature to name these species: introduced, non-indigenous, non-native, invasive, exotic, naturalized, immigrant and alien species, just to name a few. Sometimes these terms are used as synonyms but more often they are (explicit or implicit) defined quite differently (Colautti and MacIsaac 2004a; Occhipinti-Ambrogi and Galil 2004). Hence unfortunately no common terminology currently exists, making it difficult to compare studies on biological invasions. The problem is further complicated by the fact that biological invasions always happen in certain steps, which make an analysis of each appropriate (Colautti and MacIsaac 2004a). However, for the purpose of this contribution, we employ a very basic definition of introduced species: Introduced species are those that have been – intentionally or unintentionally – transported by human activities into an area where they did not occur before and where they now reproduce successfully.

How does human-mediated transport happen? One possibility is of course the intentional introduction of a species by man for purposes like aquaculture and agriculture. Another possibility is the unintentional introduction as a by-product of human activities. In aquatic systems, the transport of ballast water by ships is a major vector for introduced species globally. Larvae and also some adult stages often survive transit in ballast tanks and colonize when released into new waters (Carlton and Geller 1993). Ships can also serve as vectors by carrying fouling organisms on their hulls and anchors. This happens not only on large commercial vessels but also on smaller boats for recreational purposes (Minchin and Gollasch 2003). Fouling organisms can also be transported by drilling platforms, dry docks, navigation boys, floating debris and amphibious planes. Canals are another important pathway for invaders, connecting formerly separated biogeographical regions (Bij de Vaate et al. 2002). In addition, pet, water-garden and horticulture industries are sources of (un)intended introductions by animal releases and garden escapes (Courtenay and Stauffer 1984; Ruiz et al. 1997; Maki and Galatowitsch 2004; Rixon et al. 2005; Duggan et al. 2006).

The number of introduced species can be related to the number of introduction events and to the number of individuals introduced per event. This is why wetlands, especially freshwater, estuarine and coastal environments, are amongst the most invaded ecosystems worldwide, particularly in temperate regions (Moyle 1996; Ricciardi and MacIsaac 2000; Ricciardi 2001).These ecosystems are particularly vulnerable to invasions because they are subjected to numerous introduction vectors and activities that facilitate invasions (e.g. water diversions, shipping, recreation). Primary introductions often result from the accidental transport of species, mostly by one or a limited number of vectors. Subsequently, secondary expansion of the introduced species-range follows from the initial place of establishment and normally includes a wider range of vectors acting separately or together.

4.3 Impacts of Biological Invasions

Impacts of biological invasions can be ascribed to four different classes: ecological, evolutionary, economic and human health.

4.3.1 Ecological Impacts

Impacts of biological invasions can be found on all levels of the ecological organisation: individuals, populations, species, communities and ecosystems. The mechanisms by which an introduced species can exert impacts on native species are competition, herbivory, predation, parasitism, vectoring of pathogens and through physical or chemical modifications of habitats. Especially dramatic impacts can occur when introduced species function as keystone species or ecosystem engineers, thus affecting the functional diversity and food web structure of communities and ecosystems (Vitousek 1990; Crooks 2002). A prominent example for such a dramatic effect is the Asian clam *Potamocorbula amurensis*, which caused strong changes in phytoplankton production in San Francisco Bay and thus altered the entire food web in the bay (Alpine and Cloern 1992; Kimmerer et al. 1994; Cloern 1996). Another dramatic effect is the extinction of native species due to an introduced one. For example, the introduction of Nile perch (*Lates niloticus*) into Lake Victoria is believed to have caused up to 200 extinctions of native cichlid fishes (Kaufman 1992). Similarly, Blackburn et al. (2004) showed that the probability of native bird extinction on oceanic islands was positively correlated with the number of introduced predatory mammal species present. Indeed, introduced species are the second-leading cause of species extinction, following habitat destruction (e.g. Wilcove et al. 1998; Ricciardi and Rasmussen 1999). While the severity of introduction-induced species extinctions is hotly debated

(Gurevitch and Padilla 2004; Ricciardi 2004), it is clear that introduced species can have strong ecological impacts on native biota.

4.3.2 Evolutionary Impacts

Evolutionary impacts arise mainly by removing or lifting natural dispersal and reproduction barriers that prevented hybridization between closely related invading and native species (Mooney and Cleland 2001; Grosholz 2002; Olden et al. 2004). This prevents allopatric speciation and could restrict future biodiversity. It also alters genetic diversity and could result in outbreeding depression, one consequence of which could be reduced disease and parasite resistance. Besides these direct effects of invaders on the genetic architecture of native species, indirect effects may occur through changes in selection pressure on native populations.The introduction of predators can also affect selection for size, growth, distribution and behaviour of native species (Vermeij 1982; Trussel 1996). As an example, introduced predatory fishes often result in altered habitat use by prey species.

4.3.3 Economic Impacts

Economic impacts result from the effects of introduced species on native biota, as well as funds expended to control or mitigate invasion effects (Perrings et al. 2001). Expenses may include quarantine, control or eradication programmes. In aquaculture and agriculture, introduced species can reduce harvest quantity or quality and require costly prevention or eradication methods. Introduced species can also affect technical constructions: preventing the clogging of power plant water pipes by fouling zebra mussels *Dreissena polymorpha* costs about \$ 8×10^6 per year in Canada (Colautti et al. 2006a). The wood-boring bivalve *Teredo navalis*, the shipworm, causes severe problems with harbour constructions and ships in ports worldwide and causes about U.S.\$ 205×10^6 per year damage in the USA (Cohen and Carlton 1994). In addition, in one of the most thorough economic assessments ever, Knowler and Barbier (2000) estimated that the invasion of North American comb-jelly (*Mnemiopsis leidyi*) into the Black Sea resulted in an annual 'rent' (i.e. profit) decline from U.S.\$ 17×10^6 to U.S.\$ 0.3×10^6 for the anchovy fishery of the region. In the United States, the purple loosestrife (*Lythrum salicaria*) caused an estimated damage of U.S.\$ 45×10^6 per year, exotic aquatic weeds (53 species!) cost U.S.\$ 110×10^6 per year, the European green crab (*Carcinus maenas*) U.S.\$ 44×10^6 per year, the zebra mussel (*D. polymorpha*) U.S.\$ 1000×10^6 per year and the Asian clam (*Corbicula fluminea*) U.S.\$ 1000×10^6 per year (Pimentel et al. 2004). Extrapolations of damage and control costs for introduced species typically run into billions of dollars per year for individual countries (Pimentel et al. 2000, 2001, 2004; Colautti et al. 2006a).

4.3.4 Human Health Impacts

Human health can be affected by introduced species either directly by introduced nuisance species or indirectly by introduced species acting as vectors for human pathogens. For example, the Yellow fever (*Aedes aegypti*) and Asian tiger (*Aedes albopictus*) mosquitoes were introduced to North America and also to other localities worldwide. Asian tiger mosquitoes entered via residual water contained in scrap tyres from Japan. Besides being a nuisance species, they also serve as a vector for a range of human diseases like mosquito-borne encephalitis viruses as well as the dengue virus and possibly the West Nile virus (Moore et al. 1988; Mitchell et al. 1992; Gratz 2004; Juliano and Lounibos 2005). Another example is the unintentional introduction of the fresh water snail *Bulinus truncatus* into Ghana and Jordan, together with the trematode parasite *Schistosoma haematobium* which utilises these snails as first intermediate host and which causes schistosomiasis in humans (Burch et al. 1989; Hunter et al. 1993). Introduced species may operate in concert with other introduced or native species to impact humans, wildlife or ecosystems (Juliano and Lounibos 2005).

4.3.5 Measuring Impacts

A large gap exists in our knowledge of impacts for the majority of introduced species. Most reports on impacts are anecdotal and experimental studies are extremely rare. Part of the problem, of course, is that interest in an introduced species typically arises only after it is already abundant and has caused notable impacts. Hence, before–after, control–impact (BACI) study designs, are difficult to employ (Osenberg et al. 1994; Underwood 1994). Nevertheless, solutions for this methodological problem could be gained through comparisons between already-affected and not-yet-affected local areas and through experimental introductions in contained or quarantine areas. Better spatial and temporal considerations in biological introductions studies are required specifically; and impact studies on larger spatial and temporal scales from studies that involve more than one invading species are also required (Grosholz 2002).

When conducting impact studies, it is important to clarify what constitutes an impact. This is especially helpful if we want to compare the severity of impacts by different species or impacts of the same species at different localities. So far, no common measurement of impact is available. Parker et al.

(1999) provided a formula (Eq. 4.1) to calculate the overall impact (*I*) of a species, as:

$$I = R \times A \times E \tag{4.1}$$

with *R* being the range size, *A* the abundance and *E* the effect per individual or biomass of the invader.

Although helpful in conceptualizing factors that influence impact, the formula is not explicit with respect to the currency to measure effects or how one could compare effects of different species. This field of research needs to be explored in more detail.

The most crucial point with biological invasions is the evaluation of observed impacts. Although science can assess changes to native biota from introduced species, the evaluation of this impact is not a matter of science alone. For example, it is not the role of the scientist to determine whether invasions are good or bad, nor whether introduced species are 'evil'. Such prejudgments have been related to xenophobic and racist motives (e.g. Simberloff 2003). Whatever the motives, scientists can only report changes induced by invaders. Whether these changes are good or bad is a matter of public and political debate. Introduced species are not by definition negative and unwanted, as most crops grown today could be classified as introduced. There is common agreement that they are beneficial. The scientist thus focuses attention principally on negative impacts of introduced species.

4.4 Examples of Biological Invasions

Typical pathways of invasions and impacts of introduced species in wetlands are exemplified by four examples from an internationally important coastal wetland, the Wadden Sea in Europe (Fig. 4.1). *Crassostrea gigas*, the Pacific oyster, was introduced to the Wadden Sea for aquaculture decades ago, but rapidly spread from culture plots to native mussel beds; and this pattern of spread now threatens native communities (Reise 1998). Since oysters are effective filter feeders and occur locally in high abundance, they exert strong trophic effects. With oyster aquaculture, the associated epifauna and infauna species, including algae and tunicates which themselves are non-native species, have unintentionally been introduced (Wolff and Reise 2002). Another prominent example of unintentional introduction *via* aquaculture is provided by *Crepidula fornicata,* the American slipper limpet, which arrived in Europe with early imports of American oysters in the 1870s. This species is now superabundant in parts of its range, causing high economic impacts on oyster and scallop fisheries, especially in France (Blanchard 1997; Thieltges et al. 2003). Ecological impacts are still disputed but in blue mussels (*Mytilus*

Fig. 4.1 Four examples of introduced species and their invasion pathways from a single coastal wetland ecosystem, the European Wadden Sea. **A** An intentional aquaculture introduction, *Crassostrea gigas*, the Pacific oyster, forming dense oyster beds. **B** An unintentional co-introduction with aquaculture, *Crepidula fornicata*, typical stack of American slipper limpets on their blue mussel (*Mytilus edulis*) substrate. **C** An introduction as larvae in ship ballast water, *Ensis americanus*, shells of the American razor clam. **D** An intentional introduction for coastal protection, *Spartina anglica*, cordgrass

edulis), a major substrate for slipper limpets in the Wadden Sea, the species strongly reduces survival and growth and thus affects native mussel beds and the mussel fishery (Thieltges 2005). Besides aquaculture, ballast water of large vessels is another major vector for introduced species in wetland ecosystems. By this mode of translocation *Ensis americanus*, the American razor clam, arrived as larvae in 1978 in the German Bight. From there it spread very rapidly along European coasts, where it now constitutes the major fraction of subtidal biomass in many areas of the Wadden Sea. Impacts on native biota, however, are difficult to assess since sampling of the very fast burrowing clams is difficult. However, since it is a very abundant filter feeder, trophic effects can be expected (Armonies and Reise 1999; Armonies 2001). Other human activities, like land reclamation and coastal protection, can also serve as introduced species vectors to wetlands. Originally introduced for coastal protection in the 1920s, the cord grass *Spartina anglica* (a fertile hybrid between *S. maritima* and *S. alterniflora*) spread on the upper tidal zones of tidal flats and in the seaward ends of native salt marshes, often forming an almost monotypic belt. The plants retain sediment and displace native species like *Salicornia stricta*, glass word, *Zostera noltii*, seagrass, and *Arenicola marina*, the lugworm, and its associated species (Reise et al. 2005).

All four species co-occur in the Wadden Sea, indicating the high abundance of invaders in a single wetland ecosystem and the multiple pathways of introduction. Along North Sea coasts, more than 80 introduced species have been identified, of which about 40 occur in the Wadden Sea (Reise et al. 2002). The full array of impacts of the invaders remains to be investigated.

4.5 Understanding and Predicting Biological Invasions

Before we can predict invasions, we have to understand the underlying mechanisms of invasion and determine whether there are general patterns or whether every invasion is a unique event. Three different questions can be asked in predicting biological invasions: (1) which species will successfully establish? (2) when, where and how fast will a species invade? and (3) what will be the impact of an introduced species? Different approaches have been utilized to address these questions:

1. Determine characteristics of the invading species that make it invasive.
2. Determine characteristics of the ecosystem that make it susceptible to invasion.
3. Investigate the relationship between invader and invaded ecosystem (key-lock approach). In this approach, it is assumed that the invading species is successful when its characteristics are suited to the specific conditions in the ecosystem that is to be invaded.
4. Explore the invasion process by addressing different stages that an introduced species pass through in order to successfully invade an ecosystem. This approach is based on the premise that during every phase, the characteristics of the invader have to fit the specific conditions of the ecosystem.
5. Comparative historical approach: finding similarities and differences in invasion dynamics across ecosystems and extrapolating them across species or systems. This remains an important objective for invasion ecology, since experimental approaches are often impractical or unethical (Unmack and Fagan 2004).

4.5.1 Invading Species Approach

Early research on invasions assumed that introduced species which spread must have certain characteristics making them successful (Rejmanek and Richardson 1996; Mack et al. 2000; Richardson et al. 2000). For example, among the attributes that have been ascribed to a successful plant invader are a high dispersal rate, early fruiting, high fecundity, high growth rate, vegetative reproduction and a high tolerance of a wide range of physical conditions (Lodge 1993). For animal invaders, features of successful invaders that have been frequently cited include: early sexual maturity, asexual reproduction, large size, ability to colonize a wide range of habitat types high genetic variability, high phenotypic plasticity, gregarious behaviour, omnivory, high reproductive capacity, high dispersal rates and an association with humans. Theoretical arguments to detect invasiveness characteristics in species that establish or proliferate in new habitats are given by many authors (e.g. see Colautti et al. 2006b and literature therein).

Using these attributes to explain why certain introduced species overrun natives while others do not is not always possible. None of the characteristics is by itself suited to discriminate between invasive and non-invasive species. These attributes could instead be interpreted as those of an 'ultimate invader' and it may well be assumed that species that possess more of these characteristics are more likely to be successful invaders. The 'invading species approach' is still a frequently used method to predict invasion.

Williamson (1999) pointed out serious disagreements among three studies since 1995 that sought common traits among Britain's invasive plants. One study found that large seeds favoured invasions, in a second the opposite was found, while the results of a third indicated that the size did not matter. Likewise, Colautti et al. (2006b) found little support for the concept that invasive species possess greater physiological tolerance than their non-invasive counterparts.

Propagule Pressure

A species' invasiveness can also be increased by merely repetitive and extensive introduction. Robert Pemberton, a weed scientist with the USDA Agricultural Research Service in Fort Lauderdale, Florida, recently examined old catalogues from the Royal Palm Nurseries, a well known, trend-setting company that bred and sold plants in Manatee County, Florida, from 1881 to 1937. He discovered that plants sold for just one year had only a 2 % chance of establishing in the wild, while popular species that were in the catalogue for over 30 years had a 69 % chance of establishing. These results are consistent with an introduction effort (or propagule pressure) hypothesis to explain invasion establishment success (Andow 2003).

Before physiological factors and positive or negative biological interactions can affect invasion success, viable propagules must arrive at the novel habitat (Lockwood et al. 2005). Colautti et al. (2006b) found that propagule pressure as well as anthropogenic activities were the best predictors of invasion success. Propagule pressure (measured as the number of introductions and the number of propagules introduced) was considered in only 29 % of studies, yet was a consistent predictor of both invasiveness and invasibility in 55 out of 64 cases. Considering that non-indigenous species are almost always introduced non-randomly, propagule biases caused by a failure to consider propagule pressure may confound current paradigms in invasion ecology (Colautti et al. 2006b). Colautti et al. (2006b) concluded that propagule pressure should serve as a null model for studies of biological invasions when referring process from patterns of natural invasions. Duggan et al. (2006) state that propagule pressure is frequently mentioned as an important determinant of invasion success for terrestrial taxa, but its importance for aquatic species is unclear. They demonstrated, however, that clear relationships exist between the frequency of occurrence of aquarium fish in shops and the likelihood of introduction and establishment. Introduced and established fish caught in the wild were also typically larger than in aquaria, in accordance with the fact that these fishes were mostly released when they outgrew their aquaria. Reducing the number of introductions is the most simple way to reduce the number of successful invasions. Lonsdale (1999) found that the number of introduced plant species established in reserves was positively correlated with the number of visitors, implicating human-mediated propagule pressure. Introduction effort was also the best predictor of invasion success for introduced birds in New Zealand (Forsyth and Duncan 2001). Given the importance of propagule pressure to invasion success for so many different taxa, it would be reasonable to conclude that the best way to reduce future invasions would be to control the number of introductions.

4.5.2 Invaded Ecosystem Approach

Species Richness: Biotic Resistance Model

A long-standing theory to predict invasion success is that ecosystems with high native diversity – with their dense, interconnected webs of ecological relationships – can resist invasions better than those with fewer species (Elton 1958; Lodge 1993; Fig. 4.2). Subsequent works on species packing in a community by Robert MacArthur (1970) and others were consistent with this concept. Why should diversity affect invasibility? One might expect more diverse assemblages more fully utilize the available resources, thus leaving less space for individuals of new species (Levine and D'Antonio 1999). This is an extension of the 'empty niche' concept in ecology.

Stachowicz et al. (1999) provided support for this theory. He created artificial marine epifaunal ecosystems with anywhere from zero to four North Atlantic sessile suspension-feeding invertebrates, with all systems beginning with the same initial cover of native species. Space was the limiting factor in these systems. He then introduced a known invader, the Pacific tunicate

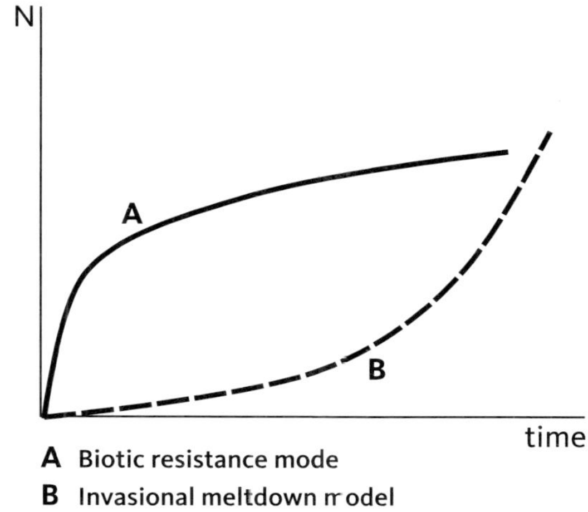

Fig. 4.2 Temporal trend in the cumulative number of successful invasions, as predicted by the biotic resistance model and the invasional meltdown model (after Ricciardi 2001)

Botrylloides diegensis into these communities. The more species initially present, the lower was the tunicate's subsequent survival (Stachowicz et al. 1999). In addition, Kennedy et al. (2002) found in their grassland plots that invasion success increased in plots with decreasing diversity. Species diversity appeared to enhance invasion resistance by increasing crowding and species richness in localized plant neighbourhoods. Both the establishment (number of invaders) and success (fraction of invaders that are large) of invading plants were reduced. These results suggest that local biodiversity represents an important defence mechanism against the spread of invaders. However, the resulting effects on biodiversity depend on the spatial scale considered: while local species richness increases with species introductions, globally the opposite effect can be observed (Sax and Gaines 2003).

The invasion processes on islands, which usually have fewer species than comparable areas of mainland, offer some support for this theory (Elton 1958). Islands are often the most heavily invaded. However, a large problem with studies that find that invaded communities are less diverse than their non-invaded counterparts is that the results are likely to reflect the impact of the invader on the native diversity and not that of diversity on the invader. In other words, cause and effect may be confounded. More diverse communities are also more likely to contain species with a strong impact on invasibility, whether deterring or facilitating invasions. How invasibility varies with diversity in natural systems is further complicated by the fact that variation in diversity is controlled by and thus covaries with disturbance, resource availability, physical stress, competitors and consumers. Also, low-diversity patches may often be the least invasible when the low diversity results from competitive dominance by one or a few species (Levine and D'Antonio 1999). Some recent studies contradict the island:continent paradigm. For example, Lonsdale (1999) determined that island:mainland nature accounted for only 13 % of the variance in introduced species richness, while native species diversity and habitat area accounted for much more variation (23 % and 22 % respectively). Similarly, Blackburn and Duncan (2001) found that island:mainland status were unable to explain patterns in the invasion success of introduced birds on islands; and they concluded that there was little support for the biological resistance hypothesis. Simberloff (1995) argued that islands were more heavily impacted by invasions (rather than being more invaded) than comparable mainland areas because the invasions often involved certain taxa (e.g. predatory mammals) that were missing entirely from islands and that these absences predisposed the islands to greater impact once those species were introduced.

Invasion theories about resistance have always been made on the basis of island-like or very small-scale invasions. Island-like models cannot be generalized to systems in which species arrive in greater frequencies or densities similar to the case on the mainland (Hewitt and Huxel 2002). Many assumptions are made when building assembly models which try to explain and predict invasions. Some of these assumptions are:

1. The Lotka–Volterra equilibrium is present in the invaded ecosystem, which assumes well mixed populations, so that all species interact with a potential invader. This restricts the applicability to the neighbourhood scale at which all component species interact.
2. The interactions of the invader are similar to those of the resident species.
3. A community is invasible if the invader persists at equilibrium. Invaders that fail to persist in communities at equilibrium may still strongly impact these systems.
4. In all of the models, the communities have reached equilibrium prior to the invasion, a state rarely reached by natural systems (Levine and D'Antonio 1999).

Hewitt and Huxel (2002) utilized assembly algorithms similar to those used in previous studies (the Lotka–Volterra model from Drake 1990), which made several more assumptions including: (5) only one species invaded per time-step and (6) invasions occurred at low densities relative to Lotka–Volterra equilibrium densities. In this study, they allowed either two species to invade per time-step (assumption 5 tested), or single species invasions at relatively high inoculation rates (assumption 6 tested). Their results showed that invasion resistance only occurred when the invasion process was restricted to single species invading at low densities (Hewitt and Huxel 2002). Moore et al. (2001) simulated plant community assemblages and tested four different mechanisms of co-existence. Their results suggest that species richness does not necessarily correlates with saturation (number of niches occupied) and is a poor predictor of invasion resistance.

Levine and D'Antonio (1999) reviewed theoretical and empirical work linking species diversity and invasions. They found that it was not possible to prove experimentally the existence of biotic resistance; and empirical studies have not produced consistent results. Similarly, Ricciardi (2001) found that there is no convincing evidence that rich aquatic communities resist invasion when organisms are introduced frequently. If abiotic conditions are suitable and dispersal opportunities exist, aquatic species will likely invade regardless of the composition of the resident community (Ricciardi 2001). Therefore, as human activities and transports continue to decrease the isolation of regional floras and faunas from each other, the numbers of species and densities of introduced invaders will increase and result in higher susceptibilities to invasion in recipient ecosystems (Hewitt and Huxel 2002). For example, the toxic dinoflagellates (including *Alexandrium catenella* and *Gymnodinium catenatum*) were spread across Australia by ballast water, while the mussel *Mytilus galloprovincialis* and even some fish species also used ships for travelling around the world (Apte et al. 2000; Lockett and Gomon 2001). Carlton and Geller (1993) termed the mass transfer of species globally in ballast tanks 'ecological roulette'.

Facilitating Invasions: 'Invasional Meltdown' Model

Up to this point, the focus has been on biotic resistance: negative interactions between native and non-indigenous species, mostly resource competition and interference. It is also possible, however, for the opposite process, facilitation, to increase invasion rate as communities become more diverse (Bruno et al. 2003). This means that a species already established in a community (native or non-native) facilitates invasion by a second species. If this is the case with two non-native species, invasional meltdown may occur, with every new introduced species making it easier for subsequent ones to establish. This results in an accelerating accumulation of introduced species rather than a deceleration, as envisioned in the biotic resistance model (Simberloff and von Holle 1999; Fig. 4.2).

There is a great variety of ways in which different non-native species can facilitate one another. For example, animals pollinate and disperse plant seeds; and animals or plants may modify habitats in ways that prove beneficial to invading species (Simberloff and von Holle 1999). Introduced species frequently interact with each other and facilitative interactions are at least as common as negative ones (Simberloff and von Holle 1999). Ricciardi (2001) found that direct positive (mutualistic and commensal) interactions among introduced species are more common than purely negative (competitive and amensal) interactions. In addition, many exploitative (predator–prey) relations appear to be strongly asymmetric in benefiting one invading species at a negligible cost to another. Combined with an increasing invasion rate in the Great Lakes, this information supports the invasional meltdown theory. The invasion history of the Great lakes is explained better by the invasional meltdown model than by the biotic resistance model. Further work is required to identify the generality of meltdown phenomena (Simberloff and Von Holle 1999). If biotic resistance to establishment exists only at a low frequency of attempted introductions, then invasional meltdown might have a threshold effect of inoculation pressure, beyond which the phenomenon may occur often. Such a case may apply with vectors that transport large numbers and large populations of species, as is the case with ballast water. In addition, biotic resistance may operate only at small spatial scales (Levine and D'Antonio 1999), while invasional meltdown could operate at multiple scales, ranging from communities to ecosystems (Simberloff and Von Holle 1999). One caution is warranted when considering the possibility of 'invasional meltdown'. Serial introductions of species from the same source pool may produce the same pattern in new areas, even with non-interacting species, as produced by invasional meltdown. Thus investigators ought to demonstrate that the species interact in a facilitative manner prior to accepting this hypothesis.

Abiotic Resistance Model

Until now, the focus has been on biotic resistance (or facilitation) rather than abiotic resistance. Moyle and Light (1996) argued that, in the case of fish invasions, abiotic resistance plays a much larger role in invasion processes than biotic resistance. They predicted that the most successful invaders in streams will be those adapted to the local hydrologic regime and mentioned several examples in North America. They also stated that all fish assemblages are invasible regardless of the diversity of the resident fish assemblages, suggesting an important role for propagule pressure. Invasive fish species have become established in a range of streams, from those with initially no fish, to those with complex fish assemblages (Moyle and Light 1996).

Abiotic resistance was a focus of Lennon et al. (2003), who investigated the impact of the supply rate of limiting nutrients (trophic state) on the invasion success of an exotic zooplankton species (*Daphnia lumholtzi*) in native plankton communities. This species is an aggressive invader in North America, primarily in reservoirs in the southern United States, which tend to be more eutrophic than lakes and reservoirs in the north. The authors initially tried to link the establishment success of *D. lumholtzi* with the trophic state of the system, assuming that a higher nutrient concentration increases the chances of successful establishment. Surprisingly, *D. lumholtzi* was associated with systems that had low nutrients, low zooplankton biomass and high zooplankton species diversity. The authors proposed that invasion success of *D. lumholtzi* was inhibited by another cladoceran species, *Chydorus sphaericus*, thereby implicating an interaction between trophic state, community composition and invasion success. Havel et al. (2005) proposed that a combination of factors, including physicochemical properties, connectivity and high levels of disturbance, have facilitated the spread of *D. lumholtzi* through reservoirs in the United States.

Disturbance (Human or Natural)

Disturbance is the process that leads to partial or total destruction of the individual or group of individuals. Disturbance is a broad term with many components and magnitudes and with much temporal variation. Often, it is the process leading to bare ground or areas of altered light conditions; and consequently disturbance is another factor that may influence invasibility. Many researchers have found, for example, that non-native species colonize disturbed areas more than pristine ones. Harper (1969) suggested that disturbance is the main determinant of plant community structure, with the important element of time influencing the level of diversity able to establish after the perturbation.

Moyle and Light (1996) predicted that successful invasions in aquatic systems are most likely to occur when native assemblages of organisms have been temporarily disrupted or depleted; as such a much wider range of species can invade in aquatic systems with high levels of human disturbance than in systems with a low level of disturbance. They mention that reservoirs (high-disturbance systems) on California's rivers often support three to four species of predatory black bass (*Micropterus* spp), while rarely more than two species are found together in natural systems. Although Lozon and MacIsaac (1997) found no support for disturbance facilitating animal invasions, there was strong support for plant communities. Colautti et al. (2006b) found that disturbance was significantly associated with invasibility of plant communities. Pollution can be considered as a disturbance leading to underexploited resources. Van der Velde et al. (2000, 2002) noticed that invasions by alien macroinvertebrates in the River Rhine occurred after a period of pollution followed by water quality improvement and these invasions accelerated again after a chemical spill followed by further water quality improvement. Strayer et al. (2005) found also that restoration efforts of large-river ecosystems led to systems heavily invaded by alien species, based on a study in the Hudson River.

Fluctuating Resource Availability

The "fluctuating resource availability" hypothesis states that a community becomes more susceptible to invasion whenever there is an increase in the amount of unused resources (Davis et al. 2000). This increase in resources occurs either by reduced consumption and/or by increased availability (Fig. 4.3).

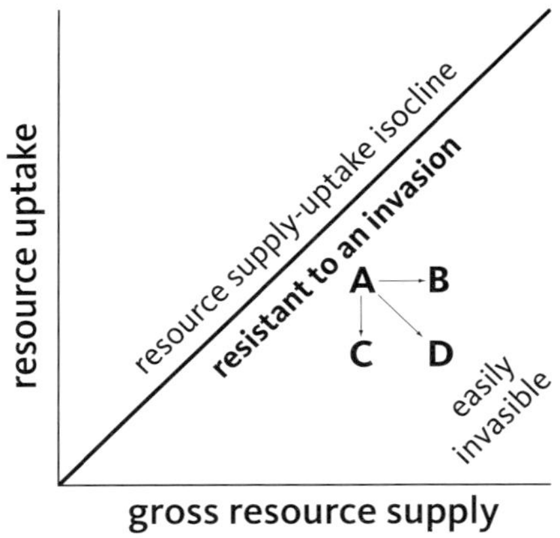

Fig. 4.3 The theory of fluctuating resource availability holds that a community's invisibility or susceptibility to invasion increases as resource availability (the difference between gross resource supply and resource uptake) increases. Resource availability can increase due to a pulse in resource supply (A→B), a decline in resource uptake (A→C) or both (A→D). In the plot shown, resource availability, and hence invasibility, increases as the trajectory moves further right and/or below the supply/uptake isocline (where resource uptake = gross resource supply). Modified after Davis et al. (2000)

Fluctuation in resource availability is identified as the key factor controlling invasibility. The theory is mechanistic and quantitative in nature, leading to a variety of testable predictions, some of which are mentioned here.

1. Environments subject to pronounced fluctuations in resource supply will be more susceptible to invasions than environments with more stable resource supply rates.
2. Invasibility will increase following disturbances, disease and pest outbreaks that increase resource availability by increasing resource supply and/or reducing the rate of resource capture by the resident vegetation.
3. There will be no necessary relationship between the species diversity of a plant community and its susceptibility to invasion. Since complete exploitation and very incomplete exploitation of resources can occur in both species-rich and species-poor communities, there is no reason to expect any consistent pattern between species richness or diversity of a community and its susceptibility to invasion.
4. There is no general relationship between the average productivity of a plant community and its susceptibility to invasion because competition intensity has not been found positively correlated with community productivity (Goldberg et al. 1999).

Resource availability may be variously high or low in either productive or unproductive habitats, with invasibility also then being either high or low in these habitats. It is not yet clear how this theory can be applied to animals and to what extent behaviour reduces the dependence of animal invaders upon a supply of unused resources. Davis et al. (2000) concluded that the elusive nature of the invasion process is caused by the fact that it depends upon resource enrichment of the habitat or release of resource usage by the inhabitants of the habitat. These two reasons occur only intermittently. Whether or not invasion actually occurs is a function of propagule pressure and the characteristics of the invading species. If a species is totally unfit for a certain environment, then the invasibility of that environment may be great but the species will never invade. Characteristics in an invading species that make it suited for the environment will support its invasiveness in a correspondingly vulnerable ecosystem (Heger and Trepl 2003). This interaction needed for a successful invasion is discussed in the next paragraph.

4.5.3 Relationship Between Invader and Invaded Ecosystem (Key-Lock Approach)

Both the invading species approach and the invaded ecosystem approach are concerned with the properties of only one component of a biological invasion. There seem, however, to be cases in which neither of these general characteristics sufficiently explains the success of an invasion. In these cases, the relationship between features of the ecosystem and characteristics of the invader is crucial for the invasion success. When there are empty niches in an ecosystem, an invader with the right characteristics to fit into such a niche has the possibility to be successful, assuming a sufficient supply of propagules. The concept of the niche was developed decades ago as an explanation for how species coexist in ecological communities.

It is difficult to determine exactly the range of a certain species' niche (including all possible niche axes) and it is still harder to quantify niche overlap between species (Lohrer 2001). Even when these niches are quantified, community composition cannot be predicted. There are other factors influencing community composition, including competition, disturbance and environmental variability, predation and timing and number of invading propagules. Even with these limitations, niche theory has value when applied to modern ecological investigations. Lohrer (2001) created a hierarchical niche-based framework (NBF) that seeks to explain *a postiori* why particular species successfully invade habitats (Fig. 4.4). The models are differentiated hierarchically by: (1) the amount of niche overlap in the invaded habitat, (2) the type/amount of change in niche breadth and (3) the nature of resource limitation. Ideally, every new invasion should be explained by only one of the models in the framework. Lohrer (2001) proposed that, when the niche-based framework is widely applied, invasion ecologists may begin to elucidate patterns necessary for making predictions. Binggeli (1997) suggested that most of the highly successful invaders do not have an ecologically equivalent species in the ecosystem invaded. Thus, ecosystem characteristics may be important to the success of the invader, who in turn must possess the 'right' characteristics to invade a particular ecosystem. This is what the relational key-lock approach comprises (Heger and Trepl 2003).

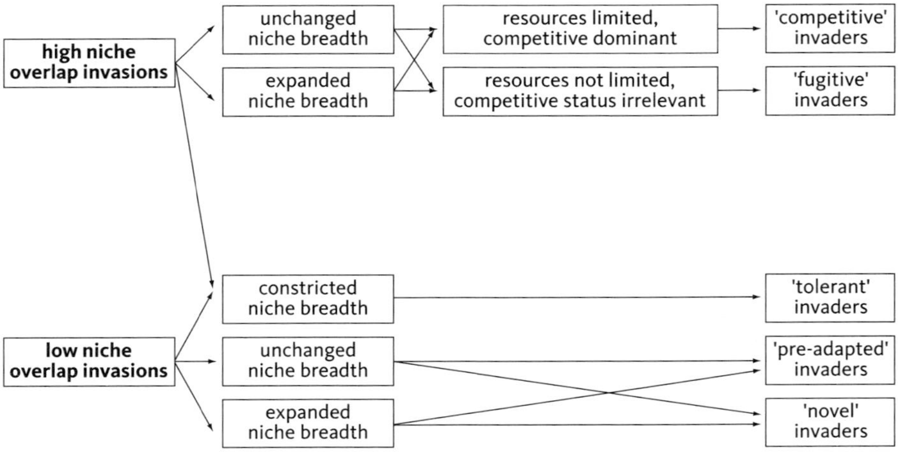

Fig. 4.4 Hierarchical niche-based framework (NBF). The models are differentiated according to: (1) the amount of niche overlap in the invaded habitat – the uppermost level, (2) the type/amount of change in a niche breadth between native and invaded habitats – the second level and (3) the nature of resource limitation – the final level. There are ten possible paths, each representing a unique invasion scenario. Modified after Lohrer (2001)

A major disadvantage of this approach is that it does not take into account factors for community composition other than the niche of a species, such as the increase in probability of establishment with number of release events or number of individuals released (Grevstad 1999). Demographic factors whose influence depends on population size or density such as demographic stochasticity, Allee effects and genetics also play an important role in the establishment of invading populations (Grevstad 1999). In addition, escape from natural enemies (predators, parasites) may influence invasion success (Binggeli 1997), although authors must be careful in ascribing invasiveness to 'enemy release' (Colautti et al. 2004b).

What happens, however, if a species manages to invade in a food web which already contains an equivalent species? Would one be redundant and driven out by competition or would both manage to coexist? In computer terms, redundancy means: the provision of multiple interchangeable components to perform a single function in order to cope with failures and errors. Translated to ecosystems, it could be the co-existence of equivalent competitors having the same or largely overlapping niches. Loreau (2004) suggested that neutral co-existence of equivalent competitors, non-linear per capita growth rates and lack of correlation between functional impact and biomass may provide the basis for the existence of functional redundancy in natural ecosystems. Difficulty of determining the existence of functional redundancy is that species may appear to perform the same function under a restricted set of conditions, yet their functional roles may vary in naturally heterogeneous environments (Wellnitz and LeRoy Poff 2001).

Duncan and Williams (2002) compiled a list of 1511 native plant species of New Zealand that occur in genera with at least one introduced species. They separated the introduced genera into those containing at least one naturalized (introduced from another region and persisting without cultivation) species and those without and also into those containing at least one native species and those without. A noticeably smaller number of naturalized genera were found to contain native species but the naturalization rate was much higher for these same genera (45 % in genera with native species, 18 % in genera without natives). Within families, genera containing at least one native species showed a significantly higher rate of naturalization than genera without native species. This contradicts Darwin's naturalization hypothesis. A possible explanation also considered by Darwin is that introduced species with native congeners are more likely to share features with them that allow them to survive in New Zealand, compared with introduced species that lack those relatives here. These shared traits may pre-adapt the plants to their new environment, helping to outweigh the potential disadvantage of stronger competition from close relatives (Duncan and Williams 2002). This might also be the case for animals. An example comes from Declerck et al. (2002), who researched patterns of diet overlap between populations of introduced and native fishes in shallow ponds. Chironomid larvae were the most important food source for all species but, by differential consumption of chironomid size classes, niche differentiation between introduced fishes was attained. In contrast, high diet overlap was found between the functional groups of one of the indigenous species and several introduced fishes, indicating a high potential for interspecific exploitative competition (Declerck et al. 2002). The introduced species established a successful self-sustaining wild population in a place with congeneric native species, which is opposite to that expected from Darwin's naturalization hypothesis.

4.5.4 Invasion Processes Differentiated in Time

The approaches described in the previous paragraphs explain several aspects of invasions, but they fail to regard invasions as processes. The characteristics of invading species, the invaded ecosystems and the interplay of both are in most cases derived from the analysis of some aspects of the invasion process, ignoring that various steps of an invasion may differ in quality and thus every step may require different characteristics from the invader (Heger and Trepl 2003).

The model of steps and stages (Heger and Trepl 2003) visualizes the invasion process as a staircase, with stages that are derived from a chronological differentiation of an idealized invasion process (Fig. 4.5A). The steps differ in

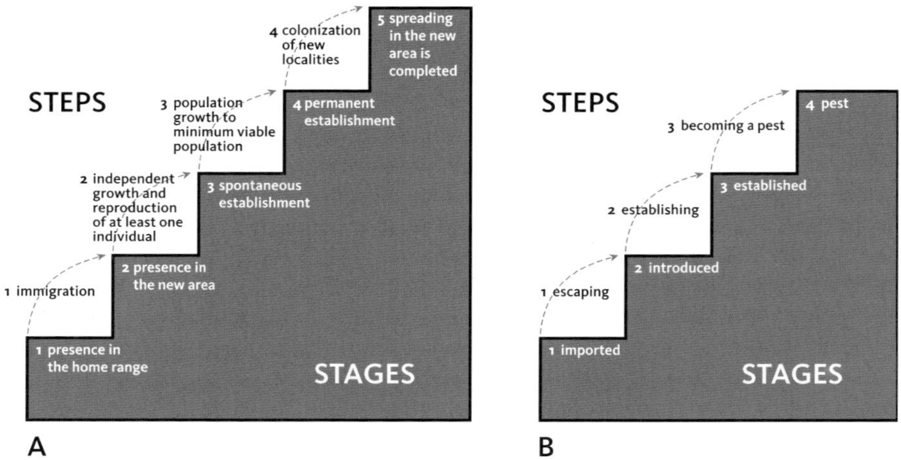

Fig. 4.5 A Chronological discrimination of an idealized invasion process into steps and stages. Different stages are reached by overcoming a sequence of steps in the course of the invasion. The height of each step depends on the ability of the species to overcome environmental limitations. Modified after Heger and Trepl (2003). **B** Chronological discrimination of an invasion process into steps and stages. *Imported* Species found in collections or accidentally brought into the country, *introduced* (or casual) found outside control or captivity as a potentially self-sustaining population, *pest* with a negative economic effect. Based on Williamson and Fitter (1996)

perspective: the first and second steps consider single individuals and investigate the possible difficulties facing an individual and the third step considers the founder population, which is confronted with problems on different levels. The fourth step concerns all new populations of the invading species and therefore considers problems of yet another quality (Heger and Trepl 2003). A successful species must pass all the different stages, taking all the steps and facing different problems at each step. Many introductions fail at the lower steps, never reaching the upper steps. The model aims to serve as a tool to define the difficulties facing a plant invader, but also pertains to animal invaders. However, the possible threats and characteristics which increase the probability of a successful invasion must be adapted for animals. It is very much focused on the invader and less on the invaded ecosystem. This is not necessary and can easily be changed: for instance, the model of biotic resistance can be incorporated into the steps and stages model. Biotic resistance is hypothesized to occur at two stages of invasion: at the establishment stage (stage 2: independent growth and reproduction) and, if establishment is successful, during subsequent population growth, when the abundance of the invader (and thus its community-wide impact) is limited by resident species richness (step 3: population growth to minimum viable population; Levine and D'Antonio 1999).

Kolar and Lodge (2001) defined four transitions (steps in the invasion sequence) in their model of chronological discrimination: transportation, release, establishment and spread. To begin the invasion process, a species must be moved away from its home range and transported into a new environment by a (human) vector. With each sequential transition, a large proportion of species fails (Kolar and Lodge 2001). When those barriers have been overcome, the species starts interacting with the invaded ecosystem. These interactions, along with other factors, determine whether that species will establish itself in the new environment (attaining a self-sustaining population). When the species does not spread from its point of establishment, it is called non-invasive; when it does, it is called invasive. This distinction is however quite imprecise because spread is partly a function of time since establishment, which is not always known. Small population size is important both to endangered species and invasive species. In both cases, consecutive periods of high mortality can drive the population (species) to extinction; this effect can be exacerbated by (and contributed to) demographic stochasticity (Mack et al. 2000). For example, having inappropriate sex ratios (for dioecious species) or age classes dominated by non-reproductive individuals could doom an introduced population even if environmental conditions were highly suited for the species' survival. However, even if sufficient propagules of appropriate sex ratio and reproductive status are introduced, successful establishment is still not guaranteed. Environmental stochasticity also affects colonizing populations and may cause invasion failures (Crawley 1989). The combined effects of demographic and environmental stochasticity may render invasion success very difficult to predict using classic, deterministic approaches. Indeed, stochastic population models may perform better in predicting invasion success than can deterministic models when inoculum sizes are very small (Crawley 1989). The model of Kolar and Lodge (2001) is meant to distinguish between stages in the invasion process because they are likely to differ in quality. Simultaneous examination of all transitions may hinder the detection of patterns. The definition of the transitions and of associated stages shown in the scheme remains unclear. The scheme does not explicitly discriminate between stages and transitions (Heger and Trepl 2003). The stage: 'survives transport and introduction' may be split up into two substages: 'survives transport' and 'survives introduction'; and, since these two stages both face different threats, a species may be adapted for transport and not for introduction.

The model of steps and stages by Williamson and Fitter (1996; Fig. 4.5B) is slightly simplified compared to that of Heger and Trepl (2003) but serves well as a 'bottleneck' model for predicting where most species will fail in the invasion process. The word 'pest' however can be replaced by 'invasive' and is corresponding with the step 'spreading in the new area is completed'. In the same way, 'imported' corresponds with 'presence in the new area', 'introduced' corresponds with 'present in the new area' and 'established' corresponds with 'permanent establishment'. It seems only one step from Heger and Trepl (2003) is missing: step 2, which is spontaneous establishment.

Richardson et al. (2000) developed a model with six major barriers that a non-native species must overcome in order to become invasive (Fig. 4.6). This model has the advantage of distinguishing between the settlement of non-native species in environments made by humans and in natural environments. However, since only barriers are described, the stages in between are less clear. In this model, a species is called casual ('alien plants that may flourish or even reproduce occasionally in an area, but which do not form self-replacing populations and which rely on repeated introductions for their persistence') when it has crossed a geographical barrier. Only when the next barrier is overcome (abiotic or biotic barriers at the site of introduction) can a species may be called casual. The first barrier 'geographic' corresponds with the first step in Heger and Trepl (2003): 'immigration'. The second barrier 'environmental (local)' corresponds with the second step 'independent

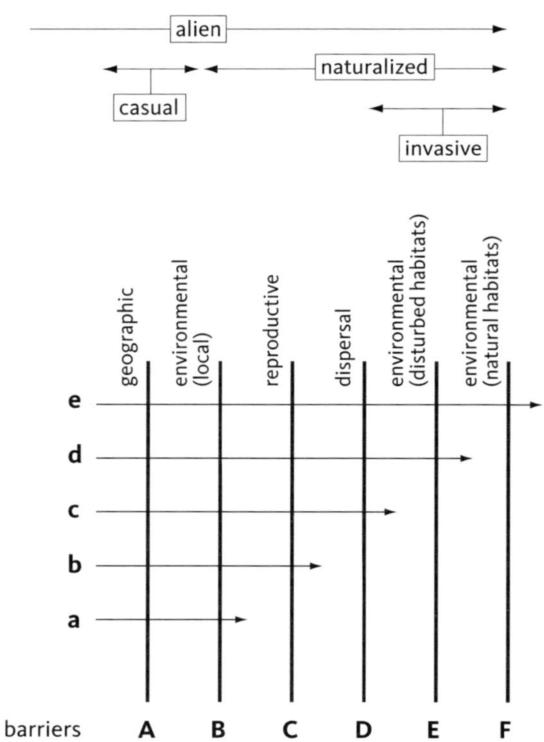

Fig. 4.6 A schematic representation of major barriers limiting the spread of introduced plants. The barriers are: *A* major geographical barriers (intercontinental and/or infra-continental, approximate scale >100 km), *B* environmental barriers (abiotic and biotic) at the site of introduction, *C* reproduction barriers (prevention of consistent and long-term vegetative and/or generative production of offspring), *D* local/regional dispersal barriers, *E* environmental barriers in human-modified or alien-dominated vegetation, *F* environmental barriers in natural or seminatural vegetation. *Arrows a–e* The paths followed by taxa to reach the different states from introduced to invasive in natural vegetation. Crossing the barriers is not irreversible: for example, climatic fluctuations can either pose new barriers, or enable the new taxon to spread. Modified after Richardson et al. (2000)

growth and reproduction of at least one individual', the third barrier 'reproductive' corresponds with 'population growth to minimum viable population' and the fourth barrier 'dispersal' corresponds with 'colonization of new localities'. In Richardson et al. (2000), however, there are two more barriers – a natural and a disturbed environmental barrier – which are actually different parts of barrier four (dispersal). Many species are better equipped to invade disturbed habitats, while others more easily invade natural habitats.

4.5.5 Comparative Historical Approach

To what extent do patterns of invasion in one region generalize to other regions? Answering this question is a key goal of invasion biology because it underlies whether we can make progress via comparative studies or must instead cope with a large set of anecdotal evidence or unrelated case studies. Prediction would be virtually impossible if the latter were true. However, historical reconstructions of the progress of invasions show remarkable similarities between diverse taxa and (eco)systems.

To illustrate this, Unmack and Fagan (2004) have investigated the Gila and Yaqui river basins (western North America) with respect to fish communities, comparing the native/non-native ratio of both basins from 1850 until today. The two basins are similar in size, physiography and ecology, but differ in the timing of regional development. Trends observed in the native/non-native ratio in both basins resemble each other (both are roughly linear), but a time lag of about 41 years exists between them. The Gila basin contains more alien fish species than the Yaqui basin; it is in a further invasion stage. Based on those data, the predictions can be made that the future of the Yaqui will echo the present-day situation of the Gila. Unfortunately, lack of this type of data and the differences in ecosystems limits the frequency that the comparative historical approach can be used.

It is questionable whether ecologists are able to identify mechanistic explanations for the phenomena observed and then be able to make predictions on the basis of universal laws. Ecological phenomena always have the component of uniqueness, particularity and historicity. Hence, a probably more suitable approach could be the so-called 'method of case studies' successfully employed in applied issues of ecology (Shrader-Frechette and McCoy 1993). The comparison of case studies allows one to identify certain patterns or analogies to be made on the ground of good knowledge on the biology and ecology of the species in question. This method seems also highly suitable to research on biological invasions and might be much more rewarding than purely mechanistic approaches (Shrader-Frechette 2001).

4.6 Shadows on the Prospects of Prediction

Several phenomena have until now escaped from prediction models. One of these is the 'time-lag' phenomenon; non-native species living inconspicuously in their new habitat for decades, until the population suddenly explodes and spreads like wildfire. Two types of lag phases may be recognized: (1) a period between the first introduction and the first spread and (2) the period preceding the switch to a significantly higher rate of population growth according to Kowarik (1995). Time-lags between the introductions of exotic plants, their establishment and subsequent spread are commonly observed in many species (Binggelli 2000). In woody plants, major changes in a biotic factor (e.g. grazing, pollinator) or an abiotic factor (e.g. hurricane, flood, logging) usually determine the duration of time-lags and there is little published evidence to support the importance of other factors and genetic adaptations in particular. Human activities and detection (e.g. interest in and ability to detect the spread of an invasive) are also common explanations for some observed time-lags (Binggelli 2000). The zebra mussel (*Dreissena polymorpha*) is another case in point. Scientists predicted its arrival into the Great Lakes from Europe via ballast water in 1981, yet no invasion occurred. It is still not clear why the animal finally invaded when it did in 1986. One possibility is a change in the environment which makes it more suitable to non-native species, although it is often unclear what that is. In the case of the zebra mussel, ironically, improving water quality may have allowed the species to survive in key European source regions and/or in the Great Lakes. Alternatively, the species may not have had appropriate dispersal opportunities from Europe until shortly before it established in the Great Lakes. Therefore, it is not clear whether the species benefited from a change in introduction effort or environmental conditions, or both (Andow 2003; MacIsaac et al. 2001). To predict time-lags, it is necessary to draw a list of species for which time-lags have been recorded elsewhere (Kowarik 1995).

Another strange phenomenon is the occasional collapse of substantial populations of invasive non-indigenous species. For example, the widespread invasive snail *Achatina fulica* and pondweed *Elodea canadensis* appear to be characterized by a rapid expansion followed by rapid decline. For the snail, disease might be the cause but, for the pondweed, the decline is yet unexplained (Simberloff and Gibbons 2004). Zebra mussel (*D. polymorpha*) populations in European systems may be differentiated into those with relative stable, long-term population dynamics and those that exhibit great variance in interannual population density (Ramcharan et al. 1992). Possible reasons for the periodic crashes include density-dependent endogenous cycles, predators, parasites or pathogens.

4.7 Conclusion

Are biological invasions unique events or can we find certain patterns that make predictions possible? Invasion biology is so intricately linked with human history, global trade and ever-decreasing transport barriers that a historical perspective is essential to understand the ecological process at hand. One should also learn from the physiology, behaviour and previous invasions of a possible invader to estimate the invasibility of valuable ecosystems. Chronological discrimination could be a useful tool in predicting invasions, especially when there is more research into the different threats posed by the different stages and more use of existing models on animal invaders. The key-lock approach is very useful as an explanatory tool. Predictions are still quite inaccurate and have an observational nature. Recent studies on propagule pressure generally meet with success in explaining vulnerability of ecosystems to invasion, although in reality it is very likely that more than one mechanism simultaneously affects invasion success. More experimentation would enable greater use of theoretical models with lower false + or – error rates. As pointed out by Rose (1997) 'nothing in biology makes sense except in the light of history'. For the prediction of future invasions, we should learn from history; and history has left us with a great deal to study.

References

Alpine AE, Cloern JE (1992) Trophic interactions and direct physical effects control biomass and production in an estuary. Limnol Oceanogr 37:946–955

Andow DA (2003) Biological invasions: assessment and management of environmental risk. Food & Fertilizer Technology Center. Available at http://www.fftc.agnet.org/library/article/eb538.html

Apte S, Holland BS, Godwin LS, Gardner JPA (2000) Jumping ship a stepping stone event mediating transfer of a non-indigenous species via a potentially unsuitable environment. Biol Invasions 2:75–79

Armonies W (2001) What an introduced species can tell us about the spatial extension of benthic populations. Mar Ecol Prog Ser 209:289–294

Armonies W, Reise K (1999) On the population development of the introduced razor clam *Ensis americanus* near the island of Sylt (North Sea). Helgol Wiss Meeresunters 52:291–300

Bij de Vaate A, Jazdzewski K, Ketelaars HAM, Gollasch S, Velde G van der (2002) Geographical patterns in range extension of Ponto-Caspian macroinvertebrate species in Europe. Can J Fish Aquat Sci 59:1159–1174

Binggeli P (1997) How relevant are ecologically equivalent species and vacant niches to the invasive potential of introduced species? In: TU Berlin (ed) Fourth International Conference on the ecology of invasive alien plants. Institut für Ökologie und Biologie der TU, Berlin, 28 pp

Binggeli P (2000) Time-lags between introduction, establishment and rapid spread of introduced environmental weeds. Proc Int Weed Sci Congr 3 2–14

Blackburn TM, Duncan RP (2001) Determinants of establishment success in introduced birds. Nature 414:195–197

Blackburn TM, Cassey P, Duncan RP, Evans KL, Gaston KJ (2004) Avian extinction and mammalian introductions on oceanic islands. Science 305:1955–1958

Blanchard M (1997) Spread of the slipper limpet *Crepidula fornicata* (L. 1758) in Europe: current state and consequences. Sci Mar 61:109–118

Bruno JF, Stachowicz JJ, Bertness MD (2003) Including positive interactions in ecological theory. Trends Ecol Evol 18:119–125

Burch JB, Bruce JI, Amr Z (1989) Schistosomiasis and malacology in Jordan. J Med Appl Malacol 1:139–163

Carlton JT, Geller JB (1993) Ecological roulette: the global transport of nonindigenous marine organisms. Science 261:78–82

Censky EJ, Hodge K, Dudley J (1998) Evidence of over-water dispersal of lizards due to hurricanes. Nature 395:556

Cloern JE (1996) Phytoplankton bloom dynamics in coastal ecosystems: a review with some general lessons from sustained investigations of San Francisco Bay, California. Rev Geophys 34:127–168

Cohen AN, Carlton JT (1994) Nonindigenous aquatic species in a United States estuary: a case study of the biological invasions of the San Francisco bay and delta. United States Fish and Wildlife Service, Washington, D.C.

Colautti RI, MacIsaac HJ (2004a) A neutral terminology to define invasive species. Div Distr 10:135–141

Colautti RI , Ricciardi A, Grigorovich IA, MacIsaac HJ (2004b) Does the enemy release hypothesis predict invasion success? Ecol Lett 7:721–733

Colautti RI, Bailey SA, Overdijk C van, Amundsen K, MacIsaac HJ (2006a) Characterised and projected costs of nonindigenous species in Canada. Biol Invasions 8:45–59

Colautti RI, Grigorovitch IA, MacIsaac HJ (2006b) Propagule pressure: A null model for biological invasions. Biol Invasions (in press)

Courtenay WR, Stauffer JR (eds) (1984) Distribution, biology, and management of exotic fish species. John Hopkins University Press, Baltimore

Crawley MJ (1989) Chance and timing in biological invasions. In: Drake JA, et al (eds) Biological invasions: a global perspective. Wiley & Sons, Chichester, pp 407–423

Crooks JA (2002) Characterizing ecosystem-level consequences of biological invasions: the role of ecosystem engineers. Oikos 97:153–166

Davis MA, Grime JP, Thompson K (2000) Fluctuating resources in plant communities: a general theory of invasibility. J Ecol 88:528–534

Declerck S, Louette G, De Bie T, De Meester L (2002) Patterns of diet overlap between populations of non-indigenous and native fishes in shallow ponds. J Fish Biol 61:1182–1197

Drake JA (1990) The mechanisms of community assembly and succession. J Theor Biol 147:213–233

Duggan IC, Rixon CAM, MacIsaac HJ (2006) Popularity and propagule pressure: determinants of introduction and establishment of aquarium fish. Biol Invasions 8:393–398

Duncan RP, Williams PA (2002) Darwin's naturalization hypothesis challenged. Nature 417:608

Elton CS (1927) Animal ecology. Sidgwich and Jackson, London

Elton CS (1958) The ecology of invasions by animals and plants. Methuen & Co., London

Forsyth DM, Duncan RP (2001) Propagule size and the relative success of exotic ungulate and bird introductions in New Zealand. Ann Nat 157:583–595

Goldberg DE, Rajaniemi T, Gurevitch J, Stewart-Oaten A (1999) Empirical approaches to quantifying interaction intensity: competition and facilitating along productivity gradients. Ecology 80:1118–1131

Gratz NG (2004) Critical review of the vector status of *Aedes albopictus*. Med Vet Entomol 18:215–227
Grevstad FS (1999) Probability of beetle population establishment increased with release size of population. Biol Invasions 1:313–323
Grosholz E (2002) Ecological and evolutionary consequences of coastal invasions. Trends Ecol Evol 17:22–27
Gurevitch J, Padilla DK (2004) Are invasive species major cause of extinctions? Trends Ecol Evol 19:470–474
Harper J (1969) The role of predation in vegetational diversity. In: Woodwell GM, Smith HH (eds) Diversity and stability in ecological systems. Brookhaven National Laboratory, Upton, N.Y., pp 48–62
Havel JE, Lee CE, Vander Zanden MJ (2005) Do reservoirs facilitate invasions into landscapes? BioScience 55:518–525
Hebert PDN, Cristescu MEA (2002) Genetic perspectives on invasions: the case of the Cladocera. Can J Fish Aquat Sci 59:1229–1234
Heger T, Trepl L (2003) Predicting biological invasions. Biol Invasions 5:313–321
Hewitt CL, Huxel GR (2002) Invasion success and community resistance in single and multiple species invasion models: do the models support the conclusions? Biol Invasions 4:263–271
Hunter JM, Rey L, Chu KY, Adekolu-John EQ, Mort KE (1993) Parasitic diseases in water sources development. World Health Organization, Geneva
Juliano SA, Lounibos LP (2005) Ecology of invasive mosquitoes: effects on resident species and on human health. Ecol Lett 8:558–574
Kaufman LS (1992) Catastrophic change in species-rich freshwater ecosystems, the lessons of Lake Victoria. BioScience 42:846–858
Kennedy TA, Naeem S, Howe KM, Knops JMH, Reich P (2002) Biodiversity as a barrier to ecological invasion. Nature 417:636–638
Kimmerer WJ, Gartside E, Orsi IJ (1994) Predation by an introduced clam as the likely cause of substantial declines in zooplankton in San Francisco Bay. Mar Ecol Prog Ser 113:81–93
Knowler D, Barbier EB (2000) The economics of an invading species: a theoretical model and case study application. Economic evaluation in classical biological control. In: Perrins C, Williamson M, Dalmazzone S (eds) The economics of biological invasions. Edward Elgar, Cheltenham, pp 70–93
Kolar CS, Lodge DM (2001) Progress in invasion biology: predicting invaders. Trends Ecol Evol 16:199–204
Kowarik I (1995) Time lags in biological invasions with regard to the success and failure of alien species. In: Pysek P, Prach K, Rejmánek M, Wade PM (eds) Plant invasions – general aspects and special problems. SPB Academic, Amsterdam. pp 15–38
Lennon JT, Smith VH, Dzialowski AR (2003) Invasibility of plankton food webs along a trophic state gradient. Oikos 103:191–203
Levine JM, D'Antonio CM (1999) Elton revisited: a review of evidence linking diversity and invasibility. Oikos 87:15–26
Lockett MM, Gomon MF (2001) Ship mediated fish invasions in Australia: two new introductions and a consideration of two previous invasions. Biol Invasions 3:187–192
Lockwood JL, Cassey P, Blackburn T (2005) The role of propagule pressure in explaining species invasions. Trends Ecol Evol 20:223–228
Lodge DM (1993) Biological invasions: lessons for ecology. Trends Ecol Evol 8:133–137
Lohrer AM (2001) A framework for empirical research on alien species. Proc Int Conf Mar Bioinvasions 2:88–91
Lonsdale WM (1999) Global patterns of plant invasions and the concept of invisibility. Ecology 80:1522–1536

Loreau M (2004) Does functional redundancy exist? Oikos 104:606–611
Lozon J, MacIsaac HJ (1997) Biological invasions: are they dependent on disturbance? Environ Rev 5:131–144
Lyell C (1832) Principles of geology, vol 2. John Murray, London
MacArthur RH (1970) Species packing and competitive equilibrium for many species. Theor Pop Biol 1:1–11
MacIsaac HJ, Grigorovich IA, Ricciardi A (2001) Reassessment of species invasions concepts: the Great Lakes basin as an example. Biol Invasions 3:405–416
Mack RN, Simberloff D, Lonsdale WM, Evans H, Clout M, Bazzaz FA (2000) Biotic invasions: causes, epidemiology, global consequences, and control. Ecol Appl 10:689–710
Maki K, Galatowitch S (2004) Movement of invasive aquatic plants into Minnesota (USA) through horticultural trade. Biol Conserv 118:389–396
Minchin D, Gollasch S (2003) Fouling and ships hulls: how changing circumstances and spawning events may result in the spread of exotic species. Biofouling 19:111–122
Mitchell CJ, Niebylski GC, Smith GC, et al (1992) Isolation of eastern equine encephalitis virus from *Aedes albopictus* in Florida. Science 257:526–527
Mooney HA, Cleland EE (2001) The evolutionary impact of invasive species. Proc Natl Acad Sci USA 98:5446–5451
Moore GC, Francy DA, Eliason DA, Monath TA (1988) *Aedes albopictus* in the United States: rapid spread of potential disease vector. J Am Mosq Control Assoc 4:356–361
Moore JL, Mouquet N, Lawton JH, Loreau M (2001) Coexistence, saturation and invasion resistance in simulated plant assemblages. Oikos 94:303–314
Moyle PB (1996) Effects of invading species on freshwater and estuarine ecosystems. In: Sandlund OT, Schei PJ, Viken A (eds) Proceedings of the Norway/UN conference on alien species, Trondheim, 1–5 July 1996. Directorate for Nature Management/Norwegian Institute for Nature Research, Trondheim, pp 86–92
Moyle PB, Light T (1996) Fish invasions in California: do abiotic factors determine success? Ecology 77:1666–1669
Occhipinti-Ambrogi A, Galil BS (2004) A uniform terminology on bioinvasions: a chimera or an operative tool? Mar Pollut Bull 49:688–694
Olden JD, Poff NL, Douglas MR, Douglas ME, Fausch KD (2004) Ecological and evolutionary consequences of biotic homogenization. Trends Ecol Evol 19:18–24
Osenberg CW, Schmitt, RJ, Holbrook SJ, Abusaba KE (1994) Detection of environmental impacts: natural variability, effect size, and power analysis. Ecol Appl 4:16–30
Parker IM, Simberloff D, Lonsdale WM, Goodell K, Wonham M, Kareiva PM, williamson MH, Von Holle B, Moyle PB, Byers JE, Goldwasser L (1999) Impact: toward a framework for understanding the ecological effects on invaders. Biol Invasion 1:3–19
Perrings C, Williamson MH, Dalmazzone S (eds) (2001) The economics of biological invasions. Edward Elgar, London
Pimentel D, Lach L, Zuniga R, Morrison D (2000) Environmental and economic costs of nonindigenous species in the United States. BioScience 50:53–65
Pimentel D, McNair S, Janecka J, Wightman J, Simmonds C, O'Connell C, Wong E, Russel I, Zern J, Aquino T, Tsomondo T (2001) Economic and environmental threats of alien plant, animal and microbe invasions. Agric Ecosyst Environ 84:1–20
Pimentel D, Zuniga R, Morrison D (2004) Update on the environmental and economic costs associated with alien-invasive species in the United States. Ecol Econ 52:273–288
Ramcharan CW, Padilla DK, Dodson SI (1992) A multivariate model for predicting population fluctuations of *Dreissena polymorpha* in North American Lakes. Can J Fish Aquat Sci 49:150–158
Reise K (1998) Pacific oysters invade mussel beds in the European Wadden Sea. Senckenbergiana Mar 28:167–175

Reise K, Gollasch S, Wolff WJ (2002) Introduced marine species of the North Sea coasts. In: Leppäkoski E, Gollasch S, Olenin S (eds) Invasive aquatic species of Europe. Kluwer, Dordrecht, pp 260–266

Reise K, Dankers N, Essink K (2005) Introduced species. In Essink K, Dettlann C, Farke H, Laursen K, Luerssen G, Marencic H, Wiersinga W (eds) Wadden Sea quality status report 2004. (Wadden Sea Ecosystems 19. Trilateral monitoring and assessment group) Common Wadden Sea Secretariate, Wilhelmshaven, pp 155–161

Rejmanek M, Richardson DM (1996) What attributes make some plant species more invasive? Ecology 77:1655–1661

Ricciardi A (2001) Facilitate interactions among aquatic invaders: is an "invasional meltdown" occurring in the Great Lakes? Can J Fish Aquat Sci 58:2513–2525

Ricciardi A (2004) Assessing species invasions as a cause of extinction. Trends Ecol Evol 19:619

Ricciardi A, MacIsaac HJ (2000) Recent mass invasion of the North American Great Lakes by Ponto-Caspian species. Trends Ecol Evol 15:62–65

Ricciardi A, Rasmussen JB (1999) Extinction of North American freshwater fauna. Conserv Biol 13:1200–1230

Richardson DM, Pyšek P, Rejmánek M, Barbour MG, Panetta FD, West CJ (2000) Naturalization and invasion of alien plants: concepts and definitions. Divers Distrib 6:93–107

Rixon CAM, Duggan IC, Bergeron NMH, Ricciardi A, MacIsaac HJ (2005) Invasion risks posed by the aquarium trade and live fish markets on the Laurentian Great Lakes. Biodiv Conserv 14:1365–1381

Rose S (1997) Lifelines. Penguin, London

Ruiz GM, Carlton JT, Grosholz ED, Hines AH (1997) Global invasions of marine and estuarine habitats by non-indigenous species: mechanisms, extent, and consequences. Am Zool 37:621–632

Sax DF, Gaines SD (2003) Species diversity from global decrease to local increase. Trends Ecol Evol 18:561–566

Shrader-Frechette K (2001) Non-indigenous species and ecological explanation. Biol Philos 16:507–519

Shrader-Frechette K, McCoy ED (1993) Method in ecology – strategies for conservation. Cambridge University Press, Cambridge

Simberloff D (1995) Why do introduced species appear to devastate islands more than mainland areas? Pac Sci 49:87–97

Simberloff D (2003) Confronting introduced species: a form of xenophobia? Biol Invasions 5:179–192

Simberloff D, Gibbons L (2004) Now you see them, now you don't! – population crashes of established introduced species. Biol Invasions 6:161–172

Simberloff D, Holle B von (1999) Positive interactions of non-indigenous species: invasional meltdown? Biol Invasions 1:21–32

Stachowicz JJ (1999) Species diversity and invasion resistance in a marine ecosystem. Science 286:1577–1578

Strayer DL, Blair EA, Caraco NF, Cole JJ, Findlay S, Nieder WC, Pace ML (2005) Interactions between alien species and restoration of large-river ecosystems. Arch Hydrobiol Suppl 155:133–145

Thieltges DW (2005) Impact of an invader: epizootic American slipper limpet *Crepidula fornicata* reduces survival and growth in European mussels. Mar Ecol Prog Ser 286:13–19

Thieltges DW, Strasser M, Reise K (2003) The American slipper limpet *Crepidula fornicata* (L.) in the northern Wadden Sea 70 years after its introduction. Helgol Mar Res 57:27–33

Trussel GC (1996) Phenotypic plasticity in an intertidal snail: the role of common crab predator. Evolution 50:448–454

Underwood AJ (1994) On beyond BACI: sampling designs that might reliably detect environmental disturbances. Ecol Appl 4:3–15

Unmack PJ, Fagan WF (2004) Convergence of differentially invaded systems toward invader-dominance: time-lagged invasions as a predictor in desert fish communities. Biol Invasions 6:233–243

Van der Velde G, Rajagopal S, Kelleher B, Muskó IB, Bij de Vaate A (2000) Ecological impact of crustacean invaders: general considerations and examples from the Rhine River. Crustacean Iss 12:3–33

Van der Velde G, Nagelkerken I, Rajagopal S, Bij de Vaate A (2002) Invasions by alien species in inland freshwater bodies in western Europe: the Rhine delta. In: Leppäkoski E, Gollasch S, Olenin S (eds) Invasive aquatic species of Europe. Kluwer, Dordrecht, pp 360–372

Vermeij GJ (1982) Phenotypic evolution in a poorly dispersing snail after arrival of a predator. Nature 299:349–350

Vermeij GJ (1991) When biotas meet: understanding biotic interchange. Science 253:1099–1104

Vitousek PM (1990) Biological invasions and ecosystem processes: towards an integration of population biology and ecosystem studies. Oikos 57:7–13

Wellnitz T, LeRoy Poff N (2001) Functional redundancy in heterogeneous environments: implications for conservation. Ecol Lett 4:177–179

Wilcove DS, Rothstein D, Dubow J, Phillips A, Losos E (1998). Quantifying threats to imperiled species in the United States. BioScience 48:607–616

Williamson M (1996) Biological invasions. Chapman & Hall, London

Williamson M (1999) Invasions. Ecography 22:5–12

Williamson M, Fitter A (1996) The varying success of invaders. Ecology 77:1661–1665

Wilkinson DM (2004) The long history of homogenization. Trends Ecol Evol 19:282–283

Wolff WJ, Reise K 2002. Oyster imports as a vector for the introduction of alien species into northern and western European coastal waters. In: Leppäkoski E, Gollasch S, Olenin S (eds) Invasive aquatic species of Europe. Kluwer, Dordrecht, pp 193–205

Section II:

Conservation and Management of Wetlands

5 Wetland Conservation and Management: Questions for Science and Society in Applying the Ecosystem Approach

EDWARD MALTBY

5.1 Introduction

Wetland loss and degradation continues worldwide despite increasing evidence from the scientific community of the significant benefits these diverse ecosystems provide to humans, wildlife and the maintenance of environmental conditions. The impact of this increased knowledge and awareness is confirmed by the unprecedented interest currently in the restoration, creation and protection of wetlands. We are then confronted by the contradictory situation of ecosystem destruction and re-establishment both featuring prominently in society's agenda. The challenge is to manage the processes of change so that we do not irretrievably lose assets difficult or impossible to replace and that we obtain the most appropriate balance between wetland conservation and socio-economic development.

5.2 Wetlands at the Interface

One of the exceptional features of wetland ecosystems is that they are closely interfaced with both natural and human systems. This is reflected in their own transitional character captured in definitions such as that of Cowardin et al. (1979) which recognised wetlands as "lands transitional between terrestrial and aquatic systems where the water table is usually at or near the surface or the land is covered by shallow water". Wetlands are often considered separately as ecotones across which one natural system gives way to another. This is particularly true of river floodplains, lake margins and coastal environments, all foci of human settlement since prehistoric times and of intense

Ecological Studies, Vol. 191
R. Bobbink, B. Beltman, J.T.A. Verhoeven, and D.F. Whigham (Eds.)
Wetlands: Functioning, Biodiversity Conservation, and Restoration

development pressure more recently. More emphatically than any other ecosystem type, the interface position places wetlands in zones of maximum potential natural and human interaction and impact. They are zones of tension where the need for conflict resolution among different sectoral interests is increasingly necessary. Examples include floods, storm surges and tsunamis, agricultural intensification, aquaculture, urbanisation, industrial and other forms of economic development. Wetland interfaces are thus highly relevant to issues of human welfare, health, safety and protection, to biodiversity and to concerns for environmental quality and sustainable development.

The relationship between human society and wetlands has changed significantly over time. It is likely that key steps in human evolution itself took place among ancestral groups occupying wetland margins of rivers, lakes and the sea, leading to the development of *Homo sapiens*. One school of thought (the so-called aquatic ape theory) attributes this to the source of fatty acids from the fish caught by these early communities (Morgan 1982). Prehistoric communities worldwide depended more or less on wetlands for food, shelter, safety and mobility (Coles and Coles 1989). Such dependence still exists today in many communities of the developing world (Maltby 1986, 1988a, b). Nevertheless human endeavour has concentrated throughout the historic period on severing the link of wetland dependence, supported by technological innovations which have enabled the natural constraints of the wetland environment to be modified and removed thereby allowing conversion to non-wetland uses. Significant social and economic gains have been achieved but we are still learning more about the full and long-term costs of such changes. It is an irony of development that inappropriate occupation, use and alternative development of wetlands have often resulted in catastrophic human, economic and environmental consequences. The intensive settlement of river floodplains and replacement of the stabilising mangrove fringe of tropical coastlines by agriculture or tourist developments are just two examples which have substituted the natural protective functions of wetlands with high-risk conditions for human populations. This has been exemplified only too clearly by the devastating effects of the December 2004 tsunami in the Indian Ocean and even more so by the flooding of New Orleans in August 2005 resulting from hurricane Katrina. Both events are stark reminders of the importance of natural wetland buffers which, if managed as such, can avoid both human misery and major economic losses.

There is increasing recognition that sustainability is one of the key challenges facing society. Ecosystem management is at the core of achieving sustainable development (Maltby 1997a, b). This recognises the need to balance considerations of human needs and desires with the capacity or limits made possible by the natural environment and ecological relationships together with the possibilities presented by technical innovation and changing social and economic circumstances. The need for protection and sustainable management of wetland ecosystems is an essential part of the Ramsar Convention on Wetlands of International Importance. It is inextricably linked with their 'wise use' (*sensu* Ramsar) and recognises the requirement for a more integrated yet holistic approach to their management. The global Convention on Biological Diversity (CBD) is fundamentally based on this premise and considers the 'ecosystem approach' (EA) as the primary framework for achieving sustainable development based on maintaining fully functional ecosystems. The EA is a methodology which emphasises the need for management to interface with ecological, economic, social, cultural and political considerations within a single framework. The EA "is a strategy for the integrated management of land, water and living resources that promotes conservation and sustainable use in an equitable way" (CBD 2003). The objective of balancing the three key objectives of the CBD is to achieve environmental sustainability, economic prosperity and social well-being (Fig. 5.1).

Wetlands, as the archetype of a natural interface and highly integrated ecosystems, are examined in the context of actually applying the ecosystem approach and against some relevant priority concerns often raised. A selection is given of major concerns, the rationale/questions underpinning the particular issue and a distillation of the challenge to wetland science and society (Table 5.1). The concerns exemplify the scope and scale of future challenges of wetland management faced by both scientists and those influencing

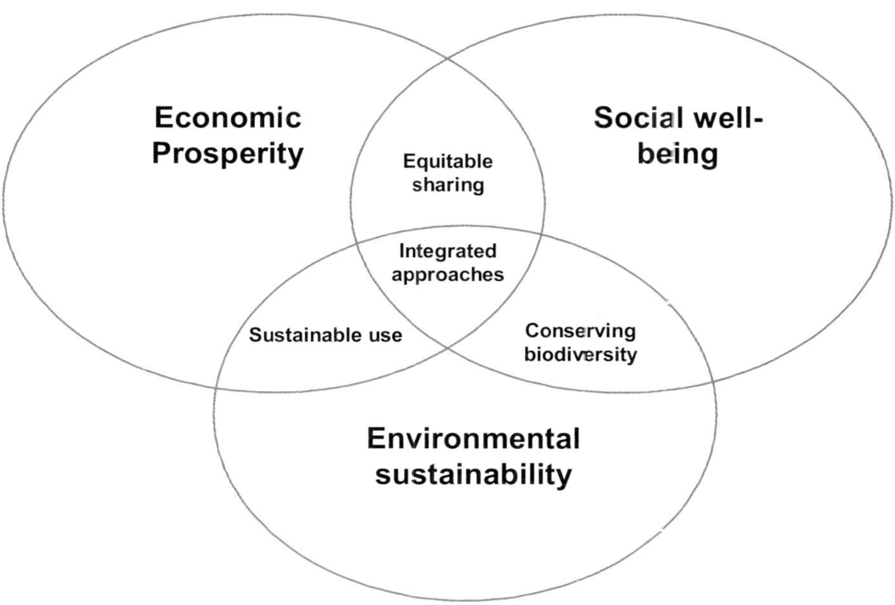

Fig. 5.1 Overlapping objectives of the Ecosystem Approach and desired outcomes (from a UK government training programme developed by E. Maltby and R.Crofts)

Table 5.1 Some priority concerns for society. Based partly on Raskin et al. (1997), Cosgrove and Rijsberman (2000), Falkenmark and Rockström (2004)

Concern	Rationale/questions	Challenge to wetland science
Competition for limited freshwater	• At least 30 % world population facing water scarcity • 10^9 people lack access to adequate drinking water • 2×10^9 people lack access to adequate sanitation • Increasing human use of freshwater has major impacts on wetland ecosystems	Can an integrated approach meet human needs as well as maintaining vital ecosystems functioning?
Development pressures, economic growth and drive for sustainability	• Poverty alleviation and health sectoral approaches • Separation of economics and ecology inappropriate • Vulnerability of water-dependent ecosystems • Use vs conservation of wetlands	Need to see wetlands as a means to achieving sustainable development not an impediment
Scientific understanding	• Rapid acceleration • Often fragmented • Insufficient linkages to policy • Inadequate translation for public/decision-makers	Need for integrated knowledge in usable forms
Global change	• Climate–wetland interactions • Wetlands and greenhouse dynamics • Wetlands as buffers	Can wetland management help deal with a changing world?
Security concerns	• Destruction to exercise control • Degradation to reduce environmental risk • Conversion to improve food security	Significance of overriding/autocratic/political actions

or delivering policies. The primary purpose of this paper is to examine how the methodological framework offered by the EA might assist in our better understanding and management of wetlands. Assessment is made by means of testing a number of the underlying principles of the EA and illustration by reference to case studies.

5.3 Recognising a New Paradigm in Ecosystem Management

Natural systems are rarely if ever static or unchanging. This is particularly true of wetlands. Geologically they are ephemeral features of the planetary landscape. Ecologically they are highly dynamic. The processes of succession ensure that ecosystem structure and functioning evolves due to autochthonous changes quite apart from the effects of any allochthonous forces (climate change is especially important here). Management must recognise the facts of change and work with it and not necessarily oppose it, notwithstanding the range of negative feedback mechanisms that might exist to assist in the maintenance of some dynamic equilibrium condition (no matter how transient that may be). Effective ecosystem management requires the widest possible view of the ecological systems themselves. Thus river ecosystem management is as much dependent on the management of catchment scale land-based processes, on farm economies and subsidies as it is on the instream dynamics of aquatic ecology.

There are at least five elements for consideration in a more holistic view of ecosystem management.

1. Different ecosystems with different functional and biodiversity characteristics can occupy the same space.
2. Ecosystems are dynamic and respond to environmental as well as human-induced changes.
3. Human societies have been a key determinant of change since prehistoric times.
4. Invasive species and unforeseen plant and animal movements are likely to be increasingly important especially given global warming.
5. Increasing recognition of links between ecosystem functioning, natural resource use and economic and conservation strategies.

What these elements recognise is that change is normal and society needs to manage or cope with it, that society itself is one of the major determinants of change and that the divisions between ecology, environment and economics are more apparent than real.

5.4 The Ecosystem Approach

Formalised under the CBD, the EA provides a methodological framework for achieving the most appropriate balance of the three objectives of the Convention: conservation of biological diversity, sustainable development and equitable sharing of the benefits of genetic resources. It acknowledges 12 underlying principles (Table 5.2) as well as five elements of guidance (Table 5.3).

Table 5.2 Principles of the Ecosystem Approach (for further background, see Maltby 1999a, b; Schei et al. 1999; UNEP 2000; Smith and Maltby 2003; and the Convention on Biological Diversity)

1. The objectives of management of land, water and living resources are a matter of societal choice.
2. Management should be decentralised to the lowest appropriate level.
3. Ecosystem managers should consider the effects (actual or potential) of their activities on adjacent and other ecosystems.
4. Recognising potential gains from management, there is usually a need to understand and manage the ecosystem in an economic context. Any such ecosystem-management programme should:
   a. Reduce those market distortions that adversely affect biological diversity;
   b. Align incentives to promote biodiversity conservation and sustainable use;
   c. Internalise costs and benefits in the given ecosystem to the extent feasible.
5. Conservation of ecosystem structure and functioning, in order to maintain ecosystem services, should be a priority target of the ecosystem approach.
6. Ecosystems must be managed within the limits of their functioning.
7. The ecosystem approach should be undertaken at the appropriate spatial and temporal scales.
8. Recognising the varying temporal scales and lag-effects that characterise ecosystem processes, objectives for ecosystem management should be set for the long term.
9. Management must recognise that change is inevitable.
10. The ecosystem approach should seek the appropriate balance between, and integration of, conservation and use of biological diversity.
11. The ecosystem approach should consider all forms of relevant information, including scientific and indigenous and local knowledge, innovatons and practices.
12. The ecosystem approach should involve all relevant sectors of society and scientific disciplines.

Development of the scientific rationale behind the approach and in formulation of the principles is elaborated further by Maltby et al. (1999) and Maltby (1999a, b). Its European context is examined by Korn et al. (1999, 2003) and Laffoley et al. (2003, 2004) and a range of international case studies are examined by Smith and Maltby (2003). Although accepted by the Conference of Parties of the CBD, the need now is to develop practical ways of implementing the EA on the ground and in different circumstances. It is intended that this will be facilitated through a web-based 'sourcebook' (http://www.biodiv.org/programmes/cross-cutting/ecosystem/sourcebook/home.shtml).

All of the EA principles are listed in Table 5.2; and they should be considered as a whole when applying the EA. However, only some are highlighted subsequently for more detailed examination related to issues of wetland protection and management. The case studies are intended to provide illustration of practical applications of the EA with particular emphasis on the particular

Table 5.3 Five points of operational guidance in applying the ecosystem approach

1. Focus on the functional relationships and processes within ecosystems
2. Enhance benefit-sharing
3. Use adaptive management practices
4. Carry out management actions at the scale appropriate for the issue being addressed, with decentralisation to lowest level, as appropriate
5. Ensure intersectoral cooperation

principle(s) highlighted. Much more comprehensive analysis is required to further the aim of mainstreaming the EA more clearly into the sectoral policies of government.

5.4.1 Principle 1: The Management of Land, Water and Living Resources is a Question of Societal Choice

The first principle of the EA is by far the most tantalising because historically at least societal choice often has been to alter and degrade wetlands. This has been the result of intended actions such as drainage for agriculture, levee raising and embankments to separate rivers from their natural floodplain and coastal flood-control structures, or has been the unintended consequences of indirect actions such as eutrophication and diffuse pollution effects from fertiliser application or sediment erosion from agricultural fields smothering the spawning grounds of migratory fish. One of the most important aspects of this principle is how to deal with the divisions with society which represent a wide range of different sectoral interests. This is recognised with the increasing emphasis on stakeholder engagement in decision-making which underpins the new water legislation in Europe embodied in the Water Framework Directive (WFD) together with the further guidance provided on its implementation with respect to wetlands (WFD CIS 2003). Whilst recognising the obvious need for conflict resolution among different interest groups there is, however, still much to be done to institutionalise the widest necessary and appropriate engagement of society in decisions affecting wetlands. It is necessary also to recognise wide variations in standpoints across the globe depending on socio-economic context and cultural views. Thus, whilst appreciation of the nature conservation values of wetlands may be a strong factor in the protection of wetlands in Europe, North America, Australia and New Zealand, such a perspective in developing countries is highly subordinate to interests such as water availability, food production and materials for shelter. It remains to be seen how the public perspective of wetlands becomes influenced by the recent concerns over the transmission of avian influenza by migrating bird populations.

Case Study: Towards Sustainable Peatlands

Peatlands comprise more than half of the globe's wetland resources. They are arguably the most hydrologically sensitive of wetland ecosystems, with water saturation a principle factor in maintaining highly concentrated stores of carbon. Even minor hydrological alteration can result in major oxidation losses and produce significant changes in vegetation and fauna. Peatlands store an estimated 329–528 × 10^{12} t of carbon (about 24 % of the world terrestrial soil carbon pool) on just over 3 % of the planetary land area. Once thought to be solely and at least primarily a high-latitude phenomenon, it is now established that tropical peatlands may store 70 Gt or 15 % of the peat carbon pool over some 10 % of the world land area (Maltby and Immirzi 1993). These tropical carbon stores are particularly vulnerable to loss as a result of hydrological change because of the effect of high temperatures which produce high rates of oxidation after drainage.

Worldwide, peatlands are being developed for mining, agricultural development, afforestation and other sectoral activities which fundamentally alter their structure and functioning. The decision to exploit the ecosystem for other purposes is invariably driven by short-term economic factors and sectoral interests. Rarely if ever is it the result of a widely informed society concensus. Yet the adverse effects of inappropriate development can affect all elements of society, such as by sea level increase resulting from the warming effect of the additional load of carbon dioxide to the atmosphere.

Immirzi and Maltby (1992) estimated that, if all peat-forming systems had escaped human alteration globally, there would have been a sink for some 367 Mt CO_2 year^{-1}. However, the past 200 years witnessed major human modification, especially by drainage for agriculture, forestry and peat mining for energy or horticultural products. Mining can lead to the emission of 180–225 t CO_2 ha^{-1} year^{-1}. Arable cropping may yield 41 t CO_2 ha^{-1} year^{-1} in temperate regions but 154 t CO_2 ha^{-1} year^{-1} in the tropics (Immirzi and Maltby 1992). Conversion to forestry and pasture has a much smaller impact. The impact of conversion of natural mires to non-wetland uses results in: (1) loss of any carbon sink capacity and (2) release to the atmosphere of accumulated carbon as CO_2. The second is progressive and rapid, often resulting in the release within a few years of carbon which has built up over millenia. The result is a significant shift in carbon balance which in the case of agriculture alone has been estimated at 426–730 Mt CO_2 year^{-1}, equivalent to about 3.5 % of fossil fuel emissions and more than 12 % of that resulting from tropical deforestation (Immirzi and Maltby 1992). In spite of the growing evidence of the size of the carbon shift and its increasing scale, especially in the tropics, there has been no explicit treatment of peatlands in global carbon and climate change models (Maltby and Proctor 1996). Their consideration in the global warming debate is underrepresented by the Kyoto Protocol; and guidelines relevant to their sustainable or 'wise' use (Safford and Maltby 1998; Ramsar

2002; Joosten and Clark 2002) appear to have had limited if any political or wider societal impact.

The developmental pressures on tropical peatlands is particularly problematic. This is partly because once disturbed they are especially vulnerable to decomposition resulting in rapid subsidence and degradation and partly because there is still a relatively poor knowledge base about their genesis and potential to support sustainable economic development. Table 5.4 gives some insight into the range of function, value or attributes of south-eastAsian peat swamp forests which may assist in the delivery of sustainable development through an integrated planning and management process. The alternative of highly sectoral development, such as conversion to agriculture, production forestry or peat extraction, might yield more immediate financial returns but inevitably results in the degradation or loss of ecosystem functions vital for sustainable land and water use (Maltby et al. 1996b; Safford and Maltby 1998). Notwithstanding the significance of any ethical argument to support the retention of biological diversity in the highly threatened areas of remaining topical peat, there is also a potentially immense wealth of pharmaceutical products based on the evidence of traditional medicinal uses (Table 5.5).There may be far-reaching effects of alternative forms of peatland management on the carbon cycle as feedback to climate change and hydrology, together with many other aspects of ecosystem functioning significant to human communities as well as wildlife. Peatland ecosystem functioning is a vital research area which

Table 5.4 Examples of functioning, values and attributes of south-east Asian peat swamp forest. For sources, see Maltby et al. (1996a, b) and Safford and Maltby (1998)

Function, value or attribute	Source
Flood mitigation	Andriesse (1988), Sim and Balamurugan (1989)
Prevention of intrusion of saline waters	Andriesse (1988), Sim and Balamurugan (1989)
Sediment removal	Sim and Balamurugan (1989)
Nutrient removal	Andriesse (1988)
Toxicant removal	Andriesse (1988)
Wildlife habitat	Giesen and Sukotjo (1991), Janzen (1974), Prentice and Aikanaithan (1989), Whitmore (1984)
Fish habitat	Johnson (1967), Davis & Abrullah (1989), Giesen and Sukotjo (1991), Ng et al. (1994)
Toxicant export	Andriesse (1988)
Biological diversity	Prentice (1990)
Water supply	Ong and Yogeswaren (1992)
Production	Lee and Chai (1996), Dwiyono and Rachman (1996), Silvius and Geisen (1996)

Table 5.5 Medicinal plants from the peat swamp forests (after Chai et al. 1989)

Family	Species	Plant form	Medicinal value
Annonaceae	*Fississtigma rigidum*	Woody climber	Drink for treating fever
Annonaceae	*Mitrella kentii* (Bl.) Miq.	Woody climber	For treating gonorrhoea (poultice from ash of stem)
Apocynaceae	*Alstonia spathulata* Bl.	Tree	For shingles (leaves made into poultice)
Araceae	*Scindapsus perakensis* Hook. F.	Herb	For easing pain caused by stings (leaf paste used)
Davalliaceae	*Nephrolepis hirsutula* (Forst.) Presl.	Fern	To stimulate lactation (drink made from boiled young shoots)
Guttiferae	*Cratoxylum arborescens* (Vahl) Bl.	Tree	For treatment of chicken pox (apply latex on rashes or skin disease)
Leeaceae	*Leea* sp.	Shrub	For scorpion and centipede bites, wasp and bee stings (apply paste made from young twigs and leavers to wound)
Leguminosae	*Sindora leiocarpa* Backer ex K. Hyne	Tree	Tonic (boil tap root and drink fruits boiled and mixed with other spices as a drink)
Moraceae	*Ficus crassiramea* Miq.	Strangling fig	For snake bite (apply paste on wounds – leaves, bark, roots)
Myrtaceae	*Eugenia cerina* Hend.	Small tree	Tonic (leaves)
Myrtaceae	*Eugenia paradoxa* Merr.	Small tree	For treating diarrhoea (leaves infused)
Myrtaceae	*Eugenia zeylanica* (L.) Wight	Small tree	Tonic
Piperaceae	*Piper arborescens* Roxb.	Climber	For treating rheumatism (plant boiled and drunk)
Simaroubaceae	*Quassia* spp	Medium-sized trees	To cure impotence and hypertension (boil tap root and drink)

must be linked more effectively to future policy initiatives that take fuller account of the role of peatlands as a biosphere resource of common, rather than a single sectoral interest.

5.4.2 Principle 3: Ecosystem Managers Should Consider the Effects of Their Activities on Adjacent and Other Ecosystems; and Principle 7: The Ecosystem Approach Should be Undertaken at the Appropriate Scale

Both these principles invite similar considerations. In particular it is necessary that there is awareness of the potential problems which can arise due to the implementation of particular actions and that the outcome is consistent with a vision shared by all of those so affected. Thus it is inequitable if the benefits resulting from actions of communities or sectoral interests upstream produce losses downstream or off-site, especially if there is no way of adjusting for these costs to other communities or interest groups. The issue is especially well illustrated in the challenges of management of river basins and trans-boundary ecosystems. Delta wetlands and adjacent waters in particular, lying at the end of often large and trans-national basins, are particularly vulnerable to activities which do not respect these principles of the ecosystem approach. Thus, the expansion of the hypoxia area of the Gulf of Mexico adjacent to the Mississippi Delta has much to do with the discharge of increasing quantities of nutrients from the vast drainage network of the Mississippi–Missouri rivers, collecting agricultural runoff from some 40 % of the coterminous United States. Agricultural intensification together with the loss of natural wetland filters and buffer zones are all likely contributory factors to the problem (Rabalais and Turner 2001; Justic et al. 2002; Rabalais et al. 2002a, b).

Case Study: The Demise and Restoration of the Mesopotamian Marshes

Probably the most instructive example of the need to apply these two principles of the ecosystem approach is provided by recent events which have impacted one of the world's most significant wetland complexes – those of Mesopotamia. The example illustrates also one of the most practical scales of wetland management – that of the whole river basin.

The Mesopotamian marshlands occupy a major part of a region heralded as one of the great 'cradles of civilisation'. Until recently they extended over an area of some 25 000 km^2. The area is steeped in cultural richness, historical value and ecological diversity. Nine thousand years ago the delta and its associated wetland ecosystems were home to the earliest of known civilisations, supporting the development of what is sometimes referred to as 'hydraulic civilisation', founded upon knowledge and sustainable utilisation of the region's water resources. It produced some of the earliest known examples of written language; and it is speculated that the roots of the biblical accounts of 'Creation' and the 'Great Flood' originate from ancient texts of the region. On the shores of the marshes, the legendary Epic of Gilgamesh was enacted. Consequently, the cultural and philosophical debt owed to the region is enormous. The recent marsh dwellers, the Marsh Arabs or 'Madan', are a link with this rich cultural past and nowhere else can a stronger relationship between water, life, economy and civilisation be demonstrated (Table 5.6).

This once extensive wetland area has suffered major desiccation and environmental, social and economic degradation from direct drainage and upstream hydrological impacts. Numerous engineering structures have been established upstream which have greatly reduced water flows and the magnitude of flood pulses. Both of these are vital elements for the sustainable functioning of the Mesopotamian Marshes. Within the past two decades alone, 20 000 km^2 of former marshland has been converted to an arid landscape

Table 5.6 Importance of the Mesopotamian marshlands (after Maltby 1994)

Unique Human Community Supported by Natural Resources
- Productive traditional agriculture
- Sustainable utilisation of wetland and adjacent land
  - Buffalo and cattle
  - Fishing and birds
  - Reeds and other plants
  - Cultivation
  - Integrated transport

Habitat for Important Populations and Species
- Intercontinental migration
- Regional biodiversity
- Rare and endemic species
  - Globally threatened birds, mammals, invertebrates
  - Cyprinid marsh species of high evolutionary significance

Linkage to Gulf
- Hydrological interface between catchment and marine ecosystem
  - Discharge
  - Water quality (nutrients, salinity, contaminants)
  - Sediment
  - Temperature
- Continuum for movement of economically important fish/shrimp, e.g.
  - *Metapenaeus affinis*
  - Pomphret
  - Saboor

Microclimate

Environmental reconstruction
- Peat/sedimentary deposits

(Maltby 1994; UNEP 2001). The reduction in flood discharge of both the Tigris and Euphrates, resulting primarily from dam construction in Turkey but reinforced by those upstream in Iraq, enabled the Saddam regime to more easily complete drainage works in the marshes. This resulted in not only displacement and misery for tens of thousands of people but also the loss of important wetland habitat, ecosystem functions and possible loss of many species (probably including some not even recorded).

One aspect of marshland loss which has not yet received sufficient attention is the question of linkage with the marine environment and the issues which arise from both disruption and subsequent re-establishment of the hydrological and ecological continuum from the river basin headwaters and flow network to the open sea. The possible effects of downstream drainage of the marshes, for whatever reason, were never considered by the former Iraq administration. This is not an unusual situation in other cases of wetland loss. As more evidence is gathered around the world it is clear that the scale of off-site impacts may be considerable and should always be an influencing factor in development decisions (Table 5.7).

Regime change in Iraq has now provided an opportunity to restore the marshes for the benefit of people, wildlife and environmental quality. However, the solution is not simple and is constrained by a wide range of considerations which demand an approach which is catchment-wide and includes numerous sectoral interests. The catchment scale is essential because the actual quantity, quality and timing of water flows available to the marshlands is controlled by upstream countries as well as areas in Iraq. Different sectoral interests include the oil industry, agriculture, transport, health and the security forces. It is still not clear how in practice the new Iraq government will engage with both local communities of marsh dwellers who have returned to

Table 5.7 Effects of closure of Aswan High Dam (after El-Sayed and van Dijken 1995)

Aswan High Dam completed 1965

Nile flow into Mediterranean
- rich in sediment/nutrients
- Nile bloom → sardines/other pelagic fish/shrimp

Catastrophic effects of decline in fertility due to hydrological change
- Sardine fishery: 35 000 t (1962/1963) → 600 t (1969)
- Shrimps: 8300 t (1963) → 1128 t (1969)
- Closure of canning factories

Some recovery
- Sardines → 8590 t (1992)
  Most landings coincide with maximum discharge from coastal waters in winter

Accelerated land loss (>100 m year^{-1})

Fig. 5.2 A MODIS satellite image of Southern Iraq acquired 15 March 2005 (copyright UNEP). Total extent of marshlands 4730 km^2 (53 % permanent marsh cover in 1973). For earlier imagery showing desiccation, see Maltby (1994) and UNEP (2001). **B** Electrofishing in re-flooded wetlands provides essential income but indiscriminant catches. **C** The traditional mudhif built of Phragmites reed has quickly reappeared in the re-flooded landscape. **D** Use of the reed as feed on island in re-flooded marshes. Photos by E.Maltby (Feb 2004) whilst on US AID-funded mission to Iraq

recently re-flooded areas as well as other stakeholders (local, national, international) to realise an appropriate balance between wetland restoration and alternative development conditions. Important steps have been taken, however, through the establishment of an Interministerial Centre for the Restoration of Iraqi Marshes (CRIM) and in bi-lateral as well as regional-level discussions with neighbouring countries. In the most recent assessment by UNEP there was, in Spring 2005, recovery of over 50 % of the former marshland extent present in the 1970s (http://imos.grid.unep.ch/; Fig. 5.2). This followed winter rains and snow melt in the headwaters of the Tigris and Euphrates rivers; and their extent by August 2005 was some 37 % of their former extent due to the effects of the high summer evaporation rates. The supply of water to the marshes is increasingly at the mercy of upstream control structures both outside as well as within Iraq. The distribution of water within the marshlands themselves will also continue to be influenced by different sectoral pressures. The outcome of decisions affecting the magnitude, timing and distribution of flows has inevitable consequences beyond the wetland boundaries (Maltby 2005). Rarely has there been such a clear need for an integrated solution to the management of land, water and living resources over a scale covering from an entire river basin to marine waters.

5.4.3 Principle 4: There is a Need to Understand the Ecosystem in an Economic Context

The failure to recognise or to be unable to properly account for the economic value of wetland ecosystems has been a major contributory factor in their progressive degradation. Often the value of wetlands is appreciated only after they have been lost or altered. The decline in the sardine fishery of the eastern Mediterranean is a much quoted example related to the closure of the Aswan Dam and the change in the nutrient and flow dynamics of the Nile Delta resulting in reduced marine productivity (Table 5.7; El-Sayed and van Dijken 1995). The wider importance of wetlands to society can be estimated through the contribution they make to 'natural capital'. Costanza et al (1997) have analysed the scale of the contribution of ecosystem services and variations among different ecosystem types to the natural capital of the planet. They estimated the average overall value of ecosystem services at U.S.$ 33×10^{18} year^{-1}, approximately twice global GNP. Their work illustrated two key points:

1. Wetlands, especially floodplains, coastal, estuarine, river and lakes, make disproportionately high contributions to ecosystem services.
2. The economic worth of wetland functioning has yet to feature prominently or consistently in decision-making on their status and management.

Environmental and ecological economists have applied considerable recent effort to demonstrating techniques for the valuation of wetland functioning (Barbier et al. 1997; Turner et al. 2000, 2003a, b). They have also analysed the damaging effects of alternative economic policies such as through conflicting policy objectives, perverse incentives (such as crop subsidies) and other financial mechanisms (such as regional development packages; e.g., Turner et al. 2003a, b). Often, however, simple market pressures can lead to wetland loss such as the rapid rise in soya bean prices for the manufacture of animal feedstuffs, resulting in part from the collapse of the anchovy harvest off the Peruvian coast in the early 1970s (Clark 1977). This encouraged clearance, drainage and cultivation of extensive areas of Pocosin (Richardson 1983) and bottomland hardwood wetlands in the United States, notwithstanding regulations to control such activity. Lack of consideration of 'natural' wetland functioning beyond that of delivering 'free' goods and services together with the immediate (but not necessarily sustainable) profit from agricultural development contributed to the loss of many thousands of hectares of wetlands in the Carolinas and Louisiana. The disastrous losses of property and jobs together with the immense human misery in New Orleans resulting from the 2005 flooding was exacerbated by the inappropriate development of wetlands which would otherwise have operated as a natural flood storage reservoir. The challenge now is to use the full economic benefits and context of wetlands as a tool for their protection, rehabilitation and sound management.

Case Study: Making it Happen. River Restoration in South-West England

An approach to wetland management which recognises the overriding significance of human, economic and social welfare has been attempted in south-west England through the pioneering work of the West Country Rivers Trust (WRT)'s Tamar 2000 project. The rationale is based on the premise that farmers and landowners, especially those on relatively low incomes, are likely first and foremost to respond to advice and opportunities which will generate cost savings and increases in profit. By gearing activities which will result in environmental benefits to actions that produce clear financial yields not only is the interest of the farmer secured but the chances of success are significantly improved. This has been the underlying philosophy of a successful sequence of collaborative projects executed by the WRT. Establishment of the Rivers Trust itself has been a key element of the model. Advisors from the Trust were able to work one to one with landowners, quickly building up the landowners' trust and confidence. This is an advantage over the relationships possible with the regulatory water agency or other statutory bodies responsible for conservation where risk of prosecution or at least severe criticism of poor management practices is a constant perceived threat. Advisors have worked with landowners to develop integrated farm management plans contributing not only to cost savings for the farmer but also significant environmental benefits. For example fencing, drinking troughs, blockage of wetland drains and reduced fertiliser applications are practical actions which reduce stock losses, improve animal health, enhance summer grazing and improve the farm balance sheet, while simultaneously reducing bank erosion and in-stream sedimentation, restoring river marginal wetland ecosystems and reducing eutrophication. The result is improvement of habitat, water quality and recovery of migratory fisheries in additional to realising significant economic benefits to the farming community.

The direct benefits derived from implementation of the Tamar 2000 project have averaged £ 2300 per farm, contributed variously from optimising farm inputs, water separation and leak reduction, improved stock health and the result of diversification (including revenue from increased 'bed and breakfast' places). Beyond this are indirect benefits to the wider community, tourists and anglers. Notoriously difficult to value, examples include improved water quality, hydrological flow regime, enhanced wildlife habitats and fisheries. The clear conclusion is that an economically driven rationale may actually be the most effective means of delivering environmental sustainability through ecological improvements including ecosystem restoration to recover degraded functions (see also www.wrt.org.uk).

5.4.4 Principle 9: Management must Recognise that Change is Inevitable

Case Study: Dealing with Sea-Level Rise: Managed Realignment at a Site in the Estuary of the River Torridge, UK

Sea-level rise is inextricably linked to the current period of global warming. Historically the threat of rising sea-levels has been met by building higher and stronger coastal defence structures. These are increasingly expensive to build and maintain. An alternative approach gaining support is to allow coastal and estuarine marginal areas to again become inundated during the tidal cycle. This is achieved by the controlled lowering or breaching of flood protection embankments and has the effect of reducing tidal peaks, absorbing floodwaters and protecting inland areas. Such managed realignment of the coastline has the added benefit of restoration of increasingly rare coastal wetland habitats. Whilst there is considerable information on the nature and rate of vegetation change in these areas, there is still little known about functional changes and especially those related to biogeochemistry (e.g., French 2004).

In July 2000 a flood defence embankment protecting abandoned agricultural land in the estuary of the River Torridge, Devon, UK, was breached as part of a managed realignment (MR) project. The objective at this site was to restore the agricultural land, reclaimed from saltmarsh some 200 years previously, to its former habitat as well as to alleviate flooding (exacerbated by both sea-level rise and land use change in the catchment) in the nearby town of

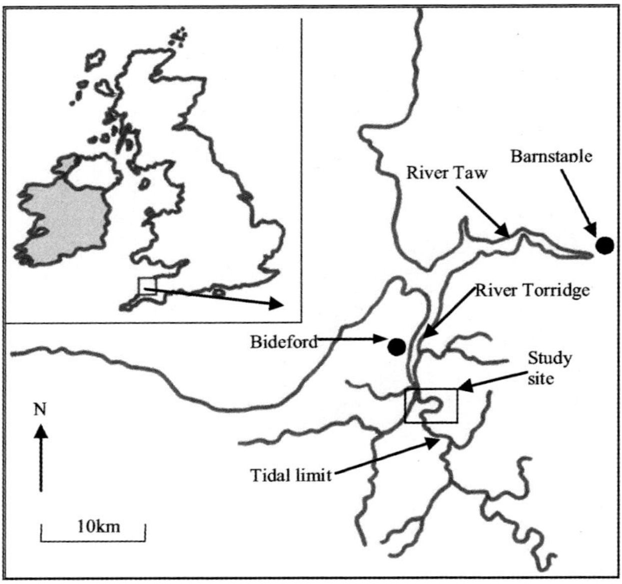

Fig. 5.3 A Location of the Pillmouth managed realignment site, Torridge Estuary (Devon, UK) and **B** the Pillmouth and Landcross study sites, showing location of flood defences, breaches and monitoring stations (after Blackwell et al., 2004)

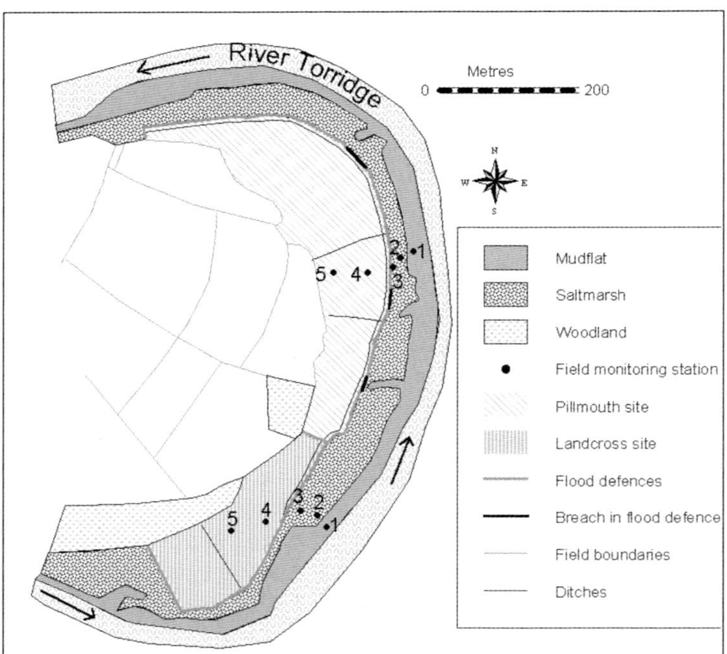

Bideford. Two contiguous sites, referred to as Pillmouth (PILM) and Landcross (LANX) were studied (Fig. 5.3), the PILM site being the subject of an MR project, with the LANX site being studied as a control.

At the PILM site permanent changes to the soil hydrological regime, plant communities and short-term changes to the soil environment at the site were observed. Blackwell et al. (2004) report these changes, the most significant observed impact being the change in flooding regime, with regular tidal inundation occurring to a maximum depth of 52 cm during spring tides; and a change in annual mean water table height of more than 50 cm was observed. These hydrological changes triggered a number of changes in soil biogeochemistry. Most notably, the pH of topsoil water (collected from 10 cm depth) fell from a pre-realignment range of pH 6.6–8.7, to pH<5.0 for approximately nine weeks, reaching a minimum of pH 3.3; and soil redox potential at 5 cm depth was reduced by over 700 mV. These changes were accompanied by changes in conductivity and NH_4^+ concentrations, conductivity rising from <2000 µS to >40 000 µS following realignment, while NH_4-N concentrations rose from 0.10 mg l^{-1} to 10.05 mg l^{-1}. Longer-term observations saw the pre-realignment vegetation at the site, which comprised mesophile pasture dominated by *Agrostis stolonifera* and *Juncus effusus*, die-off within the first year following realignment, to be replaced by pioneer salt-marsh species such as *Salicornia* spp. Two years after realignment, the site was dominated by *Aster tripolium* and some stands of *Spartina maritima* were establishing. It is likely that managed realignment will feature as an increasingly common response to progressive sea-level rise, certainly in Europe and probably worldwide. There is a need for more information on the effects of hydrological change on ecosystem functioning to assist in decisions on the optimum location and design criteria for such schemes. Additionally it will be necessary for civil society to become better informed of the rationale and benefits which can be achieved over more conventional 'hard' engineering solutions.

5.4.5 Principle 10: The Ecosystem Approach Should Seek the Appropriate Balance Between Conservation and Use of Biological Diversity

A fundamental element of the ecosystem approach is recognition of the importance of reaching and/or maintaining the most appropriate balance between conservation and use of biological diversity. This is as true in the developed world as it is in developing countries. Application of the principle requires more attention to be given to the multiple use of ecosystems and on the importance of both integrated land-use planning combined with best management practices. Inappropriate balances generate either dissatisfaction on the part of local communities, or denied access to natural resources, or environmental problems and unsustainable resource use.

Case Study: Ameliorating Acid Sulphate Soils in the Mekong Delta, Vietnam

Potential acid sulphate soil materials cover about 2 × 10^6 ha of the Mekong Delta alone. They have been transformed to severely and moderately acid sulphate soils by increased oxidation occurring as a result of agricultural intensification, canal excavation and drainage networks. Under the naturally flooded conditions of the delta, *Melaleuca* (paper bark tree) and marsh wetland ecosystems predominated. These were replaced in the second half of the twentieth century mainly by rice paddies with significantly longer periods of water drawdown in the soil, causing oxidation of iron sulphide (pyrite) to sulphates (jarosite) and producing high acidity. The pH of interstitial and surface waters may be less than pH 2.0, with consequent problems of aluminium toxicity. Rice yields are usually low to negligible on acid sulphate soils, fisheries are seriously impacted and there are significant health hazards to humans. (Maltby et al. 1996a).

Although large areas of *Melaleuca* forest have been previously cleared to make way for agricultural production, there is now growing realisation of the important role played by the natural ecosystem in delivering multiple benefits for the local people and wildlife as well as meeting nature conservation interests (Fig. 5.4). *Melaleuca* is flood-tolerant and can cope with the waterlogged soil conditions which inhibit the oxidation of pyrite, preventing severe acidification (Brinkman and Xuan 1991). Following pioneering work by Ni (2000) interest is now focussing on the re-establishment of *Melaleuca*, often in close association with rice paddy (Fig. 5.4) to counteract the acidity generated from the adjacent cropland. It still remains necessary to find the right balance between intact/restored wetland and agricultural development in the Mekong Delta. However, the improved scientific knowledge of how wetlands can deliver both direct and indirect benefits to people has a key role to play in reaching this balance.

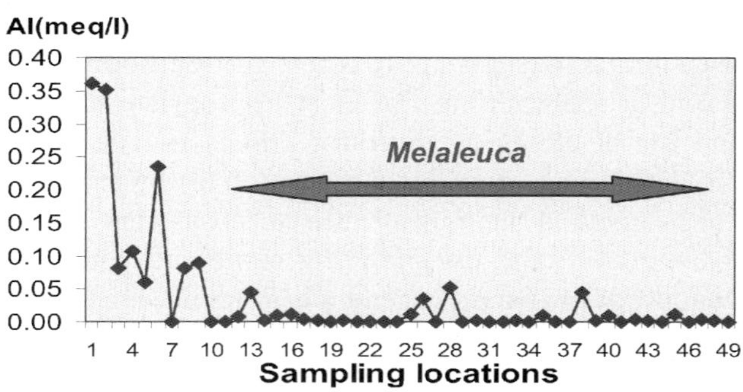

Fig. 5.4 Reduction in acidity and aluminium concentration of rice paddy drainage waters diverted through a recently planted *Melaleuca* stand on acid sulphate soil in the Mekong Delta (After Ni 2000)

5.5 Conclusion

The EA provides a methodological framework and is underpinned by principles which are highly relevant to the challenges of wetland management in today's changing world. Its application offers the possibility of balancing biodiversity conservation, sustainable use and equitable sharing of the benefits of genetic resources, whilst putting a human perspective at the centre of the process. Its implementation can be facilitated through scientific, management and societal actions.

From the scientific standpoint there is need for better understanding of the relationships between ecosystem structure, processes and functioning. There is still much uncertainty regarding the effects of human-induced and natural changes on ecosystem functioning and the delivery of ecosystem services which underpin sustainable development. More information is required on the temporal and spatial dynamics of processes and functioning at different scales, from individual wetland sites to entire river basins. It is necessary also to develop tools to enable the prediction of the effects on wetlands of different management and policy scenarios as well as those of natural changes. This requires the translation of often complex interdisciplinary science into more easy to use and understandable decision-support systems.

At the level of management, it is essential to involve all relevant stakeholders in the debate about the future of wetland resources. It is necessary to integrate the scientific research community more effectively into decision-making processes. There is a need to link more effectively the increasing knowledge of wetland ecosystem functioning into the complexities of socio-economics, recognising the often over-riding controls of short-term financial and sectoral interests which rarely if ever take into account the full values of wetlands. Wise use of the wetland resource requires stronger cross-sectoral institutional integration and/or new institutional mechanisms to overcome a long history of degradation and loss of their natural assets.

Society generally, and the political mechanisms which are empowered to deliver the appropriate balance of individual and collective interests, can undertake many key actions to promote better wetland management. One priority is to develop alternative means of redistributing the costs and benefits of land and water management. Positive measures carried out by individuals in one part of a river catchment may deny that individual certain gains but may generate wider benefits to other individuals elsewhere and help sustain overall environmental quality. Innovative financial mechanisms are required to encourage such positive measures which may achieve benefits far beyond their limits of implementation. Such actions might be linked with new types of environmental incentives recognising, for example, the wider advantages that can be achieved in reducing transboundary tensions, improving whole community welfare and achieving numerous intangible benefits. There is growing evidence that non-regulatory (advisory) incentives such as technical guidance that enables individuals, landowners or businesses to benefit from actions which also enhance wetlands or other natural ecosystems may be at least as effective as regulatory mechanisms imposed by statutory bodies. The expert advice from the research community will form a key part in elaborating these sorts of approaches.

The sustainable management of the world's wetland resources thus poses many questions to both the scientific community as well as society as a whole. The need to raise awareness that wetlands are still as important to society as they were essential to the survival of the earliest human communities is still a major challenge. It is here that the natural and social scientists must work ever more closely together. The EA is one way by which such cross-sectoral and interdisciplinary coherence may finally be achieved.

References

Barbier EB, Acreman M, Knowler D (1997) Economic valuation of wetlands, a guide for policy makers and planners. Ramsar Convention Bureau, Gland

Blackwell MSA, Hogan DV, Maltby E (2004) The short-term impact of managed realignment on soil environmental variables and hydrology. Estuarine Coastal Shelf Sci 59:687–701

Brinkman WJ, Xuan VT (1991) *Melaleuca leucadendron*, a useful and versatile tree for acid sulphate soils and some other poor environments. Int Tree Crops J 6:261–274

CBD (2003) Ecosystem approach: further elaboration, guidelines for implementation and relationship with sustainable forest management. (UNEP/CBD/SBSTTA/9/INF14) UNEP, Montreal

Chai PPK, Lee BMH, Ismawi HO (1989) Native medicinal plants of Sarawak. (Report FB1) Forestry Department, Sarawak.

Clark WG (1977) The lessons of the Peruvian anchoveta fishery. Calif Coop Oceanic Fish Invest Rep 19:57–63

Coles B, Coles J (1989) People of the wetlands. Thames and Hudson

Cosgrove WJ, Rijsberman FR (2000) World water vision: making water everybody's business. Earthscan, London

Costanza R, d'Arge R, Groot R de, Farber S, Grasso M, Hannon B, Limburg K, Naeem S, O'Neill R, Paruelo J, Raskin R, Sutton P, Belt M van den (1997) The value of the world's ecosystem services and natural capital. Nature 387:253–260

Cowardin LM, Carter V, Golet FC, LaRoe ET (1979) Classification of wetlands and deepwater habitats of the United States. US Department of the Interior, Fish and Wildlife Service, Washington, D.C.

El-Sayed SZ, Van Dijken GL (1995). The southeastern Mediterranean ecosystem revisited: thirty years after the construction of the Aswan High Dam. (Quarterdeck vol 3.1) Texas A & M University

Falkenmark M, Rockstrom J (2004). Balancing water for humans and nature: the new approach in ecohydrology. Earthscan, London

French PW (2004) The changing nature of, and approaches to, UK coastal management at the start of the twenty-first century. Geogr J 170:116–125

Immirzi CP, Maltby E (1992) The global status of peatlands and their role in carbon cycling. (A report for Friends of the Earth by the Wetland Ecosystems Research Group) University of Exeter/Friends of the Earth, London

Joosten H, Clark D (2002) Wise use of mires and peatlnads. Background and principles including a framework for decision-making. International Peat Society and International Mire Conservation Group

Justic D, Rabalais NN, Turner RE (2002) Modelling the impacts of decadal changes in riverine nutrient fluxes on coastal eutrophication near the Mississippi River Delta. Ecol Model 152:33–46

Korn H, Stadler J, Maltby E, Kerr AJ (eds) (1999) The ecosystem approach – what does it mean for european ecosystems? (International Academy for Nature Conservation, Isle of Vilm, Germany, 26–27 November, 1998) BFN, Skripten

Korn H, Schliep R, Stadler J (eds) (2003) Report of the international workshop on the "further development of the ecosystem approach". (International Academy for Nature Conservation, Isle of Vilm, Germany, 9–11 October, 2002) BFN, Skripten

Laffoley D, Burt J, Gilliland P, Connor DW, Davies J, Hill M, Breen J, Vincent M, Maltby E (2003) Adopting and ecosystem approach for the improved stewardship of the marine environment: some overarching issues. (English Nature Research Report 538) English Nature, Peterborough

Laffoley D, Maltby E, Vincent MA, Mee L, Dunn L, Gilliard P, Hamer JP, Mortimer D, Pound D (2004) The ecosystem approach. Coherent actions for marine and coastal environments. (Report to UK government) English Nature, Peterborough

Maltby E (1986) Waterlogged wealth: why waste the world's wet places. Earthscan, London

Maltby E (1988a) Wetland resources and future prospects – an international perspective. In: Zelazny J, Feierabend JS (eds) Increasing our wetland resources. National Wildlife Federation, Washington, D.C., pp 3–13

Maltby E (1988b) Global wetlands: history, current status and future. In: Hook DD, et al (eds) The ecology and management of wetlands 1. Croom Helm, pp 3–14

Maltby E (ed) (1994) An environmental and ecological study of the marshlands of Mesopotamia. (Wetland Ecosystems Research Group, University of Exeter) Amar Appeal Trust, London

Maltby E (1997a) Developing guidelines for the integrated management and sustainable utilisation of tropical lowland peatlands. In: Rieley JO, Page SE (eds) Biodiversity and sustainability of tropical peatlands. Samara, Cardigan, pp 9–18

Maltby E (1997b) Ecosystem management: the concept and the strategy. IUCN World Conserv 3/97:3–4

Maltby E (1999a) Ecosystem approach: from principles to practice. In: Schei PJ, Sandland OT, Starnd R (eds) The Norway/UN conference on the ecosystem approach for sustainable use of biological diversity, Trondheim, September 1999. Norwegian Directorate for Nature Management/Norwegian Institute for Nature Research, Trondheim, pp 30–40

Maltby E (1999b) Some European perspectives of the ecosystem approach. In: Korn H, Stadler J, Maltby E, Kerr AJ (eds) The ecosystem approach – what does it mean for european ecosystems? (International Academy for Nature Conservation, Isle of Vilm, Germany, 26–27 November, 1998) BFN, Skripten

Maltby E. (2005) Approaches to the re-establishment of the freshwater–marine continuum through the Mesopotamian marshes: regional, sectoral and transboundary challenges In: UNEP (ed) High-level meeting on the restoration of the Mesopotamian marshlands and the marine environment. Regional Organisation for the Protection of the Marine Environment/United Nations Environment Programme, Geneva, pp 7–8

Maltby E, Immirzi P (1993) Carbon dynamics in peatlands and other wetland soils: regional and global perspectives. Chemosphere 27:999–1023

Maltby E, Proctor MCF (1996a) Peatlands: their nature and role in the biosphere. In: Lappalainen E (ed) Global peat resources. International Peat Society/ Geological Survey of Finland, Helsinki, pp 11–19

Maltby E, Immirzi CP, Safford RJ (1996b) Tropical lowland peatlands of south east Asia. IUCN, Gland

Maltby E, Burbridge P, Fraser A (1996) Peat and acid sulphate soils: a case study from Vietnam. In: Maltby E, Immirzi P, Safford RJ (eds) Tropical lowland peatlands of southeast Asia. IUCN, Gland, pp 187–197

Maltby E, Holdgate M, Acreman MC, Weir A (1999) Ecosystem management: questions for science and society. Royal Holloway (University of London), Egham

Morgan E (1982) The aquatic ape – a theory of human evolution. Souvenir Press, London

Ni DV (2000) Developing a practical trial for sustainable wetland management based on environmental and socio-economic functions of *Melaleuca cajuputi* in the Mekong Delta. PhD thesis, Royal Holloway (University of London), London

Rabalais NN, Turner RE (2001) Hypoxia in the Northern Gulf of Mexico: description, causes and change. In: American Geophysical Union (ed) Coastal hypoxia: consequences for living resources and ecosystems. (Coastal and Estuarine Studies 58) American Geophysical Union, Washington, D.C., pp 1–36

Rabalais NN, Turner RE, Scavia D (2002a). Beyond science into policy: Gulf of Mexico hypoxia and the Mississippi River. BioScience 52:129–142

Rabalais NN, Turner RE, Wiseman WJ Jr (2002b) Hypoxia in the Gulf of Mexico, a.k.a. "The Dead Zone." Annu Rev Ecol Syst 33:235–263

Ramsar (2002) Global action plan for peatlands. Ramsar, available at http://www.ramsar.org/key_res_viii_17_e.htm

Raskin P, Gleick P, Kirshen P, Pontius RG, Strzepek K (1997) Water futures: assessment of long-range patterns and problems. Stockholm Environmental Institute, Stockholm

Richardson CJ (1983) Pocosin wetlands: vanishing wastelands or valuable wetlands? Bioscience 33:626–633

Safford L, Maltby E (eds) (1998) Guidelines for integrating planning and management of tropical lowlands peatlands. IUCN, Gland

Schei PJ, Sandland OT, Starnd R (1999) The Norway/UN conference on the ecosystem approach for sustainable use of biological diversity, Trondheim, September 1999. Norwegian Directorate for Nature Management/Norwegian Institute for Nature Research, Trondheim

Smith R, Maltby E (2003) Using the ecosystem approach to implement the CBD. Key issues and case studies. IUCN, Gland

Turner RK, Brouwer R, Crowards TC, Georgiou S (2003a) The economics of wetland management. In: Turner RK, Bergh JCJM van den, Brouwer R (eds) Managing wetlands: an ecological economics approach. Edward Elgar, Cheltenham, pp 73–105

Turner RK, Bergh JCJM van den, Brouwer R (eds) (2003b) Managing wetlands: an ecological economics approach. Edward Elgar, Cheltenham

Turner RK, Bergh JCJM van den, Söderquist T, Barendregt A, Straaten J van der, Maltby E, Ierland EC van (2000) Ecological-economic analysis of wetlands: scientific integration for management and policy. Ecol Econ 35:7–23

UNEP (2000) Report of the fifth meeting of the conference of the parties to the convention on biological diversity. (UNEP/CBD/COP/5/23) UNEP, Nairobi

UNEP (2001) The Mesopotamian marshlands: demise of an ecosystem. UNEP, Nairobi

WFD CIS (2003) Horizontal guidance on the role of wetlands in the water framework directive. (WFD CIS guidance document 12) Directorate General Environment of the European Commission, Brussels

6 Wetlands in the Tidal Freshwater Zone

A. Barendregt, D.F. Whigham, P. Meire, A.H. Baldwin, and S. Van Damme

6.1 Characteristics of Tidal Freshwater Wetlands

Tidal freshwater wetlands occur in the upstream reaches of many temperate estuaries. An estuary is "an inlet of the sea reaching into the river valley as far as the upstream limit of the tidal rise" (Fairbridge 1980). Within estuaries, tidal freshwater wetlands are restricted to the portion of the estuary where there is tidal action but little or no salinity. Depending on the magnitudes of tidal energy, river discharges and topography, freshwater tidal areas [practical salinity units (PSU) <0.5 parts per thousand (ppt), or <500 ppm] can be present between the highest point of tidal reach (i.e., the head of the estuary) and the oligohaline upper estuary, with PSU = 0.5–5.0 (McLusky 1993; Fig. 6.1). In some settings, tidal freshwater areas can be found as far as 150 km from the mouth of the estuary (Van Damme et al. 1999). The location of the tidal freshwater zone within the estuary depends on the balance between water volume and velocity of the incoming tide and the discharge of the out-flowing river. Towards the upstream tidal limit, the horizontal movement of tide becomes less important as the tidal wave attenuates and the vertical tide movement becomes zero partly due to the prevention of river discharge by the higher tide in the estuary. Further downstream, both the horizontal and vertical movement of the tide is important. Tidal characteristics are strongly dependent on river discharge: during high discharge, the tidal limit may be much further downstream than during low discharge. Symmetrical tides (i.e., tides with similar duration of flood and ebb) occur in the saline zone of estuaries, but in tidal freshwater zone they are not symmetrical. In the latter, there typically is a short period of intense incoming tide and a longer period of ebbing tide. Moreover, in the tidal freshwater zone, hydrologic conditions interact with the funnel-shape of the river channel to produce a tidal range that can be as high as 6 m.

In contrast to salinity, which changes in a unidirectional way, suspended matter concentrations often show a clear maximum near or in the freshwater

Ecological Studies, Vol. 191
R. Bobbink, B. Beltman, J.T.A. Verhoeven, and D.F. Whigham (Eds.)
Wetlands: Functioning, Biodiversity Conservation, and Restoration

Fig. 6.1 The three zones in the estuary (saline, brackish, fresh) with the direction of flow at low tide and at high tide

tidal zone, as illustrated in Fig. 6.2 (also in, e.g., Meade 1972; Grabemann et al. 1997; McManus 2005). Two processes are responsible for the high turbidity. First, saline water has a higher density; and it therefore remains at the bottom of the channel and forms a wedge as it moves upstream during a tidal cycle. During the change in tides, the upper freshwater flows downstream and the saline bottom-water flows upstream, facilitating the development of boundary conditions. At the tangent plane where the two flows meet, the velocity is zero, without physical disturbance, and the concentration of suspended matter is the highest (Officer 1981). Second, at the same location, physical–chemical conditions occur that result in the flocculation of suspended particles, increased by the input of particles from the brackish zone. This flocculation of suspended particles is stimulated by differences in electric charge in the particles, which is optimal at very low salinities (Eisma et al. 1994). As a result, the concentration of suspended particles is highest in the tidal freshwater zone just above the brackish zone (Meade 1972).

Tidal freshwater wetlands are thus unique ecosystems because of their physical location within river–estuarine systems. As will be described later, they have a high level of habitat and species diversity and, because they typically occur in upper parts of estuaries where human activities predominate, nutrient concentrations are also high (Mesnage et al. 2002). Accordingly, tidal freshwater wetlands are located in parts of estuaries that have had a long history of human intervention. Even though they are located in culturally important areas, they have not been studied as frequently or intensively as brackish and saline wetlands that occur nearer to the coast (Elliot and McLusky 2002). The lack of research focus on tidal freshwater wetlands has at least three explanations. First, marine ecologists have rarely investigated these systems, since the water is fresh and out of their scope. Second, river ecologists have rarely been interested in these systems because they are tidal. And third, these wetlands may be unattractive to some researchers since there are few species that are unique to freshwater tidal wetlands and, in

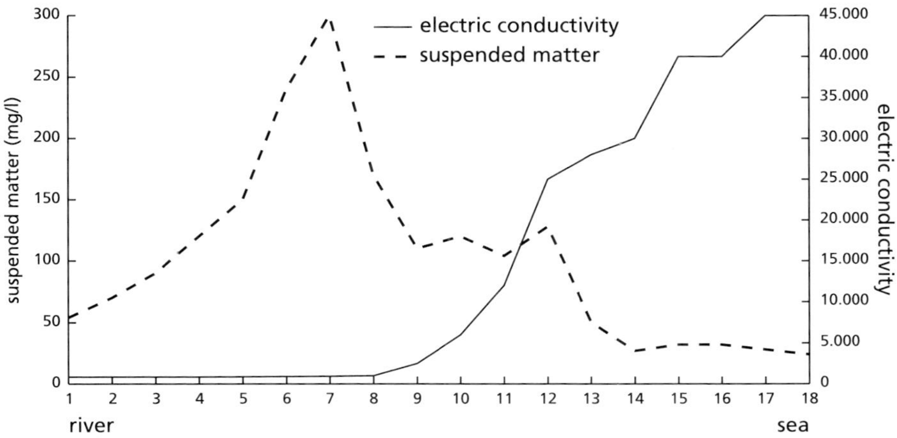

Fig. 6.2 Electric conductivity (µS/cm) and suspended matter (mg/l) in the estuary in the gradient from the river to the mouth. Data derived from river Scheldt (Van Damme et al. 1999)

addition, the system is often rich in nutrients, very muddy, and not easily accessible.

Because there have been relatively few studies of tidal freshwater wetlands (for a bibliography, see Yozzo et al. 1994), their distribution has not been fully examined throughout much of the world. In western Europe, tidal freshwater wetlands occur in, at least, the estuaries of the rivers in northwest Europe from Elbe to the Gironde (Meire and Vincx 1993), including British and Irish estuaries. On the Atlantic coast in North America, tidal freshwater wetlands occur in all major river systems from the Gulf of St. Lawrence to Georgia (Odum et al. 1984). Tidal freshwater wetlands also occur on the Pacific coast of North America (Boule 1981; Tanner et al. 2002), including Alaska, but their distribution has not been adequately documented. Junk (1983) reported that tidal freshwater systems occur on the Atlantic Coast of South America; but he offered no details on where they occurred, nor their extent.

Another type of tidal freshwater wetland may also be abundant in the upstream portions of large river deltas, such as the delta of the Mississippi River in Louisiana, USA (Penfound and Hathaway 1938; Chabreck 1972; Gosselink 1984; Mitsch and Gosselink 2000). While much research has been conducted on the ecology of Louisiana delta plain wetlands, the freshwater wetlands in this area are rarely referred to as "tidal freshwater wetlands". Many of the delta plain freshwater wetlands do have a tidal signature, but the range is small (<0.3 m). Also, there is no clear diurnal pattern of tidal flooding, but rather periodic flooding due to upstream runoff and wind effects, which can alter the water level by a meter or more. This pattern of tidal fluctuation is vastly different from those of tidal wetlands, where tides can fluctuate by several meters twice daily. While the ecological and socioeconomic importance of delta plain wetlands is unquestioned, we choose to omit detailed discussion of their ecology in this review because their flat topography and microtidal environment make them a unique case compared with the wider distribution of tidal freshwater wetlands in non-deltaic environments.

In tropical regions, coastal wetland types are replaced by the mangrove forest (Mitsch and Gosselink 2000). One of the authors (D.W.) has observed tidal freshwater wetlands in the freshwater portions of several mangrove-dominated rivers in Okinawa (Japan), suggesting that they are probably more common on a global basis than has been recognized. The presence and abundance of tidal freshwater wetlands in river systems is controlled by several factors including:

- shape of the river mouth should result in estuarine conditions in which there is a tidal flow of saline water into the river.
- The river system should include enough of a lowland area for a relatively large tidal area to exist in the river system.
- There should be a constant discharge of freshwater from the river, so that the tidal freshwater portion of the estuary occurs throughout the entire year.
- The tidal impact should dominate near at least the mouth of the river (i.e., the freshwater flow does not reach the sea without being influenced by saline and brackish water).

Consequently, it is not likely that tidal freshwater wetlands occur in tropical areas where mountains reach the sea, areas where there are deltaic plains, and areas where river flows are so large (e.g., Amazon River) that fresh water flows directly into the sea without the important saline wedge.

The following review of tidal freshwater wetlands has several objectives. First, we want to combine the relevant literature from Europe and North America, a synthesis that has not been previously attempted. Second, we want to update and summarize the literature on tidal freshwater wetlands, since the publication of the few previous syntheses in the 1980s and 1990s (Simpson et al. 1983a; Odum 1988; Meire and Vincx 1993). The only other literature-update on American tidal freshwater wetlands has been provided by Mitsch and Gosselink (2000).

6.2 Human Activities

Since human activities have modified tidal freshwater wetlands in many ways, a short introduction to the historical changes in land use and water quality is given.

6.2.1 Historical Development

Tidal freshwater wetlands have been intimately linked to the development of human cultures, especially in parts of Europe. Archaeological research west of Rotterdam in The Netherlands suggests that humans associated with the "Vlaardingen culture" were present in tidal freshwater systems as far back as 2700 BC (De Ridder 1999). Basket-worked fish-traps have been found in former streams; and dams and dikes were constructed to change the flooding frequency. By 175 BC, people were using culverts made from tree trunks to drain tidal freshwater habitats. The culverts had a one-way valve at the outer side, preventing the inflow of water at high tide and thus allowing areas to drain. Analyses of pollen and plant remnants from that period suggest that the wetlands were dominated by grasses (probably *Phragmites*) and sedges, with other frequent species such as *Typha*, *Atriplex*, and *Rumex*. No species from brackish or saline conditions were represented. The most common woody plants in or near the wetlands were alder (*Alnus*), with some oaks (*Quercus*), pines (*Pinus*), and willows (*Salix*; Brinkkemper and De Ridder 2001). Paleological reconstruction suggests that, at that time, tidal freshwater systems were part of an open landscape at the edge of peatland with distinct tidal influences. There is also strong evidence that people had agricultural fields and kept cattle in the wetlands. Humans undoubtedly used tidal freshwater wetlands in North America but the impacts of humans in pre-colonial times have not been well documented (Kiviat 1991).

Due to sedimentation and erosion processes, enormous changes occurred over the centuries in the tidal freshwater zones in European rivers. From the Middle Ages to a century ago, maps of the tidal areas in western Belgium, southwestern Netherlands and northern Germany documented drastic changes. New estuaries were formed in the delta of the Rhine system and wetland areas were changed to land by diking to prevent flooding and for purposes of reclamation to agricultural lands. During the past 200 years, technology supported this change and the former large tidal areas are at present mostly restricted to the main stream-bed of the estuary, e.g., as illustrated by the river Seine (Avoine et al. 1981). Large port cities developed in the regions of rivers where tidal freshwater systems occurred and they were exploited by societies for their economic potential (Pinder and Witherick 1990; Preisinger 1991). Major cities and harbors in Europe and North America arose in this way (e.g., Hamburg, Rotterdam, Antwerp, Dublin, London, Philadelphia, Richmond, Washington, D.C.). Odum et al. (1984, p. 88) state that "almost all of this habitat on the USA–Atlantic coast is in the 13 original colonies" and that diking, dredging and filling of these wetlands have occurred widely.

6.2.2 Water Quality Changes

In addition to physical changes and loss in area associated with agricultural and urban development, another important change was the enormous increase in the discharge of human and industrial wastewater into the rivers upstream of and within the tidal freshwater zone. As a result, the chemistry of the rivers changed in three ways. First, the discharge of sewage increased the nutrient load and the organic components, resulting in a high oxygen demand that led to a reduction of oxygen concentrations in the tidal freshwater zone (Riedel-Lorje and Gaument 1982; Marchand 1993; Van Damme et al. 1999). Second, the discharge of heavy metals and other contaminants resulted in the further degradation of tidal freshwater systems (Schuchardt et al. 1993; Khan and Brush 1994; Knight and Pasternack 2000; Middelkoop 2000; Ciszewski 2001; Walling et al. 2003). Third, the concentrations of organic micropollutants in river water increased. Due to severe sedimentation of the suspended matter including the pollutants, concentration of pollutants increased, resulting in toxicological problems in tidal freshwater systems (Heemken et al. 2000; Steen et al. 2002; Jonkers et al. 2003).

Many of these problems continue today, even though advanced wastewater treatment facilities are present and industries do a much better job of minimizing pollutant discharges. Pollution is especially problematic where sedimentation rates are very high. Consequently sediments contain high concentrations in heavy metals and organic micropollutants; sometimes the quantities are too large to be stabilized, transformed, or removed by natural processes (Ridgley and Rijsberman 1994). In The Netherlands, for example, harbor areas had to be dredged and the polluted sediments were often stored in newly created polders next to the river in the freshwater tidal areas.

6.3 Biological Variation Within the Freshwater Tidal Ecosystem

This section describes the zonation in vegetation, the temporal variation and seed dynamics in plant communities, and the presence of wildlife and other biota. The aim is to compare and contrast the biology and ecology of tidal freshwater wetlands on both sides of the Atlantic Ocean.

6.3.1 Vegetation Zonation

The primary ecological factors influencing the distribution of plant species are the impacts of tides and variations in surface elevations. The duration and frequency of flooding at any location depends on the elevation of the substrate relative to mean high tide and result in distinct zonation in vegetation (Zonneveld 1960; Simpson et al. 1983a; De Boois 1982; Preisinger 1991). North American and European tidal freshwater wetlands are comparable in processes and zonation; however, the species differ considerably. Table 6.1 summarizes the most important zones and their species. In general, the num-

Table 6.1 Some characteristic species in the vegetation of tidal freshwater wetlands in Europe and North America

Vegetation Zone	Europe	North America
Aquatic vegetation	*Potamogeton* spp, *Nuphar lutea*	*Ceratophyllum demersum*, *Elodea* spp, *Hydrilla verticillata*, *Myriophyllum* spp, *Potamogeton* spp, *Vallisneria americana*
Low marsh	*Scirpus* spp, *Phragmites australis*, *Caltha palustris*	*Nuphar advena*, *Peltandra virginica*, *Pontederia cordata*, *Sagittaria latifolia*, *Zizania aquatica*
High marsh	*Valeriana officinalis*, *Lythrum salicaria*, *Epilobium spp.*, *Bidens* spp, *Symphytum officinale*, *Solanum dulcamare*	*Acorus calamus*, *Amaranthus cannabinus*, *Ambrosia trifida*, *Aster* spp., *Bidens* spp, *Carex* spp, *Cicuta maculata*, *Cuscuta gronovii*, *Cyperus* spp, *Echinochloa* spp, *Eleocharis* spp, *Galium tinctorium*, *Impatiens capensis*, *Iris* spp, *Ludwigia* spp, *Hibiscus moscheutos*, *Leersia oryzoides*, *Peltandra virginica*, *Phragmites australis*, *Pilea pumila*, *Polygonum* spp, *Scirpus* spp, *Schoenoplectus* spp, *Typha* spp
Forests/swamps	Willow species: *Salix* spp with, e.g., *Anthriscus sylvestris*, divided in subtypes: (1) with *Callitriche stagnalis*, (2) with *Cardamine amara*, (3) with *Circea lutetiana*	Trees: *Acer rubrum*, *Chamaecyparis thyoides*, *Fraxinus pennsylvanica*, *Magnolia virginiana*, *Nyssa biflora*, *Carpinus caroliniana*Shrubs: *Clethra alnifolia*, *Ilex verticillata*, *Itea virginica*, *Leucothoe racemosa*, *Rhododendron viscosum*, *Taxodium distichum*, *Vaccinium corymbosum*, *Viburnum* spp.Vines: *Parthenocissu quiquefolia*, *Smilax* spp, *Toxicodendron radicans*Herbs: many of the species listed under high and low marsh above; *Cinna arundinacea*, *Viola cucullata*, *Osmunda* spp, *Thelypteris thelypteriodes*, *Woodwardia* spp

ber of plant species in a zone increases with decreasing frequency and duration of flooding.

6.3.2 The Vegetation of European Tidal Freshwater Wetlands

In the tidal creeks, the current is usually too strong and turbidity too high for vascular plants to become established. Tidal creeks also have dynamic sand banks and areas of mud flats that just occur above low tide (Zonneveld 1960). Only in more isolated shallow pools are some aquatic plants present, such as *Potamogeton* spp, *Sagittaria sagittifolia*, and *Nuphar lutea*. Closer to high-tide level, vascular plants become common and four basic types of vegetation are present.

Below mean high tide, plant species are represented that can stand flooding and wave activity. In areas with wave energy, *Scirpus triqueter*, *S. maritimus*, and *S. lacustris* are most common. *Scirpus triqueter* is an endangered species (Deegan and Harrington 2004), being unable to persist in areas impacted by lack of wave energy (after human control). Just above mean high tide level, the vegetation dominated by *Scirpus* spp is replaced by a marsh vegetation that has many herbaceous plants, such as species in the genera *Lythrum*, *Phalaris*, *Epilobium*, *Typha*, *Symphytum*, *Valeriana*, and *Sparganium*. As the marshes are eutrophic (Verhoeven et al. 2001), there is a dominance of *Urtica dioica* and *Calystegia sepium*. In habitats with a similar tidal regime but with a high concentration of organic material in the substrate, rare characteristic species are *Leucojum aestivum* and *Equisetum fluviatile*.

At other locations, reed (*Phragmites australis*) is the dominant plant but many other species, e.g., *Veronica anagallis-aquatica* also occur. Very characteristic at these locations is *Caltha palustris* var. *araneosa*, which produces special roots, at leaf nodes, that can be transported by the tides (Van Steenis 1971). However, the majority of locations in higher tidal zones are covered by willow (*Salix*) forests, either managed as plantations (osier beds) or as natural vegetation (Fig 6.3A, B). At present, three sub-types of this forest occur (Barendregt 2005). At or just below the mean high water line, where anaerobic conditions prevail and soil development is just starting, the conditions are still too wet for many terrestrial species; and *Callitriche stagnalis* and a species of *Vaucheria* occur. Just above high tide, *Cardamine amara* is abundant; and 50 cm above mean high water level *Circea lutetiana* occur. Within this wetland–upland transition zone, there is often evidence of succession (including species such as *Carex remota*) and soil genesis that results in the accumulation of organic matter. European freshwater tidal wetlands do not suffer for invasive species, although *Angelica archangelica* and *Impatiens noli-tangere* might be represented in the vegetation.

Biomass varies across different zones within tidal freshwater wetlands and the general pattern is for biomass to increase with increasing elevation within

Fig. 6.3 **A** (*top*) Characteristic zonation in European tidal freshwater wetlands, with a tidal creek between the zones, close to the river a zone with low marsh (left side) dominated by bulrush (*Scirpus lacustris*), in the center a reed bed (*Phragmites australis*), and to the right side a willow forest (*Salix* spp). Photo from river Oude Maas, The Netherlands, by A. Barendregt. **B** (*bottom*) A tidal creek at low tide, entering the willow forest (both sides of the creek) and a small reed bed (center), illustrating the muddy soil in the creek. At high tide, the water table is up to the willow trees. Photo from the river Oude Maas, The Netherlands, by A. Barendregt

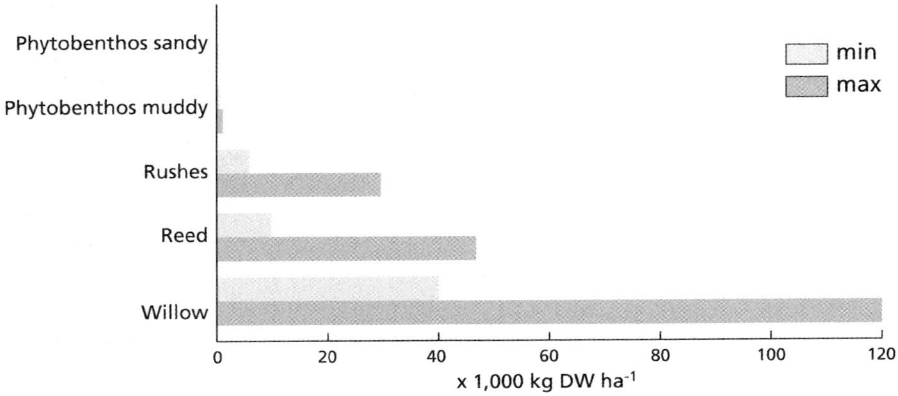

Fig. 6.4 Biomass (minimum and maximum values) in some types of vegetation in tidal freshwater wetlands in Western Europe (unpublished data from Scheldt catchment)

the tidal zone. This pattern is illustrated for a European tidal freshwater wetland (Fig. 6.4). At the lowest elevation within the tidal zone, the biomass of non-vascular plants (e.g., phyto-benthos) is low on both sandy and muddy substrates. At higher elevations, the biomass of vascular plants such as bulrushes (*Scirpus* spp) and reed (*Phragmites)* is intermediate and the highest biomass is associated with stands of willow (*Salix* spp) at the highest elevations within the intertidal zone.

6.3.3 The Vegetation of North American Tidal Freshwater Wetlands

In North America, tidal freshwater wetlands include swamps, which are dominated by trees and shrubs, and emergent wetlands dominated by herbaceous plants. Along the Atlantic Coast, tidal freshwater swamps may abut rivers or be separated from the river by a band of tidal freshwater marsh of varying width. Swamps often have a complex microtopography created by fallen logs that results in elevated "hummocks" interspersed with lower "hollows" (Rheinhardt 1992; Peterson and Baldwin 2004). The hummocks may extend 30 cm or more above the hollows, resulting in marked differences in species composition of herbaceous plants over short distances.

Tidal freshwater wetland habitats that are dominated by herbaceous species can be divided into two broad categories: high marsh and low marsh (Simpson et al. 1983a). The two habitats are distinguished by differences in hydro-period along an elevation gradient. In many locations, the pattern of zonation proceeding away from the river is low marsh–high marsh–swamp (Fig. 6.5A, B). Interestingly, the elevation of the high marsh may be above that of the swamp hollows due to a natural levee effect where river sediment

Fig. 6.5 A (*top*) Characteristic zonation in North American tidal freshwater wetlands. Visible on the right of the river at the lowest elevations is the low marsh, here dominated by *Nuphar advena*. At a slightly higher elevation is the high marsh, here visible as a line of lighter shade above the low marsh, with a diverse mixture of herbaceous annuals and perennials. Further to the right, but at a slightly lower elevation, lies the tidal freshwater swamp forest, with species-rich communities. Photo from the Nanticoke River, Maryland, by A. Baldwin. **B** (*bottom*) A tidal creek at low tide. In this late summer (August) scene, the dominant species are both annuals. Along the creek is *Polygonum punctatum* and the tall plant behind it is wild rice (*Zizania aquatica* var. *aquatica*). Photo from the Hamilton Marshes located near Trenton, New Jersey, by D.F. Whigham

settles out primarily directly along the banks of the river during high tides or flooding events. Moreover, a latitudinal climate gradient results in large variation in species composition and distribution of types of tidal freshwater wetlands.

Tidal freshwater swamps are typically dominated by hardwood species, such as *Nyssa* spp, *Fraxinus pennsylvanica*, and *Acer rubrum*; they have a diverse shrub layer, containing species such as *Vaccinium corymbosum*, *Rhododendron viscosum*, *Viburnum* spp, and *Magnolia virginiana*. Herbaceous communities also occur in swamp habitats if the tree and shrub strata are sparse enough to allow adequate light penetration. Hummocks may be populated by species of *Carex* spp as well as grasses such as *Cinna arundinacea* and ferns such as *Osmunda cinnamomea*, *O. regalis*, and *Thelypteris palustris*. Other herbaceous species like *Viola cucullata* occur on hummocks. In low areas between hummocks (i.e., hollows) one finds species that also occur in the high marsh in full sunlight. Common species in hollows are *Peltandra virginica*, *Leersia oryzoides*, *Typha latifolia*, and *Zizania aquatica*.

High marsh vegetation is often more diverse than low marsh vegetation, containing mixtures of annual such as *Bidens* spp, *Impatiens capensis*, *Polygonum arifolium* and perennials like *Panicum virgatum*, *Peltandra virginica*, and *Leersia oryzoides*. Low marshes, in contrast, are dominated by perennials such as *Nuphar advena* and *Pontederia cordata*. The only annual species that is common to many low marsh habitats along the Atlantic Coast is *Zizania aquatica*.

Along the Atlantic Coast, only one species has been recognized as being rare, *Aeschynomene virginica* (Griffith and Forseth 2003); and it appears to be a fugitive species that persists by colonizing disturbance patches on the levee of tidal creeks. Invasive species have also received attention in tidal freshwater wetlands because of the potential impacts that they may have on native species and ecosystem processes (Findlay et al. 1990). Especially noteworthy is the increased abundance of European genotypes of *Phragmites australis* that invade and expand more readily than native North American genotypes (Saltonstall 2002). The expansion of *Phragmites* has been noted in several estuarine systems (Chambers et al. 1999; Meyerson et al. 2000), including freshwater areas (Rice et al. 2000), but the impact on native species remains undetermined and there are few efforts to eradicate or control it (Findlay and Groffman 2003; Teal and Peterson 2005). *Lythrum salicaria* and *Murdannia keisak* are two Eurasian species that occur in some tidal freshwater marshes in the United States (Baldwin and DeRico 1999; Baldwin and Pendleton 2003).

Temporal Variation in Plant Communities

Most of the ecological studies of plant communities of tidal freshwater wetlands are from the United States Atlantic coast; and marshes have been studied more than swamps. While species composition has been described in many systems, seed bank and vegetation dynamics have been studied primarily in high marsh habitats. One of the most dramatic features of plant communities of tidal freshwater wetlands is the tremendous variation in relative abundance of species throughout the growing season. Many of the species of tidal freshwater wetlands do not persist as dead culms or shoots above the soil surface over the winter; and so the wetland has a flat, barren appearance, with a layer of decomposing plant material that decreases in thickness during the winter months. Early in the spring (e.g., April in the mid-Atlantic), perennial species begin to send up shoots. Simultaneously, the high-light environment, increasing photoperiod, and warming temperatures stimulate germination of primarily obligate or facultative annual species that are part of soil seed bank, often yielding high densities of established seedlings (Leck 2003). The leaves and shoots of the emerging perennials (e.g., *Acorus calamus*, *Peltandra virginica*, *Leersia oryzoides*, *Hibiscus moscheutos*) typically grow more rapidly than the seedlings, reaching their maximum height and biomass and forming a canopy early in the summer (e.g., June in the mid-Atlantic; Odum 1988; Whigham and Simpson 1992). The marshes during this time of year have a deep green, lush appearance, perhaps broken by occasional splotches of flower color. The annuals continue to grow, experience thinning, and eventually overtop the perennials and reach their maximum biomass later in the season (e.g., September) as dominant perennials are senescing. Examples of annuals that dominate by late summer are *Polygonum arifolium*, *P. sagittatum*, and *Bidens laevis*. By the end of the season, the marshes have a red-brown appearance, enhanced by the yellow flowers of *Bidens*. However, this pattern does not hold for all species; some perennials also flower late in the season (e.g., *Symphyotrichum puniceum*, *Cicuta maculata*, *Helenium autumnale*).

In addition to seasonal changes in community structure, which are primarily a result of life history and phenology of the various species, vegetation of tidal freshwater wetlands changes in composition between years (Leck and Simpson 1995). Much of this variation may be due to changes in the abundance of annual species between years. Annual species are a major component of wetland vegetation, often comprising half or more of the species, number of individuals, and biomass (Parker and Leck 1985; Whigham and Simpson 1992). Because annuals must be recruited each year from the seed bank or recently dispersed seeds, conditions that inhibit germination and seedling establishment and growth (which occur early in the growing season) will result in lower abundance of those species throughout the growing season (Baldwin et al. 2001).

Seed–Vegetation Dynamics

Because of the abundance of annual species in tidal freshwater wetlands, seed banks, seed dispersal, and seed germination are important aspects of their plant community dynamics. Just as vegetation fluctuates between years in these systems, so does the relative abundance of species in the seed bank (Leck and Simpson 1995). However, the seed bank may not mirror the standing vegetation. While many species do occur in both seed bank and vegetation (e.g., *Bidens* spp. *Impatiens capensis*, *Polygonum* spp, *Typha* spp), some of the dominant vegetation species (e.g., *Nuphar advena*, *Phragmites australis*) are rarely found in the seed bank, while some common species (e.g., *Juncus effusus*, *Cyperus erythrorhizos*) in the seed bank occur rarely in undisturbed vegetation (Leck et al. 1988).

Differences between seed bank and vegetation have been noted in other ecosystems and are likely due to differences in evolved life history strategy. For example, seed banks provide a mechanism for species with persistent seeds to emerge following a disturbance such as ice scouring or fire that removes dominant canopy vegetation. The seed bank species in this case can be considered "fugitive species" that are recruited in patches following disturbances but are otherwise rare in matrix vegetation. Similarly, many perennial marsh species rely on clonal propagation as their primary mode of reproduction and may have lost their capacity to survive for long or even short periods as buried seeds due to genetic drift. Others may be incapable of sexual reproduction: *Acorus calamus*, a dominant perennial in many United States tidal freshwater wetlands, is sterile due to polyploidy, at least in Europe (Eckert 2002). Seeds also differ between species in how long they typically exist in the seed bank. Leck and Simpson (1987) identify three strategies for dominant seed banks species in mid-Atlantic tidal freshwater wetlands: (1) transient species that have seeds that overwinter but are completely depleted during spring germination (e.g., *Impatiens capensis*), (2) high turnover with some reserve of viable seeds (e.g., *Bidens laevis*), and (3) large long-term seed bank (e.g., *Ranunculus sceleratus*). Because species with different strategies are dependent to different degrees on annual renewal of the seed bank, variation in seed bank strategy contributes to interannual variation in species composition of vegetation. For example, if *Impatiens capensis* (a transient species) is not abundant in vegetation one year due to wetter-than-normal conditions, it contributes fewer seeds to the seed bank, and so it has less propagules available for establishment the following year. Species with persistent seed banks, in contrast, are less impacted by the prior year's seed production.

Just as seed bank strategy varies between species, so do seed germination requirements (Leck 1996). Some species require light for germination (e.g., *Bidens laevis*), while others do not (e.g., *Peltandra virginica*). Others require the presence of oxygen (*Impatiens capensis*), while others germinate under anaerobic conditions (*Pontederia cordata*). Many species of temperate tidal freshwater wetlands require cold stratification for two months or more before germination can occur; and germination percentage varies, depending on temperature regime. Flooding has been found to reduce richness and density of seedlings emerging from tidal freshwater marsh seed banks (Baldwin et al. 2001); and salinity was found to reduce seedling emergence from seed banks of oligohaline tidal marshes (PSU <5; Baldwin et al. 1996).

6.3.4 Wildlife

In Europe most mammals have been reduced in abundance and none of them seem to impact tidal freshwater wetlands in a significant way, with the exception of beaver (*Castor fiber*) which was reintroduced in one large area in the Netherlands (Biesbosch) about a decade ago. The beaver now has a stable population but its impacts are relatively restricted. In the United States two herbivores, the native muskrat (*Ondatra zibethicus*) and the introduced nutria (*Myocastor coypus*), are known to impact vegetation and nutrient cycling (Connors et al. 2000) in tidal freshwater wetlands, mostly by digging the rhizomes (Odum et al. 1984). Beaver are now also common and rapidly spreading in North America and they have become regular inhabitants of tidal freshwater wetlands (D.F. Whigham, personal observation).

The freshwater tidal region is important for birds, especially waterfowl such as ducks and waders, as resting and feeding places during migration (Ysebaert et al. 2000). Tidal freshwater habitats are especially important during freezing periods because the tidal influence causes open water areas to persist. These habitats are used by numerous migratory and resident species in the spring because they provide food resources early in the growing season (Odum et al. 1984). Tidal freshwater wetlands are also important breeding sites for a number of duck species and passerines (Hawkins and Leck 1977). In willow forests many passerine species breed, as well as species of herons and cormorants. In the Netherlands, for instance, there has been a sudden increase in the population of the endangered bluethroat (*Luscinia svecica*) in freshwater tidal wetlands, resulting in a stimulation of the whole national population (Meijer and van der Nat 1989).

6.3.5 Fish Species

Many fish species occur in estuaries, but no species are solely restricted to the freshwater tidal areas (Odum et al. 1984; Anderson and Schmidt 1989; Lobry et al. 2003). In contrast, most fish species occur only in the saline areas and brackish parts of the estuary and rarely move into tidal freshwater areas, except for the juvenile stage of some species (Thiel and Potter 2001; Lobry et al. 2003). In the freshwater tidal parts of estuaries, most of the common fish species also occur in the non-tidal parts of the rivers. There are also anadromous fish species that pass through the tidal freshwater areas during movement from or to the sea (Pomfret et al. 1991). Because of river regulation (e.g., dams) and pollution, some anadromous species no longer occur in some European rivers that have tidal freshwater habitats. The salmon (*Salmo salar*) and the sturgeon (*Acipenser sturio*) disappeared from most European rivers more than 50 years ago (Verhey et al. 1961) and the allis shad (*Alosa alosa*) disappeared in 1920 from the river Rhine.

An important European fish species that requires freshwater tidal habitats is theTwaite shad (*Alosa fallax*). Their spawning grounds are at the upstream border of the tidal areas and the juveniles use the freshwater tidal areas as nursery grounds. Most of the populations of theTwaite shad have been affected by over-fishing and pollution, but damming of the estuary of the Rhine system in the 1970 also negatively impacted the species. However, in recent years the recovery of theTwaite shad has been reported from Germany and Belgium as a result of reduced water pollution (Gerkens and Thiel 2001). Along the Atlantic coast of North America, the american shad (*Alosa sapidissima*) and the alewife (*A. pseudoharengus*) occupy habitats similar to those of theTwaite shad and they have suffered many of the same consequences for the same reasons (Odum et al. 1984), indicating comparable ecology in the New World.

6.3.6 Other Biota

Just as elsewhere in the world, the biodiversity of other groups is high in tidal freshwater wetlands but most of them have not been studied in detail. In Europe and North America, the diversity and ecological importance of algal species (with special diatom species: Rehbehn et al. 1993; Muylaert and Sabbe 1996) has not been examined thoroughly (e.g., Kiviat and Barbour 1996; Muylaert et al. 1997). Animal-related publications for tidal freshwater wetlands in North America can be found in the bibliography published by Yozzo et al. (1994). Here we focus on two groups of animals: benthic species and terrestrial invertebrates that live on emergent herbaceous and woody vegetation.

Compared to brackish and saline zones in estuaries (Seys et al. 1999), the diversity of benthic species is less in tidal freshwater areas due to high turbidity of unconsolidated substrates (Van Damme et al. 1999, Wolff 1973). Only Oligochaeta (*Tubifex*) have been described as being abundant from The Netherlands by Heyligers (see: Verhey et al. 1961). In North America, studies of various animal groups in Tivoli Bay (Hudson River) and Chesapeake Bay have focussed on microbenthos (Simpson et al. 1984), meiofauna (Yozzo and Smith 1995), ostractods (Yozzo and Steineck 1994), and fish-microcrustacean interactions (Yozzo and Odum 1993).

In contrast to benthic animals, terrestrial invertebrates in tidal freshwater wetlands are highly diverse and include many taxonomic groups, such as spiders, beetles, woodlice, molluscs, millipedes, worms, and springtails (Barbour and Kiviat 1986). Only few investigators, however, have examined this group of animals in tidal freshwater wetlands. Heyligers (see: Verhey, 1961) reports the flooding frequency and the weakness of the soil as the principle variables controlling the diversity of terrestrial invertebrates. Desender and Mealfait (1999) and Hendrickx et al. (2001) suggested that the characteristics of the vegetation structure were also important. In the willow forests along the river Oude Maas, where nowadays the best freshwater tidal wetlands in The Netherlands occur, 65 pitfall traps at different locations have been used to examine ground-dwelling invertebrates (Barendregt 2005). For each species captured, Barendregt calculated their range according to mean high water table as an index of their response to flooding frequency. No species occurred over the entire range of locations and three types of distribution patterns were found (Fig. 6.6). A group of semi-aquatic species (low range) needs daily flooding and they are well represented only below mean high tide. A second group of species (high range) mostly occur above mean high tide, ranging from species that starting just 20 cm above mean high tide level (= occasionally flooded) to locations more than 60 cm above high tide (= rarely flooded). The lower the elevation, the more tolerant these species have to be to flooding. This second group consists of many common wetland or wet forest species. A third group of species (middle range) occurs around or just above mean high tide, in habitats that are flooded frequently; and this group avoids the higher range. This identification

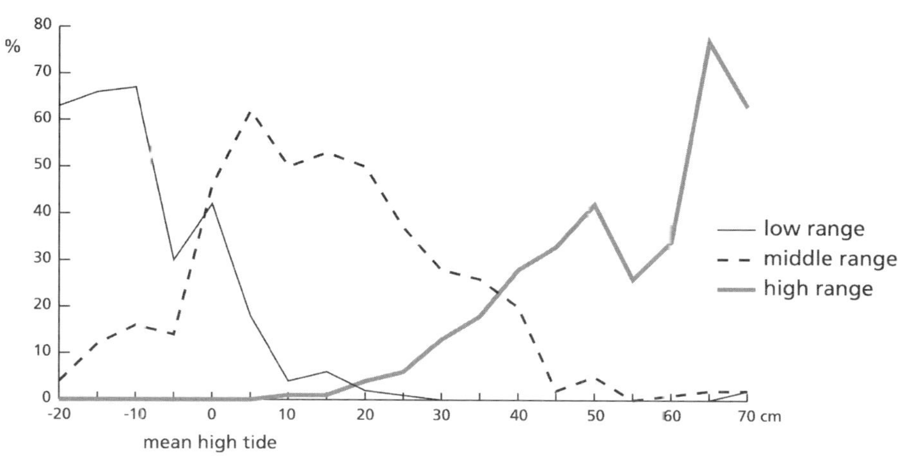

Fig. 6.6 Cumulative relative preference in Dutch tidal freshwater wetlands of some ground-dwelling invertebrate species according to the elevation related to the level of mean high tide, indicated by three groups of species: (1) four species from the low range, (2) five species from the middle range, (3) six species from the high range

of three distinct groups of invertebrates supports the hypothesis that by stress the competition from aquatic and terrestrial species is less prominent in this intermediate zone and this facilitates the presence of another group of species from the middle range, probably stimulated by this flooding stress. The evaluation of all species from this system demonstrates that most rare or infrequent species of the tidal freshwater systems of The Netherlands, such as representatives from Mollusca, Coleoptera, Isopoda, and Araneae, are found in this intermediate zone, supporting the explanation that characteristic tidal wetland species are represented in this zone (Barendregt 2005).

In addition to the pattern in Fig. 6.6, another dimension of the biota in tidal freshwater wetlands is given in the distribution of terrestrial invertebrates within the wetland. In The Netherlands, willow forests are managed and are cut, on average, every four years. Harvesting procedures result in significant physical changes; areas of bare clay soils are exposed. Following harvesting the habitat is characterized by high light intensity but after about two years of vegetation recovery, light intensities in willow stands are much lower. When areas are abandoned for a decade or more, the willow forest comes into the next phase of succession and biomass plus litter fall within the system increases. Some species of terrestrial invertebrates occur in all stages of vegetation development during the succession but about half of all invertebrate species prefer the early or the late phase in succession.

6.4 Chemical and Physical Processes: the Wetland as a Filter

Estuaries are well known to act as filters that prevent compounds transported by rivers from reaching the sea. Tidal action is not only the driving force behind the existence of fringing wetlands in estuaries, but it also helps them to play an important role in the estuarine filter function. As described in the Introduction, tidal freshwater wetlands occur near the upper limit of tide in the region of most rivers that is nutrient-rich. From a landscape perspective, the salt, brackish, and fresh types of wetlands that occur in estuaries function decidedly differently and the material presented in this section is mostly based on comparisons between freshwater tidal wetlands and brackish or saline wetlands.

A first obvious function that freshwater wetlands exert on the quality of flooding water is physical aeration. As many estuaries are heterotrophic, especially in their upstream stretches, wetlands provide a shallow distribution surface for oxygen poor water, enhancing oxygen influx from the atmosphere. As primary production in estuaries is often limited by turbidity, physical aeration often contributes more to the oxygen status of the water than primary production (Soetaert and Herman 1995). Moreover, the oxygen levels differ from neap to spring tides (Parker et al. 1994). The surplus in carbon input from the river does not reach the sea, since the majority of the organic input is broken down to CO_2 by bacteria and disappears from the system in the freshwater tidal zone. At the same time, the oxygen concentration is lowest in this zone, as a result of the metabolic activity of bacteria that interact with, for example, compounds with a high biological oxygen demand that result from human activities (Marchand 1993). Attempts to draw inferences about freshwater wetland – estuarine interactions were in the past predominantly performed by measuring nutrient concentrations of flooding and draining water over tidal cycles (classic exchange budgets; e.g., Simpson et al. 1983b). These results indicated that most of the nitrogen (N) and phosphorus (P) entering the wetland was transformed in its characteristics with nutrients mostly changed from particulate to dissolved forms. There is apparently only a relatively small net import or export of nutrients during a typical tide cycle (Fig. 6.7). Although these results may well be biased by errors in the hydrologic budget, concentration profiles show that the seepage water quality in creeks at low tide can clearly diverge from the river water quality.

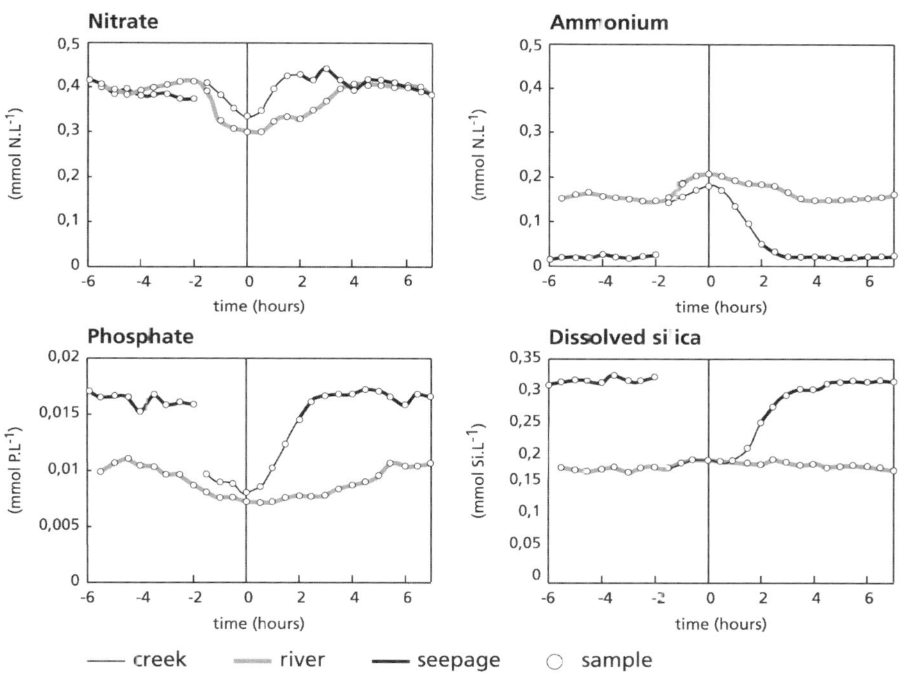

Fig. 6.7 Concentration profiles of nutrients in the main creek of a freshwater marsh and in the adjacent river channel of the Scheldt estuary (Belgium), 29 April 1998. The seepage phase consisted of the outflow of water from the creek when the water level at the creek mouth was <30 cm deep. Time 0 is the moment of high tide in the river

In general, saline wetlands are considered to be carbon and nutrient sinks (e.g., Odum 1988; Mitsch and Gosselink 2000), brackish wetlands are either carbon sinks or sources depending on the flooding regime (e.g., Jordan and Correll 1991); and tidal freshwater wetlands have been shown to be nutrient sinks during the growing season and sources of carbon and nutrients in the non-growing season (Simpson et al. 1983a; Neubauer et al. 2000; Neubauer and Anderson 2003).

Seasonal patterns of nutrient uptake and release also differ among the three types of tidal wetlands, especially between freshwater tidal wetland and brackish and saline wetlands due to the influence of saline conditions in the latter two (Simpson et al. 1983a; Odum et al. 1984; Mitsch and Gosselink 2000). Patterns of nutrient uptake and release are also influenced by differences in patterns of sediment deposition. Brackish tidal wetlands and freshwater tidal wetlands occur in the maximum turbidity zone of tidal rivers where sediment deposition rates are high (Darke and Megonigal 2003). Sediment inputs and consequently the build-up of the height of the wetland surface is typically greater in freshwater tidal wetlands, which are closer than saline wetlands to upland sediment sources and the turbidity maximum zone (Pasternack et al. 2000; Neubauer et al. 2002). Sedimentation rates are, however, highly variable within freshwater tidal wetlands and both vegetation and location of the wetland relative to the turbidity maximum of the river are important factors in determining rates of sediment deposition (Uncles et al. 1998; Pasternack and Brush 2001; Darke and Megonigal 2003).

The consequences of high sediment deposition rates and lack of salinity stress result in high productivity, typically high rates of decomposition, and higher rates of nutrient uptake during the growing season in freshwater tidal wetlands relative to non-wetland ecosystems (Whigham and Simpson 1976, 1992; Whigham et al. 1978; Simpson et al. 1983a, 1983b; Bowden 1984b; Odum et al. 1984; Findlay et al. 1990, 1991, 1998, 2002; Neubauer et al. 2000, 2002; Pasternack et al. 2000).

Patterns of nutrient cycling also differ even though all tidal wetlands are characterized by sediments that are mostly anaerobic. Anaerobic metabolism in tidal freshwater wetlands is dominated by methanogenesis compared to a dominance of sulfate reduction in saline wetlands (Odum 1988). All aspects of the nitrogen cycle (nitrification, denitrification, mineralization, fixation) are important in tidal freshwater wetlands and all components of the system play important roles in controlling the patterns and rates of N cycling (Bowden 1984b; Morris and Bowden 1986). Only the surface sediments are aerobic; and nitrate produced by mineralization is quickly assimilated (Bowden 1984a, 1986). The net transfer of N is to the sediments; and N appears to be used more efficiently than P (Bowden 1984b, 1986).

The bivalent aerobic–anaerobic characteristic of surface sediments in tidal freshwater wetlands triggers questions about which aspect dominates nutrient cycling: the aerobic surface sediments that are flooded with water that is often oxygen-poor, or the anaerobic subsurface sediments into which and from which nutrient exchange is limited because of the waterlogged nature of the sediments and the presence of dense inorganic particles such as silts and clays. The relatively new technique of in situ stable isotope enrichment allows for the examination of N flow through multiple pools simultaneously while maintaining natural hydrologic and biogeochemical gradients and ecosystem functions. Some whole ecosystem ^{15}N enrichments have been performed, showing unequivocally that nitrification is one of the most important transformation processes associated with eutrophic tidal freshwater marshes (Gribsholt et al., in preparation). Comparison of ecosystem-scale nitrification with pelagic nitrification demonstrated that the reactive surface area of the wetland is the key site for nitrification. Three key factors may potentially be important determinants of nitrification in wetland ecosystems: nitrifier biomass, oxygen dynamics, and ammonium availability for nitrifiers. The distribution, abundance, and activity of nitrifiers is known to be influenced by their attachment to particles; and the surfaces of the dense vegetation and plant litter in the marsh may provide an excellent substrate for microbial colonization.

Phosphorus cycling is mostly controlled by sediment input, phosphorus generation in anaerobic sediments through release, and interactions with iron. In freshwater tidal wetlands when FeO_x is reduced in the anoxic sediments, the resulting Fe(II) diffuses upward into aerobic layers and is converted back to FeO_x that binds PO_4^{3-} that comes from external (e.g., tidal) or internal (mineralization) sources (Cornwell 1987, Chambers and Odum 1990), resulting in a net retention of phosphorus. In brackish wetlands, the efflux of dissolved PO_4^{3-} from sediments is much faster and FeO_x sequesters much less PO_4^{3-} (Callender 1982, Callender and Hammond 1982, Hopkinson et al. 1999). The distribution of Fe(III) in the sediments of tidal wetland also varies along the salinity gradient, with higher abundance in freshwater sediments than in brackish sediments, and with a stronger tendency to be concentrated near the surface of the freshwater sediment (Phillips and Lovley 1987). Likewise, concentrations of FeO_x-bound PO_4^{3-} in sediments decline from the freshwater reaches down the estuary into brackish waters.

Although silica (Si) plays a major role in coastal eutrophication events, the cycling of Si in freshwater tidal marshes has not received much scientific attention. The lithogenic fraction of Si in tidal marsh sediments is considered inert at biological timescales. However, tidal wetlands contain large amounts of biogenic Si (BSi) in vegetation and sediments (Norris and Hackney 1999). Dissolved Si (DSi) taken up by wetland plants is stored in specific structures of various shapes, the phytoliths. Plant BSi can become available again to the estuarine ecosystem after the plants decay, through phytolith dissolution. However, phytoliths can be highly resistant to decomposition and a large part of plant BSi is buried in the sediments. Together with buried diatom BSi, buried phytoliths comprise the sediment BSi fraction. Due to the high BSi amounts available for dissolution, DSi concentrations in marsh pore-water can be several times the DSi concentration in the main river channel, providing excellent conditions for benthic diatoms on the wetland surface and providing the energetic base for secondary production (Hackney et al. 2000). Moreover, floodwater containing lower DSi concentrations, compared to pore-water, is enriched with DSi when flowing through the sediments (Fig. 6.7). Thus, large amounts of DSi can be exported from tidal wetland sites to the main river channel between tidal inundations, especially in spring and summer, when DSi in the main river is depleted by diatom communities. Tidal freshwater wetlands could therefore play a major role in estuarine silica cycling, supporting secondary production through DSi recycling.

Heavy metals and trace metals can be retained in significant quantities in freshwater tidal wetlands, especially in areas with high sedimentation rates (e.g., Khan and Brush 1994). Vegetation plays a major role in removing heavy metals from tidal waters during the growing season and the metals are mostly transferred to the substrate for long-term sequestration via litter accumulation (Simpson et al. 1983b; Dubinski et al. 1986).

6.5 Restoration and Future Outlook

We have attempted to demonstrate that tidal freshwater wetlands are widespread and both ecologically and environmentally important, and they are thus worthy of both preservation and restoration. We also hope to convey the message that they are still threatened in many areas because of human intervention. The main problems that need to be considered now and into the future are issues related to water chemistry, the extent of tidal freshwater wetlands, and direct human impacts.

6.5.1 Europe

The pollution of European rivers that include tidal freshwater areas has improved drastically in the past couple of decades. Since the first positive changes in chemistry mentioned by Van Dijk et al. (1994), water quality has continued to improve during the past decade. The concentration in heavy metals and organic micropollutants has declined; but the concentration in nitrogen and the deficit in oxygen remain difficult problems due to diffuse pollution sources. Positive changes have, however, been reported in water quality for some systems (e.g., Gerkens and Thiel 2001; Soetaert et al. in press).

As a consequence of poor water quality, sediment quality in most tidal flats and wetlands has also deteriorated. Elevated concentrations of metals and hydrocarbons can be found. In the Scheldt estuary, there are no signs that this has a negative impact on the vegetation (Van Regenmortel, personal communication) but spiders show a clear accumulation (Du Laing et al. 2002). Concentrations in the suspended sediments are however decreasing in parallel with the improving water quality, resulting in clear gradients: recent deposits have much lower concentrations than older ones. It remains unclear what the impact of improving oxygen conditions in the estuary will be on the mobilization of these pollutants. Also, polluted sediments can be brought into suspension by erosion.

The second issue that needs to be considered is the fact that many areas that once supported tidal freshwater wetlands have been destroyed or suffer from the absence of tidal influences (De Boois 1982). In many instances, the wetlands were reclaimed for other functions, such as agriculture or housing and industry (Preisinger 1991). The consequence of previous wetland losses is that there is a limited area available in tidal freshwater areas for ecosystem restoration. Restoration is an important element and it needs to be actively considered and pursued. At the same time, the closing of the tidal regions of rivers has caused serious problems in water quality. In the Netherlands, discussions have been held about the re-opening of the sluices of the estuary Haringvliet in order to restore tidal exchange between the former estuary and the sea. Changes in current hydrological management might result in a reduction of the enormous problems in blue-algal growth in some (former saline tidal) areas. This tidal flow can also restore the ecological conditions in the river and especially in the Biesbosch-region, the part of the system that historically had large areas of tidal freshwater wetlands.

At the same time, Europe fights the problem of excessive discharge of river water, heavier and more frequent storms, and a sea level rise which may occur due to climatic changes. Further heightening of dikes is possible but ultimately results in more risks. Indeed the height difference between the water levels during storms in the estuary and the adjacent polder areas becomes larger. In the case of a dike breach, the effects are then of course much more severe. In the Scheldt estuary in Belgium next to reinforcing dikes, controlled inundation areas are built to improve safety against storms. These areas are low-lying polders without human occupation. A high dike is build around the polder, and a lower one between the estuary and the polder. During storms the water overtops the lower dike and the polder is filled with water. Upstream water levels do not increase as most of the water flows into the polder. Through large sluices the polder is emptied during the next low tide and most of the storage capacity is restored before the next high-water event. These sluices can also be used to create a reduced tide in these polders, allowing the development of tidal wetlands, hence combining safety and restoring habitat (Meire et al. 2002; Van den Bergh et al. 2005). These new wetlands can also have a significant effect on the water quality. Modelling exercises show that they can substantially increase the oxygen content in the river and enhance the primary productivity and nutrient removal. A first experimental site will become functional in 2005 and a larger area of about 300 ha is under construction. Next to combining safety and wetland development in controlled inundation areas, managed realignment by reshaping the river is an important option for restoration.

The third issue that impacts tidal freshwater areas of estuaries is direct human activity, such as modification of rivers for shipping. As the size of ships increases, there is increasing pressure to increase the depth of the shipping channels (Preisinger 1991). Dredging and increasing the depth of rivers increases the tidal range (e.g., up to 6 m in Elbe and Scheldt) and erosion of the shoreline by the stronger currents. This results in very steep gradients from the stream to the higher zones in the freshwater range, with the loss of mud flats and the bordering vegetation due to erosion. The only option is to reduce erosion in a structural way, such as widening of the river through managed realignment. Other activities that use stones or other structural materials to stabilize the eroding wetland edge only reduce the natural dynamics of the wetlands by preventing the development of low marsh or subtidal vegetation and hydrologic exchange between the river and adjacent wetlands.

The preservation of the remaining freshwater tidal areas in Europe might not be the problem, since most extensive areas are preserved as nature areas or reserves, mostly incorporated in the national ecological networks of reserves. For instance, the remaining 1000 ha in Belgium or the 1000 ha in the Netherlands are both protected by national regulations; the same is true for German or UK areas. In addition to national conservation programs, EU legislation known as the Bird and Habitat Directive is valid for many areas, including areas that have tidal freshwater wetlands, such as in the estuaries of the Elbe, Ems, Rhine, and Scheldt. However, only one animal and one plant species of tidal freshwater wetlands have been incorporated into the EU Bird and Habitat Directive. TheTwaite shad (*Alosa fallax*) is restricted to these wetlands and is still seriously endangered, although it is recovering at some locations (Gerkens and Thiel 2001, Maes et al. 1998). *Oenanthe conioides* is the only endemic plant species in tidal freshwater wetlands that has been incorporated in the Bird and Habitat Directive. This plant is restricted to the Elbe catchments and preservation can only occur following the restoration of habitats in that estuary. More striking is the absence of the freshwater tidal systems as a priority ecosystem in the EU Bird and Habitat Directive, given the fact that these systems provide an important habitat for these two species.

6.5.2 United States

In the United States, federal, state, and local regulations protect coastal wetlands, and the rate of loss of estuarine areas has decreased by 82 % between 1986 and 1997 (Dahl 2000) compared to the period from the mid-1970s to the mid-1980s (Dahl and Johnson 1991). It is not possible, however, to determine the status and trends for tidal freshwater wetlands from the national assessment documents because they are not considered separately in the analysis of either estuarine or freshwater wetlands. In recent years, increased emphasis has been given to the restoration and conservation of tidal freshwater wetlands. Large freshwater tidal wetland restoration projects have been initiated in the Delaware estuary at the Hamilton Marshes (Leck 2003; Fig. 6.5B) and sites near the Philadelphia airport (D. Whigham, personal observation). In the Chesapeake Bay, restoration of tidal freshwater wetlands has been initiated in Washington (D.C.) in the Anacostia river, a tributary of the Potomac River (Baldwin 2004). Also in the Chesapeake Bay, governmental and non-profit conservation organizations have purchased and preserved tidal freshwater wetlands. The Nature Conservancy, for example, has preserved tidal freshwater wetlands on the Choptank and Nanticoke River in Maryland and the National Park Service created the Dyke Marsh Preserve near Washington, D.C. (Johnston 2000).

Several states on the Atlantic coast of North America (Maryland, Delaware, New York, Virginia) have included tidal freshwater systems as components of estuarine reserves that form the National Estuarine Reserve System (NEERS). New York NEERS sites include two tidal freshwater systems (Tivoli Bay, Stockport Flats). Three reserves in the Chesapeake Bay are freshwater tidal systems or include tidal freshwater wetland habitats. In Maryland, Jug Bay and Otter Creek represent two of the largest tidal freshwater wetland reserves in the country. In Virginia, the Virginia Coastal Reserve also includes tidal freshwater wetland habitats. The Delaware NEERS reserve system includes tidal freshwater wetlands at Blackbird Creek.

6.6 Conclusions

Tidal freshwater wetlands occur in the upper part of estuaries in Europe and North America (and likely elsewhere) and experience tides of up to several meters in amplitude twice a day. They occur at the interface between the brackish zone in the estuary and the river; and where brackish and fresh water mix is an area of maximum suspended matter (i.e., the maximum turbidity zone). The tidal freshwater zone within the estuary plays an important role in overall patterns of nutrient cycling for the whole estuary and the pattern appears to differ in the brackish and saline sections. Although tidal freshwater wetlands do not include many endemic or restricted species, they are characterized by high species and habitat diversity. There is distinct zonation in flora and fauna species, responding to the relationship between surface elevation and tidal amplitude. The dominant species are different between Europe and North America, but the structure of the system and the life strategy of the species are fully comparable. The tidal freshwater wetlands in Europe and North America also have a common history of being highly influenced by human activities, resulting in altered hydrology, losses in wetland area, and high levels of sediment and nutrient input on both sides of the Atlantic Ocean. In recent years, restoration and preservation activities have also been initiated on both sides of the Atlantic and there is hope that tidal freshwater wetlands will increasingly become important elements of estuarine systems that provide many free ecological services to man and nature.

Acknowledgements We wish to thank the participants of the Special Symposium at the Wetland Conference in 2004, since they started a basic discussion about an ecosystem that was almost forgotten. The figures in this chapter are drawn by the Cartographic Laboratory Faculty GeoSciences of Utrecht University. We thank two anonymous referees for their constructive criticisms.

References

Anderson AB, Schmidt RE (1989) Survey of larval and juvenile fish populations in water-chestnut (*Trapa natans*) beds in Tivoli South Bay, a Hudson River tidal marsh. In: Blain EA, Waldman JR (eds) Polgar fellowship reports of the Hudson River national estuarine research reserve program, 1988. Hudson River Foundation, New York, pp VI.1–VI.34

Avoine J, Allen JP, Nichols M, Salomon JC, Larsonneur C (1981) Suspended-sediment transport in the Seine Estuary, France: effects of man-made modifications on estuary-shelf sedimentology. Mar Geol 40:119–137

Baldwin AH (2004) Restoring complex vegetation in urban settings: the case of tidal freshwater marshes. Urban Ecosyst 7:125–137

Baldwin AH, DeRico EF (1999) The seed bank of a restored tidal freshwater marsh in Washington, D.C. Urban Ecosyst 3:5–20

Baldwin AH, Pendleton FN (2003) Interactive effects of animal disturbance and elevation on vegetation of a tidal freshwater marsh. Estuaries 26:905–915

Baldwin AH, McKee KL, Mendelssohn IA (1996) The influence of vegetation, salinity, and inundation on seed banks of oligohaline coastal marshes. Am J Bot 83:470–479

Baldwin AH, Egnotovich MS, Clarke E (2001) Hydrologic change and vegetation of tidal freshwater marshes: field, greenhouse, and seed-bank experiments. Wetlands 21:519–531

Barbour S, Kiviat E (1986) A survey of Lepidoptera in Tivoli North Bay (Hudson River Estuary). In: Cooper JC (ed) Polgar fellowship reports of the Hudson River national estuarine research reserve program, 1985. Hudson River Foundation, New York, pp IV.1–IV.26

Barendregt A (2005) The impact of flooding regime on ecosystems in a tidal freshwater area. J Ecohydr Hydrobiol 5:95–102

Boule ME (1981) Tidal wetlands of the Puget Sound region, Washington USA. Wetlands 1:47–60

Bowden WB (1984a) Nitrogen-15 isotope dilution study of ammonium production and consumption in a marsh sediment. Limnol Oceanogr 29:1004–1015

Bowden WB (1984b) Nitrogen and phosphorus in the sediments of a tidal freshwater marsh in Massachusetts USA. Estuaries 7:108–118

Bowden WB (1986) Nitrification nitrate reduction and nitrogen immobilization in a tidal freshwater marsh sediment. Ecology 67:88–99

Brinkkemper O, De Ridder T (2001) Het archeobotanisch onderzoek. (VLAK-verslag 2.3, Hoogstad 5.019) VLAK, Vlaardingen. Available at www.archeologie.vlaardingen.nl

Callender E (1982) Benthic phosphorus regeneration in the Potomac River Estuary. Hydrobiologia 92:431–446

Callender E, Hammond DE (1982) Nutrient exchange across the sediment–water interface in the Potomac River Estuary. Estuar Coast Shelf Sci 15:395–413

Chabreck RH (1972) Vegetation, water, and soil characteristics of the Louisiana coastal region. (Louisiana state university agricultural experiment station bulletin 664) Louisiana State University, Baton Rouge

Chambers RM, Odum WE (1990) Porewater oxidation, dissolved phosphate and the iron curtain. Iron–phosphorus relations in tidal freshwater marshes. Biogeochemistry 10:37–52

Chambers RM, Meyerson LA, Saltonstall K (1999) Expansion of *Phragmites australis* into tidal wetlands of North America. Aquat Bot 64:261–273

Ciszewski D (2001) Flood-related changes in heavy metal concentrations within sediments of the Biala Przemsza River. Geomorphology 40:205–218

Connors L, Kiviat E, Groffman P, Ostfeld RS (2000) Muskrat (*Ondatra zibethicus*) disturbance to vegetation and potential net nitrogen mineralization and nitrification rates in a fresh-tidal marsh. Am Midl Nat 143:53–63

Cornwell JC (1987) Phosphorus cycling in arctic lake sediments: adsorption and authigenic minerals. Arch Hydrobiol 109:161–179

Dahl TE (2000) Status and trends of the wetlands in the conterminous United States 1986–1997. US Department of the Interior/Fish and Wildlife Service, Washington, D.C.

Dahl TE, Johnson CE (1991) Status and trends of wetlands in the conterminous Untied States, mid-1970s to mid-1980s. US Department of the Interior/Fish and Wildlife Service, Washington, D.C.

Darke AK, Megonigal JP (2003) Control of sediment deposition rates in two mid-Atlantic coast tidal freshwater wetlands. Estuar Coast Shelf Sci 57:255–268

De Boois H (1982) Veranderingen in het milieu en de vegetatie in de Biesbosch door afsluiting van het Haringvliet. PhD thesis, Wageningen Agricultural University, Wagenigen

De Ridder T (1999) The oldest deltaworks of western Europe (in Dutch). Tijdschr Waterstaatgesch 8:10–22

Deegan BM, Harrington TJ (2004) The distribution and ecology of *Schoenoplectus triqueter* in the Shannon estuary. Proc R Ir Acad Sect B 104:107–117

Desender K, Mealfait JP (1999) Diversity and conservation of terrestrial arthropods in tidal marshes along the River Schelde: a gradient analysis. Biol Conserv 87:221–229

Dubinski BJ, Simpson RL, Good RE (1986) The retention of heavy metals in sewage sludge applied to a freshwater tidal wetlands. Estuaries 9:102–111

Du Laing G, Bogaert N, Tack FMG, Verloo MG, Hendrickx F (2002) Heavy metal contents (Cd, Cu, Zn) in spiders (*Pirata piraticus*) living in intertidal sediments of the river Scheldt estuary (Belgium) as affected by substrate characteristics. Sci Total Environ 289:71–81

Eckert CG (2002) The loss of sex in clonal plants. Evol Ecol 15:501–520

Eisma D, Chen S, Li A (1994) Tidal variations in suspended matter floc size in the Elbe river and Dollard estuaries. Neth J Aquat Ecol 28:267–274

Elliot M, McLusky DS (2002) The need for definitions in understanding estuaries. Estuar Coast Shelf Sci 55:815–827

Fairbridge RW (1980) The estuary, its definition and geodynamic cycle. In: Olausson E, Cato I (eds) Chemistry and biogeochemistry of estuaries. Wiley, New York, pp 1–35

Findlay S, Groffman P (2003) Effects of *Phragmites australis* removal on marsh nutrient cycling. Wetl Ecol Manage 11:157–165

Findlay S, Howe K, Austin HK (1990) Comparison of detritus dynamics in two tidal freshwater wetlands. Ecology 71:288–295

Findlay S, Pace ML, Lints D (1991) Variability and transport of suspended sediment, particulate and dissolved organic carbon in the tidal freshwater Hudson River. Biogeochemistry 12:149–169

Findlay S, Sinsabaugh RL, Fischer DT, Franchini P (1998) Sources of dissolved organic carbon supporting planktonic bacterial production in the tidal freshwater Hudson River. Ecosystems 1:227–239

Findlay SEG, Kiviat E, Nieder WC, Blair EA (2002) Functional assessment of a reference wetland set as a tool for science, management and restoration. Aquat Sci 64:107–117

Gerkens M, Thiel R (2001) Habitat use of age-0 Twaite shad (*Alosa fallax* Lacepede, 1803) in the tidal freshwater region of the Elbe River, Germany. Bull Fr Peche Piscicult 362/263:773–784

Gosselink JG (1984) The ecology of delta marshes of coastal Louisiana: a community profile. (FWS/OBS-84/09) US Fish and Wildlife Service, Washington, D.C.

Grabemann I, Uncles RJ, Krause G, Stephens JA (1997) Behaviour of turbidity maxima in the Tamar (UK) and Weser (FRG) estuaries. Estuar Coast Shelf Sci 45:235–246

Griffith AB, Forseth IN (2003) Establishment and reproduction of *Aeschynomene virginica* (L) (Fabaceae) a rare, annual, wetlands species in relation to vegetation removal and water level. Plant Ecol 167:117–125

Hackney CT, Cahoon LB, Prestos C, Norris A (2000) Silicon is the link between tidal marshes and estuarine fisheries: a new paradigm. In: Weinstein MP, Kreeger DA (eds) Concepts and controversies in tidal marsh ecology. Kluwer, Dordrecht, pp 543–552

Hawkins P, Leck CF (1977) Breeding bird communities in a tidal freshwater marsh. Bull NJ Acad Sci 22:12–17

Heemken OP, Stachel B, Theobald N, Wenclawiak BW (2000) Temporal variability of organic micropollutants in suspended particulate matter of the river Elbe at Hamburg and the river Mulde at Dressau, Germany. Arch Environ Contam Toxicol 38:11–31

Hendrickx F, Desender K, Mealfait JP (2001) Verspreidingspatronen en bedreigingen van de arthropodenfauna van het Schelde-estuarium. Levende Nat 102:68–69

Hopkinson CS, Giblin AE, Tucker J, Garritt RH (1999) Benthic metabolism and nutrient cycling along an estuarine salinity gradient. Estuaries 22:863–881

Johnston DW (2000) The Dyke marsh preserve ecosystem. Va J Sci 51:223–272

Jonkers N, Laane RWPM, De Voogt P (2003) Fate of nonylphenol ethoxylates and their metabolites in two Dutch estuaries: evidence of biodegradation in the field. Environ Sci Technol 37:321–327

Jordan TE, Correll DL (1991) Continuous automated sampling of tidal exchanges of nutrients by brackish marshes. Estuar Coast Shelf Sci 32:527–545

Junk WJ (1983) Wetland of tropical South America. In: Whigham D, Dykyjováand D, Hejny S (eds) Wetland of the world: inventory, ecology and management. Kluwer, Dordrecht, pp 679–739

Kiviat E (1991) Wetland human ecology. PhD thesis, Union Institute, Cincinnati

Kiviat E, Barbour JG (1996) Wood turtles in fresh-tidal habitats of the Hudson River. Can Field Nat 110:341–343

Khan H, Brush GS (1994) Nutrient and metal accumulation in a freshwater tidal marsh. Estuaries 17:345–360

Knight MA, Pasternack GB (2000) Sources, input pathways, and distribution of Fe, Cu, and Zn in a Chesapeake Bay tidal freshwater marsh. Environ Geol 39:1359–1371

Leck MA (1996) Germination of macrophytes from a Delaware River tidal freshwater wetland. Bull Torrey Bot Club 123:48–67

Leck MA (2003) Seed-bank and vegetation development in a created tidal freshwater wetland on the Delaware River, Trenton, New Jersey, USA. Wetlands 23:310–343

Leck MA, Simpson RL (1987) Seed bank of a freshwater tidal wetland: turnover and relationship to vegetation change. Am J Bot 74:360–370

Leck MA, Simpson RL (1995) Ten-year seed bank and vegetation dynamics of a tidal freshwater marsh. Am J Bot 82:1547–1557

Leck MA, Simpson RL, Whigham DF, Leck CF (1988) Plants of the Hamilton marshes. Bartonia 54:1–17

Lobry J, Mourand L, Rochard E, Elie P (2003) Structure of the Gironde estuarine fish assemblages: a comparison of European estuaries perspective. Aquat Living Resour 16:47–58

Maes J, Taillieu A, Van Damme PA, Cottenie K, Ollevier F (1998) Seasonal patterns in the fish and crustacean community of a turbid temperate estuary (Zeeschelde estuary, Belgium). Estuar Coast Shelf Sci 47:143–151

Marchand J (1993) The influence of seasonal salinity and turbidity maximum variations on the nursery function of the Loire estuary (France). Neth J Aquat Ecol 27:427–436

McLusky DS (1993) Marine and estuarine gradients – an overview. Neth J Aquat Ecol 27:489–493

McManus J (2005) Salinity and suspended matter variations in the Tay estuary. Cont Shelf Res 25:729–747

Meade RH (1972) Transport and deposition of sediments in estuaries. Geol Soc Am Mem 133:91–120

Meijer R, Van der Nat J (1989) Bluethroat *Luscinia svecica* saved by the Biesbosch? Limosa 62:67–74

Meire P, Vincx M (eds) (1993) Marine and estuarine gradients. Neth J Aquat Ecol 27:75–493

Meire P, Van den Bergh E, Ysebaert T, Nijssen D (2002) Nature development along the river Scheldt: combining ecosystem functions in the Kruibeke–Bazel–Rupelmonde polder. In: Redecker B, Finck P, Härdtle W, Riecken U, Schröder E (eds) Pasture landscapes and nature conservation. Springer, Berlin Heidelberg New York, pp 173–185

Mesnage V, Bonneville S, Laignel B, Lefebvre D, Dupont JP, Mikes D (2002) Filling of a wetland (Seine estuary, France): natural eutrophication or anthropogenic process? A sedimentological and geochemical study of wetland organic sediments. Hydrobiologia 475/476:423–435

Meyerson LA, Saltonstall K, Windham L, Kiviat E, Findlay SEG (2000) A comparison of *Phragmites australis* in freshwater and brackish marsh environments in North America. Wetland Ecol Manage 8:89–103

Middelkoop H (2000) Heavy-metal pollution of the river Rhine and Meuse floodplains in the Netherlands. Neth J Geosci 79:411–428

Mitsch WJ, Gosselink JG (2000) Wetlands. Van Nostrand Reinhold, New York

Morris JT, Bowden WB (1986) A mechanistic numerical model of sedimentation, mineralization and decomposition for marsh sediments. Soil Sci Soc Am J 50:96–105

Muylaert K, Sabbe K (1996) *Cyclotella scaldensis* spec. nov. (Bacillariophyceae), a new estuarine diatom. Nova Hedwigia 63:335–345

Muylaert K, Van Kerckvoorde A, Vyverman W, Sabbe K (1997) Structural characteristics of phytoplankton assemblages in tidal and non-tidal freshwater systems: a case study from the Scheldt basin, Belgium. Freshwater Biol 38:263–276

Neubauer SC, Anderson IC (2003) Transport of dissolved inorganic carbon from a tidal freshwater marsh to the York River estuary. Limnol Oceanogr 48:299–307

Neubauer SC, Miller WD, Anderson IC (2000) Carbon cycling in a tidal freshwater marsh ecosystem: a carbon gas flux study. Mar Ecol Prog Ser 199:13–31

Neubauer SC, Anderson IC, Constantine JA, Kuehl SA (2002) Sediment deposition and accretion in a mid-Atlantic (U.S.A.) tidal freshwater marsh. Estuar Coast Shelf Sci 54:713–727

Norris AR, Hackney CT (1999) Silica content of a mesohaline tidal marsh in North Carolina. Estuar Coast Shelf Sci 49:597–605

Odum WE (1988) Comparative ecology of tidal freshwater and salt marshes. Annu Rev Ecol Syst 19:147–176

Odum WE, Smith III TJ, Hoover JK, McIvor CC (1984) The Ecology of tidal freshwater marshes of the United States East Coast: a community profile. (FWS.OBS-83-17) US Fish and Wildlife Service, Washington, D.C.

Officer CB (1981) Physical dynamics of estuarine suspended sediments. Mar Geol 40:1–14

Parker VT, Leck MA (1985) Relationships of seed banks to plant distribution patterns in a freshwater tidal wetland. Am J Bot 72:161–174

Parker WR, Marshall LD, Parfitt AJ (1994) Modulation of dissolved oxygen levels in a hypertidal estuary by sediment resuspension. Neth J Aquat Ecol 28:347–352

Pasternack GB, Brush GS (2001) Seasonal variations in sedimentation and organic content in five plant associations on a Chesapeake Bay tidal freshwater delta. Estuar Coast Shelf Sci 53:93–106

Pasternack GD, Hilgartner WB, Brush GS (2000) Biogeomorphology of an upper Chesapeake Bay river-mouth tidal freshwater marsh. Wetlands 20:520–537

Penfound WT, Hathaway ES (1938) Plant communities in the marshlands of southeastern Louisiana. Ecol Monogr 8:1–56

Peterson JE, Baldwin AH (2004) Variation in wetland seed banks across a tidal freshwater landscape. Am J Bot 91:1251–1259

Phillips EJP, Lovley DR (1987) Determination of Fe(III) and Fe(II) in oxalate extracts of sediment. Soil Sci Soc Am J 51:938–941

Pinder DA, Witherick ME (1990) Port industrialization, urbanization and wetland loss. In: Williams M (ed) Wetlands – a threatened landscape. Blackwell, Oxford, pp 234–266

Pomfret JR, Elliott M, O'Reilly MG, Phillips S (1991) Spatial and temporal patterns in the fish communities in two UK North Sea estuaries. In: Elliott M, Ducrotoy JP (eds) Estuaries and coasts: spatial and temporal intercomparisons, Olson and Olson, Fredensborg, pp 277–284

Preisinger H (1991) Strukturanalyse and Zeigerwert der Auen- und Ufervegetation im Hamburger Hafen- und Hafenrandgebiet. (Dissertationes Botanicae Band 174) Cramer, Berlin

Rehbehn R, Schuchardt B, Schirmer M, Kirst GO (1993) The distribution of *Actinocyclus normanii* (Bacillariophyceae) in estuaries: field observations and laboratory investigations. Neth J Aquat Ecol 27:205–214

Rheinhardt R (1992) A multivariate analysis of vegetation patterns in tidal freshwater swamps of lower Chesapeake Bay, U.S.A. Bull Torrey Bot Club 119:192–207

Rice D, Rooth J, Stevenson JC (2000) Colonization and expansion of *Phragmites australis* in upper Chesapeake Bay tidal marshes. Wetlands 20:280–299

Ridgley MA, Rijsberman FR (1994) Multicriterion analysis and the evaluation of restorations policies for a Rhine estuary. Socio Econ Planning Sci 28:19–32

Riedel-Lorje JC, Gaument T (1982) A century of Elbe research – Hydrobiological conditions and fish populations from 1842 to 1943 under the influence of construction projects and sewage discharge (in German). Arch Hydrobiol Suppl 61:317–376

Saltonstall, K. (2002) Cryptic invasion by non-native genotypes of the common reed, *Phragmites australis*, into North America. Proc Natl Acad Sci USA 99:2445–2449

Schuchardt B, Haesloop U, Schirmer M (1993) The tidal freshwater reach of the Weser estuary: riverine or estuarine? Neth J Aquat Ecol 27:215–226

Seys J, Vincx M, Meire P (1999) Spatial distribution of Oligochaetes (*Clitellata*) in the tidal freshwater and brackish parts of the Schelde estuary (Belgium). Hydrobiologia 406:119–132

Simpson KW, Bode RW, Fagnani JP, DeNicola DM (1984) the freshwater macrobenthos of the main channel. Part B: biology, taxonomy and distribution of resident macrobenthid species. (Final Report to the Hudson River Foundation 008/83A) New York State Department of Health, Albany

Simpson RL, Good RE, Leck MA, Whigham DF (1983a) The ecology of freshwater tidal wetlands. BioScience 33:255–259

Simpson RL Good RE, Walker R, Frasco BR (1983b) The role of Delaware river USA freshwater tidal wetlands in the retention of nutrients and heavy metals. J Environ Qual 12:41–48

Soetaert K, Herman P (1995) Carbon flows in the Westerschelde estuary (The Netherlands) evaluated by means of an ecosystem model (MOSES). Hydrobiologia 311:247–266

Steen RJCA, Evers EHG, Van Hattum B, Cofino WP, Brinkman UAT (2002) Net fluxes of pesticides from the Scheldt Estuary into the North Sea: a model approach. Environ Pollut 116:75–84

Tanner CD, Cordell JR, Rubey J, Tear LM (2002) Restoration of freshwater intertidal habitat functions at Spencer Island, Everett, Washington. Restor Ecol 10:564–576

Teal JM, Peterson S (2005) The interaction between science and policy in the control of *Phragmites* in oligohaline marshes of Delaware Bay. Restor Ecol 13:223–227

Thiel R, Potter IC (2001) The ichthyofaunal composition of the Elbe Estuary: an analysis in space and time. Mar Biol 138: 603–616

Uncles RJ, Howland RJM, Aeston AE, Griffiths ML, Harris C, King RS, Morris AW, Plummer DH, Woodward EMS (1998) Concentrations of dissolved nutrients in the tidal Yorkshire Ouse River and Humber Estuary. Sci Total Environ 210/211:377–388

Van Damme S, Ysebaert T, Meire P, Van den Bergh E (1999) Habitatstructuren, waterkwaliteit en leefgemeenschappen in the Schelde-estuarium. (Report Instituut voor Natuurbehoud 99/24) Instituut voor Natuurbehoud, Brussels

Van den Bergh E, Van Damme S, Graveland J, De Jong D, Baten I, P Meire (2005) Ecological rehabilitation of the Schelde estuary (The Netherlands-Belgium; Northwest Europe): linking ecology, safety against floods and accessibility for port development. Restor Ecol 13:204–214

Van Dijk GL, Van Liere L, Admiraal W, Bannink BA, Cappon JJ (1994) Present state of the water quality of European rivers and implications for management. Sci Total Environ 145:187–195

Van Steenis CGGJ (1971) De zoetwatergetijdedotter van de Biesbosch en de Oude Maas, *Caltha palustris* L. var. *araneosa*, var. nov. Gorteria 5:213–219

Verhey CJ, Heyligers PC, Lebret T, Zonneveld IS (1961) De Biesbosch, land van het levende water. Thieme, Zutphen

Verhoeven JTA, Whigham DF, Van Logtestijn R, O'Neill J (2001) A comparative study of nitrogen and phosphorus cycling in tidal and non-tidal riverine wetlands. Wetlands 21:210–222

Walling DE, Owens PN, Carter J, Leeks GJL, Lewis S, Meharg AA, Wright J (2003) Storage of sediment-associated nutrients and contaminants in river channel and floodplain systems. Appl Geochem 18:195–220

Whigham DF, Simpson RL (1976) The potential use of freshwater tidal marshes in the management of water quality in the Delaware River. In: Tourbier J, Peirson JF (eds)

Biological control of water pollution. University of Pennsylvania, Philadelphia, pp 173–186

Whigham DF, Simpson RL (1992) Annual variation in biomass and production of a tidal freshwater wetland and comparisons with other wetland systems. Va J Sci 43:5–14

Whigham DF, McCormick J, Good RE, Simpson RL (1978) Biomass and primary production in freshwater tidal wetlands of the Middle Atlantic Coast. In: Good RE, Whigham DF, Simpson RL (eds) Freshwater wetlands. ecological processes and management potential. Academic, New York, pp 3–20

Wolf WJ (1973) The estuary as a habitat – an analysis of the soft-bottom macrofauna of the estuarine area of the rivers Rhine, Meuse, and Scheldt. (Zoologische Verhandelingen No. 126) Brill, Leiden

Yozzo DJ, Odum WE (1993) Fish predation on epiphytic microcrustacea in Tivoli South Bay, a Hudson river tidal freshwater wetlands. Hydrobiologia 257:37–46

Yozzo DJ, Smith DE (1995) Seasonality, abundance, and microhabitat distribution of meiofauna from a Chickahominy river, Virginia tidal freshwater marsh. Hydrobiologia 310:197–206

Yozzo DJ, Steineck PL (1994) Ostracoda from tidal freshwater wetlands at Stockport, Hudson River estuary: abundance, distribution, and composition. Estuaries 17:680–684

Yozzo DJ, Smith DE, Lewis ML (1994) Tidal freshwater ecosystems. Bibliography. (Virginia Institute of Marine Sciences, Contribution No. 1880) Virginia Institute of Marine Sciences, Gloucester Point

Ysebaert T, Meininger PL, Meire P, Devos K, Berrevoets CM, Strucker RCW, Kuijken E (2000) Waterbird communities along the estuarine salinity gradient of the Schelde estuary, NW-Europe. Biodiver Conserv 9:1275–1296

Zonneveld IS (1960) The Brabantsche Biesbosch. A study of soil and vegetation of a freshwater tidal area. (Bodemkundige Studies No. 4) PhD thesis, Wageningen University, Wageningen

7 Biodiversity in European Shallow Lakes: a Multilevel–Multifactorial Field Study

L. De Meester, S. Declerck, J.H. Janse, J.J. Dagevos, R. Portielje, E. Lammens, E. Jeppesen, T. Lauridsen, K. Schwenk, K. Muylaert, K. Van der Gucht, W. Vyverman, G. Zwart, E. van Hannen, P.J.T.M. van Puijenbroek, J.M. Conde-Porcuna, P. Sánchez-Castillo, J. Vandekerkhove, and L. Brendonck

7.1 Introduction

Shallow lakes and reservoirs are very common in Europe. They harbour a wide variety of organisms and are an important component of the landscape structure. In addition to their ecological role, they also support a wide array of socio-economic activities, such as fisheries, recreation, water storage (flood prevention) and water supply for agriculture. Shallow lakes are, however, subject to many threats and many are highly impacted by human activities. They are particularly vulnerable to nutrient loading, water level regulation and the infiltration of pesticides (Jeppesen et al. 1997; Moss 1998; Scheffer 1998).

With and since the development of the theory that shallow lakes may exist in two alternative stable states (Scheffer et al. 1993), major progress has been made in the understanding of shallow lake functioning (Jeppesen et al. 1997; Hansson et al. 1998; Moss 1998; Scheffer 1998). The concept developed by Scheffer et al. (1993) states that shallow lakes can occur in either a transparent or a turbid state and that both states are stabilized by feedback loops over a broad range of nutrient concentrations. The transparent state is characterized by clear water, submerged macrophytes, low phytoplankton biomass and concentrations of suspended matter, high densities of large zooplankton, low biomass of planktivorous and benthic fish biomass and a high proportion of piscivorous fish biomass (Jeppesen et al. 1997; Scheffer 1998). Conversely, the turbid state is typically characterized by turbid water, absence of submerged macrophytes, high concentrations of phytoplankton and suspended matter, dominance by small zooplankton and high biomasses of zooplanktivorous

Ecological Studies, Vol. 191
R. Bobbink, B. Beltman, J.T.A. Verhoeven, and D.F. Whigham (Eds.)
Wetlands: Functioning, Biodiversity Conservation, and Restoration

and benthivorous fishes. Nutrient levels determine the stability of each state (Scheffer et al. 1993): the stability of the turbid state increases with increasing nutrient concentrations, whereas the stability of the transparent state increases with decreasing nutrient concentrations. The turbidity state of a lake not only affects its ecological role, but also its economic and recreational value. In many respects, the clearwater state is to be preferred over the turbid state. The production of drinking water, for example, can be seriously impeded by the formation of algal blooms that may lead to clogging of filters or in bad taste or odour of the water. These problems are even worsened in the event of blooms of cyanobacteria, of which several genera (e.g. *Microcystis*, *Anabaena*, *Aphanizomenon*) are known to produce substances that are toxic to cattle or humans (Jackson et al. 1984; Jochimsen et al. 1998; Chorus and Bartram 1999). Also with respect to fisheries, the clearwater state is preferred: fish kills through acute anoxia are more common in phytoplankton-infested eutrophied lakes than in clearwater lakes; and the type of fish (e.g. the presence of pike, *Esox lucius*) as well as their meat quality for consumption are generally higher valued when caught from clear-water lakes. Finally, the amenity value of clear-water lakes with a well-developed macrophyte vegetation is higher than that of lakes in the turbid state, enhancing their value for recreation.

Although a number of studies on lakes have analysed the association between biodiversity of specific organism groups with key habitat characteristics (e.g. biodiversity in zooplankton in relation to lake size; Dodson 1992), or covered a number of trophic levels (Jeppesen et al. 2000), we know of no studies having considered several environmental variables simultaneously with biodiversity among organism groups at different trophic levels. It is, for example, symptomatic that the success of lake restoration measures is often evaluated in easy-to-measure variables, such as water transparency, the amount of suspended matter, phytoplankton biomass, nutrient levels or the degree of macrophyte coverage, whereas biodiversity measurements are only seldom used as evaluation criteria.

The many anthropogenic environmental alterations that threaten the integrity of shallow lakes undoubtedly affect the biodiversity of several groups of aquatic organisms. Measures that aim to conserve and restore biodiversity can only be successful if they are evaluated against scientifically sound indicators of biodiversity. Biodiversity is, however, difficult to assess, as it involves different aspects (e.g. taxon richness, evenness, rarity) that have to be measured at different levels or scales (e.g. taxonomic diversity vs population genetic diversity; α-, β- and γ-diversity). Furthermore, conservation is concerned with the biodiversity of different groups of organisms. Joint assessment of biodiversity in a wide array of taxa is generally very expensive, laborious and time-consuming, which explains the lack of data sets documenting the biodiversity of a large number of entire lake food webs. The design of a biodiversity index that estimates system-wide diversity in a relatively simple way would therefore be a big step forward. Such a tool would facilitate the monitoring of changes in lake integrity, the evaluation of restoration measures and the selection of freshwater bodies that deserve highest priority for conservation. Thus, the aims of the BIOMAN project were:

1. To develop a database on the current state of biodiversity in European shallow water bodies, covering a wide variety of organism groups and different trophic levels and involving both the classic pelagic food web as well as the microbial loop. For zooplankton, this database on taxon diversity in the active communities was completed with an analysis of taxon diversity in their resting egg banks (see Vandekerkhove et al. 2004a, b, 2005a, b) as well as with an assessment of within-species genetic diversity of key taxa (*Ctenodaphnia*, *Hyalodaphnia*).
2. To assess the degree to which biodiversity at the ecosystem-wide level is multidimensional.
3. To study the unique association between biodiversity and selected key variables (e.g. potential system productivity, water plant cover, lake surface area, degree of connectivity) that potentially affect biodiversity.
4. To develop mathematical models predicting the impact of environmental changes on biodiversity in shallow lakes.
   The purpose of the models is to predict the impact of (anthropogenically induced) changes in the environment on the lake ecosystem; and they are intended to be used as management tools.

7.2 Lake Selection

A total of 98 shallow lakes were selected, with a comparable number of lakes being located in each of three geographic regions that represent different latitudes in Europe: Denmark (DK; 32 lakes), Belgium/The Netherlands (BNL; 34 lakes) and the south of Spain (SP; 32 lakes; Fig. 7.1). A detailed written protocol was developed to standardize lake selection criteria, sampling methods and analyses (the protocol can be downloaded from www.kuleuven.ac.be/ bio/eco/bioman). Figure 7.1 shows the geographic location of the lakes. The lakes were not selected randomly but according to mutually independent gradients of four potentially important key variables: submerged vegetation cover, total phosphorus, lake surface area, degree of connectivity. In practice, we attempted to select two lakes belonging to each of 16 different categories in each geographic region. These 16 categories were defined as all possible combinations of four dichotomous criteria: (1) area of the lake covered by submerged water plants more or less than 20 % (being our criterion to differentiate between the clear-water and the turbid state), (2) total phosphorus level (TP) lower or higher than 100 $\mu g\ l^{-1}$, (3) surface area (lakes smaller and larger than 5 ha) and (4) degree of connectivity (connected vs

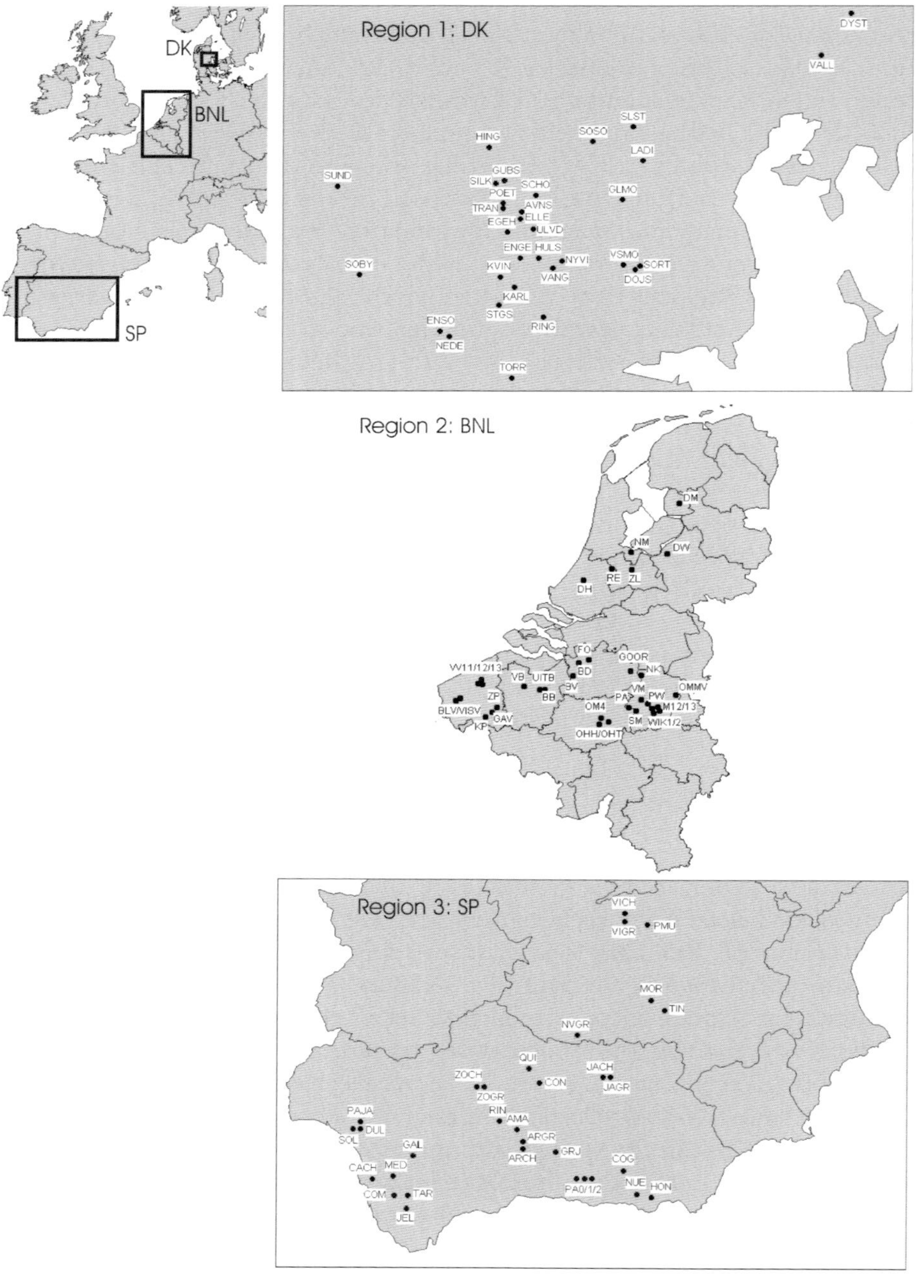

Fig. 7.1 Location of the selected model systems in Denmark (*DK*), Belgium/The Netherlands (*BNL*) and the south of Spain (*SP*). *Codes* refer to lake names (list available from the authors upon request)

non-connected). These selection criteria were defined: to ensure a good representation of a large variety of lake types in each geographic region, to allow an independent evaluation of the association between biodiversity and each of the four focal environmental variables (avoidance of collinearity; see Graham 2003) and to optimally prevent systematic biases between geographic regions with respect to the chosen gradients. The main advantage of this approach is that it allowed us to disentangle the relative importance of focal key environmental variables. For instance, in a random selection of lakes from a given region, the likelihood of a significant negative association between macrophyte cover and nutrient concentration is very high (Jeppesen et al. 2000). Such an association prevents assessment of the unique relationship between biodiversity and a variable without controlling for variation in the other variable. The drawback of our selection procedure is, however, that the selected lakes cannot be considered a random subset of lakes from a given geographic region. One can therefore argue that the we cannot extrapolate our data to the population of lakes in a region, simply because our selection no doubt has resulted in the inclusion of lake types that are not particularly common (e.g. lakes with a high nutrient load which also have an abundant macrophyte coverage). We thus have indeed to be careful in extrapolating our results: our data set allows separation of the influence of the four selected core variables on biodiversity, pinpointing at a more mechanistic understanding, but the 16 categories of lakes we identified are by far not equally abundant in nature.

7.3 Sampling and Analysis

Each lake was sampled monthly during a period of six months (May–October), either in 2000 or in 2001. Sampling followed a strict protocol (see www.kuleuven.ac.be/bio/eco/Bioman). More information on the sampling design and methods is given by Declerck et al. (2005; including electronic archives). In addition to the four mentioned core variables (macrophyte cover in summer, TP, size, connectivity), we also assessed many other local environmental variables [both abiotic: transparency as measured by Secchi disc and Snell's tube, pH, temperature, conductivity, total nitrogen (TN), total suspended matter; and biotic: chlorophyll *a* concentration], morphometry (mean depth), and land use in the drainage area. Monthly quantitative samples were taken to investigate community structure and diversity of several groups of organisms at different trophic levels of the aquatic food web: bacterioplankton, ciliates, phytoplankton, rotifers and crustacean zooplankton. Fish were sampled once (in autumn), using standardized mixed mesh-size gill-netting, supplemented with fyke nets. Macro-invertebrates were sampled during summer (July or August), using a semi-quantitative unit of effort approach (macro-invertebrate net, sampling all microhabitats during 2 min in each of six randomly selected littoral sampling stations). Depending on the surface area of the lake, planktonic organisms were sampled at eight (<5 ha) or 16 (>5 ha) randomly selected locations (including the littoral zone). These samples were pooled to one sample per sampling date. In addition, the six monthly samples were pooled so that we obtained, for each group of organisms, one sample per lake.

Bacterioplankton was analysed using molecular techniques: communities were screened with denaturing gradient gel electrophoresis (DGGE) and specific taxa were identified with reverse line blot hybridization (RLB; Zwart et al. 2003). Protist communities were investigated both with molecular techniques (DGGE) and using morphological criteria (to the genus level). Cladoceran and copepod zooplankton was identified to species level, whereas rotifers were identified to genus level. Some cladoceran taxa that are difficult to identify (e.g. *Hyalodaphnia* species complex) were screened with species-specific nuclear DNA markers (RFLP analysis) to verify species identity.

Taxonomic analyses of phytoplankton, ciliates and zooplankton were done by different researchers for the different geographic areas, but were standardized among the collaborating partners via the organization of technical workshops and the exchange of digital photographs. Data analysis involved mainly multivariate statistics testing for congruence among different organism groups and their relationships with environmental variables (see also Declerck et al. 2005).

7.4 Lake Characteristics

Initially, 16 categories of lakes were defined according to four dichotomous criteria (macrophyte cover, phosphorus content, size, degree of connectivity) and each lake category was designed to be represented by six lakes, i.e. two lakes per geographic region. Despite the efforts to standardize the selection of model systems, it proved very difficult to represent all *a priori* defined categories of lakes to an equal extent. Specific lake types were hard to find, especially clearwater/high-nutrient lakes and turbid water/low-nutrient lakes. Also, the expected values, based on earlier surveys or preliminary visits, of lake characteristics used for lake selection were not always fully corroborated by the actual analyses. As a result, not all categories were represented by exactly two lakes in all regions, although the deviation was reasonably low. Although the selection procedure imposed a dichotomous scheme with a threshold value for each of the four focal variables, the selected lakes showed a continuous variation in the different areas. As a result, data analysis was based on multivariate ordination techniques and regression analyses rather than ANOVA-type approaches (see Declerck et al. 2005). Importantly, the

Table 7.1 Average (±SD) of the three main dichotomous variables in each of the studied geographic regions. *SUBMCOV* Cover by submerged macrophytes, *TP* total phosphorus, *Area* lake size, *DK* Denmark, *BNL* Belgium/The Netherlands, *SP* Spain

Variable	DK	BNL	SP
SUBMCOV (%)	21±31	16±22	30±40
TP (μg l^{-1})	131±135	331±610	153±283
Area (ha)	19±29	19±36	18±22

selection scheme fulfilled its goal: for each geographic region, it resulted in the variance of the four focal variables to be mutually independent. There was only one exception to this: in DK, TP tended to be related to macrophyte cover (Spearman rank correlation: –0.47; P=0.007).

Table 7.1 summarizes the environmental characteristics with respect to the three main dichotomous variables of the lakes in the three regions. Thanks to the rigid lake selection, major systematic biases between geographic regions with respect to the dichotomic variables were avoided. Yet, considerable inter-regional differences were found for several other variables. Conductivity was on average very low in DK (median: 33 μS cm^{-1}), while it was very high in SP (median: 4066 μS cm^{-1}). The high conductivity of Spanish lakes was not caused by sea salt intrusion, but rather was related to strong lake-level fluctuations of many Spanish lakes (of which several dried out during summer). For some lakes, soil composition (high concentration of carbonates, sulphates, etc.) also caused variation in conductivity. Levels of TN and suspended matter tended also to be higher in SP than in DK or BNL, while levels of TP and chlorophyll *a* were substantially lower. Land use tended to differ between regions as well. Agri-cultivated land made up an average of about 45 % in DK and BNL but 63 % in SP, while the proportion of urban areas was higher in BNL (20 %) than in DK (11 %) and SP (3 %). The percentages of wetlands and uncultivated areas were overall higher in DK (42 %) and SP (34 %) than in BNL (28 %). The geographical regions studied were so different that it was impossible to select lakes that were similar in all respects. However, by selecting lakes according to specific criteria, we managed to eliminate biases in key factors that were intrinsically related to the occurrence of alternative stable states in shallow lakes.

Despite geographic differences in the average values of some abiotic variables, the patterns of association between lake characteristics showed some strong similarities among geographic regions. For each geographic region, standardized PCA analyses were performed on a set of variables, including the abiotic variables and the biomass of the major organism groups. The major axis of variation (PCA1) was in each case strongly related to variables that are typically associated to water clarity, such as the degree of cover by

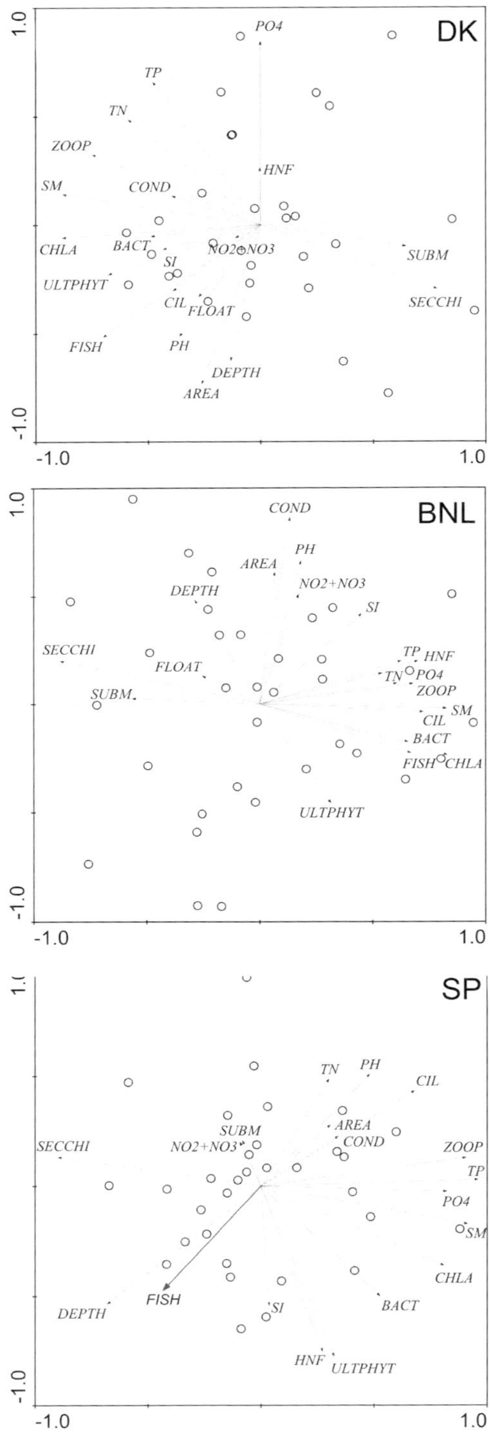

Fig. 7.2 Biplot of standardized PCA analyses performed for each geographic region (*upper* Denmark, *middle* Belgium/The Netherlands, *lower* Spain) on environmental variables (abiotic data, biomass data). The biplot represents the first and second PCA axes. *Bold labels* Environmental variables. *SECCHI* Secchi depth, *DEPTH* lake depth, *AREA* lake surface area, *SM* suspended matter, *COND* conductivity, *PH* pH, *TP* total phosphorus, *TN* total nitrogen, *PO4* phosphates, *NO2+NO3* nitrites and nitrates, *SI* silicates, *SUBM* cover by submerged vegetation, *FLOAT*, cover by floating vegetation, *CHLA* chlorophyll *a*, *BACT* density of bacteria, *ULTPHYT* density of phytoplankton, size range 0.2–20.0 μm, *CIL* ciliate densities, *ZOOP* total zooplankton biomass (rotifers, copepods, cladocerans), *FISH* fish numerical catch per unit of effort

macrophytes, the concentration of suspended matter, phytoplankton biomass and Secchi depth (Fig. 7.2). Furthermore, the first axes of the PCA analyses were strongly associated with the biomass of most groups of aquatic organisms: the biomass of bacteria, ultraphytoplankton (0.2–20.0 μm), ciliates, zooplankton (sum of rotifer, cladoceran and copepod biomass) and fish tended to be higher in turbid lakes than in transparent lakes. In each geographic region, pH was positively related to lake surface area. In BNL and DK, surface area was also positively correlated to lake depth, and in BNL to conductivity as well. In SP, surface area was uncorrelated to depth and conductivity, but lake depth was negatively correlated to conductivity and pH.

7.5 Multidimensionality of System-Wide Biodiversity

In all three geographic regions, congruence in taxon diversity between groups tended to be rather low (Declerck et al. 2005): out of all possible correlations among organism groups, only few were reasonably high and significant. The number of correlations higher than 0.5 equalled two in DK, one in BNL and two in SP (Fig. 7.3). The degree of multidimensionality in the diversity hypervolume was also evaluated by Declerck et al. (2005), by means of standardized PCA-analyses. The total amount of variation explained by the first four PCA axes only ranged between 60.0 (BNL) and 62.5% (DK); and PCA4 still

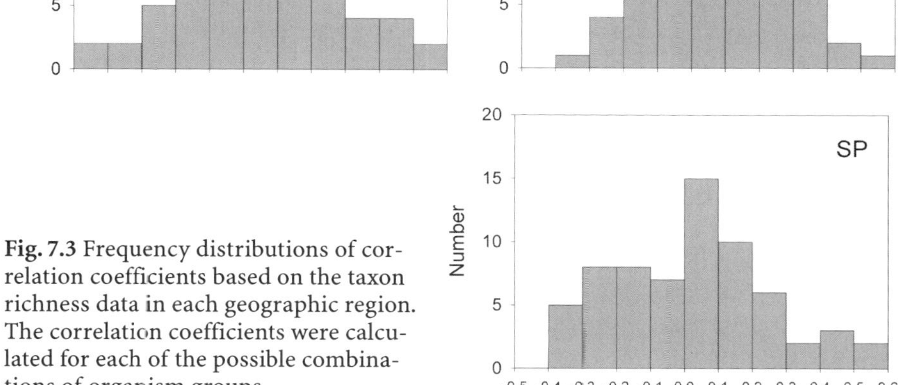

Fig. 7.3 Frequency distributions of correlation coefficients based on the taxon richness data in each geographic region. The correlation coefficients were calculated for each of the possible combinations of organism groups

explained approximately half the amount of the variation explained by PCA1. These results suggest that diversity in different groups of aquatic organisms varies independently to a large degree and that the richness hypervolume is not dominated by one or a few gradients. This leads to the important conclusion that taxon richness intrinsically is multidimensional and that taxon richness assessed for one group of organisms cannot reliably be used to predict taxon richness for other groups nor for the entire freshwater community.

7.6 Macrophytes and Nutrient Concentrations

Several studies have documented a variety of responses of taxon richness to system productivity or surrogates for productivity, such as TP (Leibold 1999; Dodson et al. 2000; Jeppesen et al. 2000; Chase and Leibold 2002). Most reported in freshwater systems are unimodal associations, with richness of different organism groups (e.g. phytoplankton, zooplankton, macro-invertebrates) peaking at intermediate levels of productivity (Mittelbach et al. 2001). In sets of randomly selected lakes, however, several environmental factors may show a high degree of interdependence. For instance, water plant cover and macrophyte richness also tend to respond unimodally to productivity: in oligotrophic lakes, water plant growth is limited by nutrients; in mesotrophic lakes, vegetations are typically species rich and well-developed; at high nutrient concentrations (highly eutrophic to hypereutrophic lakes), there is a clear-cut increase in the likelihood of a turbid state which is often characterized by the absence of any vegetation (Scheffer 1998). The presence of vegetation may on itself affect aquatic diversity, independently of system productivity, because its structure creates habitat heterogeneity, forms a substrate for a large variety of organisms (periphyton, several macro-invertebrate groups) and provides refuge for prey against predation.

Our study differs from the previous ones in that the studied lakes were selected along mutually independent gradients of water plant cover and productivity. This allowed us to disentangle the effects of TP and vegetation cover and to evaluate their unique association with the richness of different organism groups. Declerck et al. (2005) showed that variation partitioning applied on a redundancy analysis model yielded a pattern in which both TP and water plant cover tended to be independently associated with the richness of different subsets of the organism groups. The most pronounced associations between group richness and water plant cover were positive (e.g. number of DGGE bands in bacteria, species number of cyclopoid copepods and cladocerans in DK, genus richness of rotifers and number of *a priori* defined taxonomic groups of macro-invertebrates in BNL; Declerck et al. 2005). An exception to this pattern was found for genus richness in phytoplankton, which showed negative associations with water plant cover. In SP, associations with vegetation could not be disentangled from associations with conductivity. TP also explained a significant proportion of richness in DK and SP, but not in BNL. A detailed examination of the response curve of taxon richness to TP revealed six linear associations, of which three were negative (phytoplankton genus richness in DK, genus richness of rotifers and species richness of cyclopoid copepods in SP) and three were positive (number of DGGE bands in bacteria and eukaryotes in BNL, species richness of submerged vegetation in SP; Declerck et al. 2005). Six response curves showed curvilinearity, but only one of these was significantly hump-shaped. This contrasts with the conclusion of Mittelbach et al. (2001), who reported the occurrence of humped curves in 40 % of the studies that involved diversity in aquatic invertebrates. We hypothesize that this difference may partly be due to the fact that none of the earlier studies considered or explicitly controlled for the potential impact of aquatic vegetation. The unimodal responses that are often observed for taxon richness along the productivity gradient in shallow lakes may indirectly reflect the unimodal response of water plant vegetation to this gradient and need not always be directly steered by productivity itself.

The associations between TP and richness in DK and SP suggest that TP indeed may affect aquatic diversity in direct ways, as proposed by several former studies (e.g. Leibold 1999). In addition, TP may also affect aquatic diversity in a more indirect way by influencing the aquatic vegetation. We postulate that such indirect mediation of richness by TP may provide an additional explanation for unimodal richness responses along broad TP gradients in aquatic ecosystems.

7.7 Model and Expert Tools

7.7.1 Approach

As one of the objectives was to predict the impact of lake or catchment management on biodiversity, the BIOMAN results were used to extend and adapt two predictive shallow lake models [a functional model PCLake (Janse 1997) and the expert system BASIS (Best Analogous Situations Information System; Van Nes 2002)]:

- The output variables from the model PCLake were linked to predictions on biodiversity, based on the multiple regression models derived in BIOMAN. The PCLake model was chosen because it contains the main biotic groups and environmental variables.
- The data were integrated in the BASIS system, so as to allow predictions on biodiversity in various organism groups in shallow lakes, using analogies on relationships between ecosystem variables.

- Both approaches were combined by using the output from PCLake as an input for BASIS.

7.7.2 PCLake

PCLake is a functional ecological eutrophication model developed at RIVM (Janse et al. 1992; Janse 1997). It is based on the simulation of functional groups of organisms, embedded in their abiotic environment and linked by nutrient cycles in the lake ecosystem. The model describes a completely mixed water body and comprises both the water column and the upper sediment layer (Fig. 7.4); optionally, a surrounding marsh zone can be defined if applicable. The PCLake model calculates water quality parameters, such as transparency, the biomass of functional groups of organisms at different trophic levels (three phytoplankton groups, submerged macrophytes, zooplankton, zoobenthos, whitefish, predatory fish) as well as the distribution

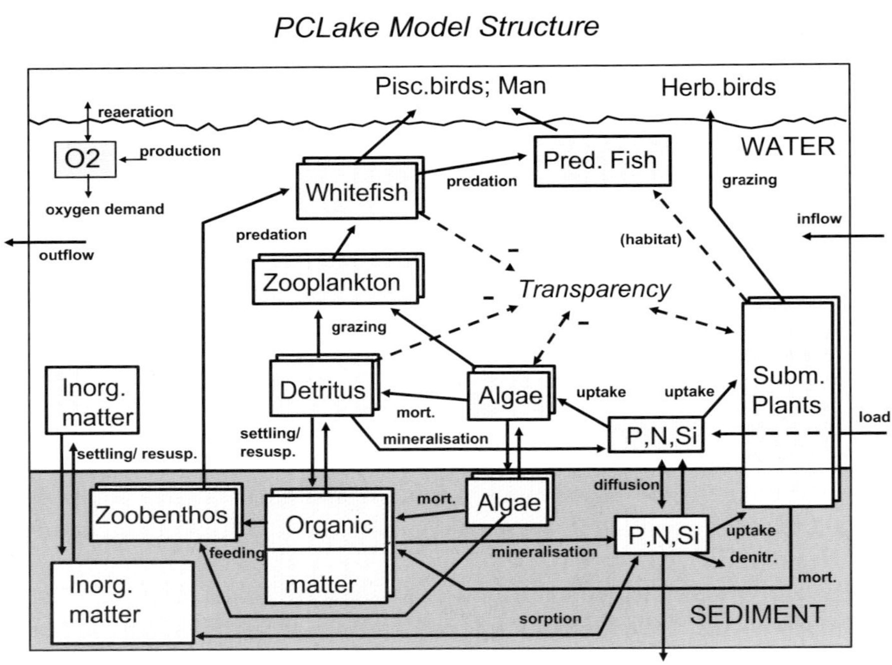

Fig. 7.4 Schematic structure of the PCLake model. *Doubled blocks* denote compartments modelled in both dry weight and nutrient units. Three functional groups of phytoplankton are distinguished: cyanobacteria, diatoms and other small edible algae. *Solid arrows* Mass fluxes (e.g. food relations), *dotted arrows* 'empirical' relations. Egestion and mortality of animal groups and respiration are not shown (adapted from Janse 1997)

and fluxes of the nutrients N and P. Main inputs to the model are: lake hydrology, nutrient loading, dimensions (lake depth, size), size of a surrounding marsh zone (if any), sediment characteristics and lake history (initial conditions). The model has been used in the past to simulate the effects of nutrient load reduction and other measures (e.g. biomanipulation) in a number of case studies. It has also been used to estimate the (threshold) nutrient levels for dominance of either phytoplankton or submerged macrophytes, depending on lake features (Janse 2005). The response showed a hysteresis effect in many cases, consistent with recent theories on shallow lakes ecology (e.g. Scheffer 1998).

In the BIOMAN project, three actions were performed with respect to PCLake: (1) validating the model over a wider range of conditions, (2) establishing a link with biodiversity and (3) developing a more user-friendly version of the combined model. Firstly, the PCLake model was validated using data from the BIOMAN data set. Output variables were calculated for 13 lakes (seven in DK, four in BNL, two in SP) for which estimates on retention time and nutrient loading were available. The lakes in the subset were highly variable in characteristics: mean depth 0.9–3.0 m (median 1.5 m), area 0.5–170 ha (31 ha), retention time 5–600 days (70 days), TP input concentration 0.002–0.5 g P m^{-3} (0.08 g P m^{-3}), TN input concentration 0.03–19.0 g N m^{-3} (3.8 g N m^{-3}). Both 'clear' and 'turbid' lakes and all three climatic regions were represented. Because information concerning initial states and soil composition was often incomplete, different scenarios were calculated for these parameters. Lakes without data concerning hydrology and nutrient loading had to be left out. Simulations were performed until a quasi-steady state was reached. Simulated and measured summer means of chlorophyll *a* and vegetation coverage are shown in Fig. 7.5; the BIOMAN data are shown together with the data on >40 lakes (four were overlapping), also widely differing in dimensions and loading, which were previously used for calibration by means of a Bayesian method. In agreement with the model's aim, this was a combined (compromise) fit on all lakes and variables together, rather than on a particular case (Aldenberg et al. 1995; Janse 2005). Although most of the BIOMAN lakes fitted reasonably well within the earlier results, at least in terms of clear/turbid, there were some lakes in which macrophyte cover was over-predicted. Some of the assumptions on soil features and suspended matter, which had to be very rough due to lack of data, were possibly not correct. The outlier was a lake under recent restoration where macrophyte development had not (and still not yet) started for unknown reasons. Nevertheless, the results were considered satisfactory to allow the model's applicability outside the calibration set.

Secondly, the model was extended to include predictions on biodiversity, in addition to the aspects of ecological functioning of a lake ecosystem (functional groups of organisms, abiotic characteristics), as covered by the original PCLake model. The extension was made using statistical relationships

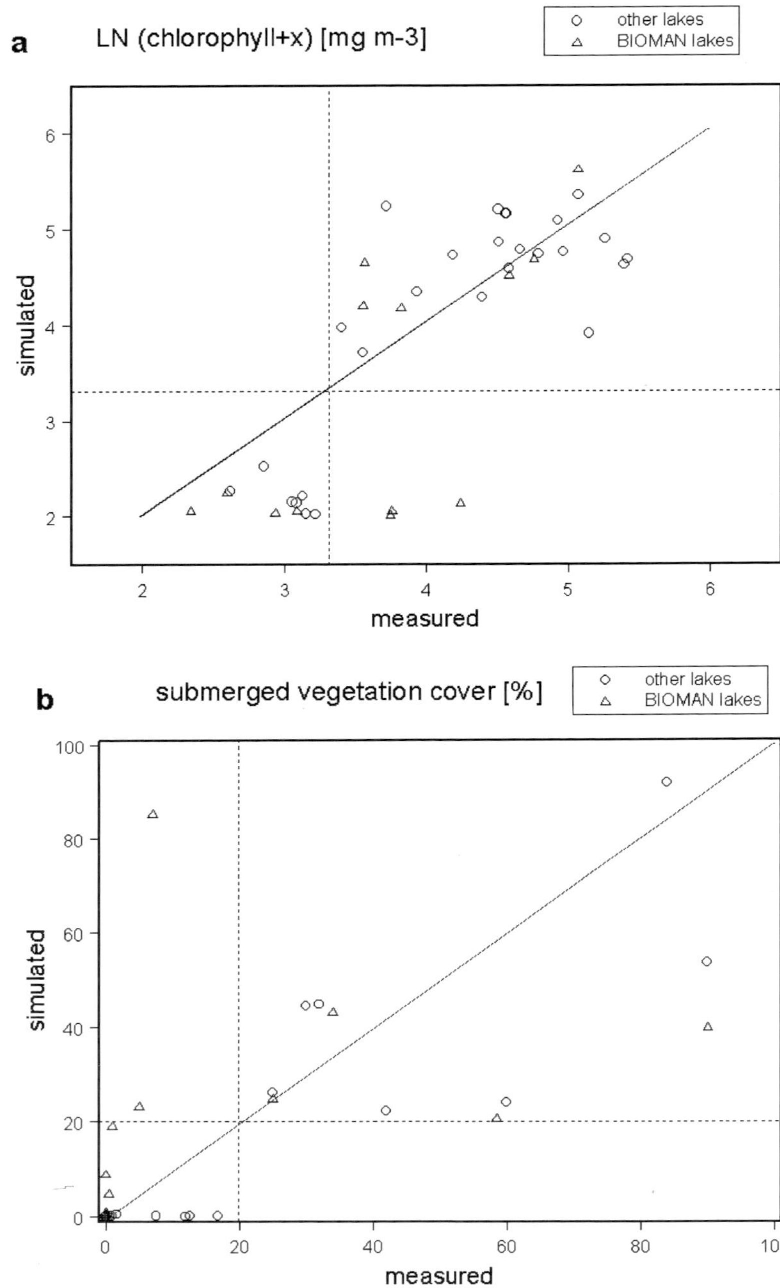

Fig. 7.5 PCLake simulations compared to measured data (summer averages). *Open squares* BIOMAN lakes, *solid circles* other lakes (previous calibration data set). **a** Chlorophyll-*a* (natural log scale, x=7.4 mg m^{-3}). *Dashed lines* indicate the limit of 20 mg m^{-3}. **b** Cover of submerged macrophytes. *Dashed lines* indicate the limit of 20 % cover

between environmental variables and biodiversity, expressed as taxon richness per biotic group. We did not attempt to further split up the functional groups in the model, as information on the extra parameters would be lacking in most lake studies and the model's predictability would probably decrease. Instead, output values calculated by PCLake for a set of abiotic (water transparency, nutrient concentrations) and biotic variables (biomass of phytoplankton, zooplankton and fish, cover by submerged macrophytes) were coupled to taxon richness via multiple regression equations established from the BIOMAN biodiversity data set. This was done for each geographical region separately (DK, BNL, SP). For a variable not included in PCLake, for instance conductivity, the measured value was used. An example of the results (for a clear, macrophyte-dominated, mesotrophic lake >5 ha in the region BNL) is shown in Fig. 7.6. For most groups, the calculated taxon richness based on PCLake output was close to the calculated value based on measurements. Some groups were underestimated by the model, for example the eukaryotic micro-organisms investigated by molecular techniques. The figure also shows the results for a combined application of PCLake and BASIS (see below).

The original model was converted into a user-friendly version, incorporating the biodiversity relations. This consists of an Excel application, making use of an already compiled C++ executable file (Mooij 2002). The model and user manual are available upon request from Jan Janse.

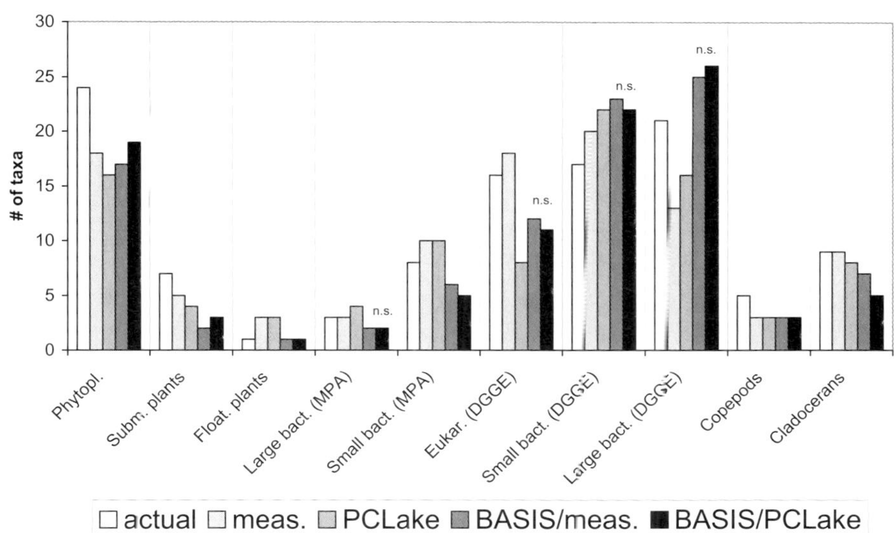

Fig. 7.6 Biodiversity (taxon richness) for different biotic groups in one of the BIOMAN lakes, estimated by different methods: (a) measured taxon richness, (b) multiple regression equations applied to measured abiotic and biomass data, (c) multiple regression equations applied to PCLake output, (d) BASIS output based on measured abiotic and biomass data, (e) BASIS output based on PCLake output. *n.s.* BASIS calculations not significant

7.7.3 The Expert System BASIS

BASIS is an expert system using analogies. It is based on a technique called case-based reasoning (CBR). It was originally developed in research on artificial intelligence (Aamodt and Plaze 1994). Similar to what experts do, the diagnosis of an existing situation and prediction of new situations is based on previous experiences in similar cases. In computerised expert systems like BASIS, these cases of previous experience are stored in a database from which the most similar cases can be extracted. Statistics on these similar cases are used for decision-making in new situations. The advantages of the CBR technique are: (1) there is no need for prior information about the nature of relations between variables, (2) it is easy to find and browse through available information on the most analogous cases in the database, (3) the system is flexible in that it can handle missing data both in the cases stored in the database as well as in the 'new' case for which the user wants to make predictions and (4) the system can improve when new information is added. This capability of learning is an important general feature of artificial intelligence and CBR. The theory behind the analogies and all the features implemented in BASIS, including a user manual of the software package, is described in detail by Van Nes (2002). Application of BASIS on the BIOMAN database allowed us: (1) to visualise the relationships between biodiversity indices and explanatory variables by several graphical options and (2) to predict biodiversity indices for any 'lake in question' as a function of (sub)sets of explanatory variables. This prediction is based on observations in those lakes in the dataset that are most similar to this 'lake in question' with respect to the chosen set of explanatory variables. The method can cope with missing data, although the uncertainty of the prediction is then increased. An example of the results is shown in Fig. 7.6. For six out of ten biotic groups, BASIS could find a significant result for taxon richness. The results were close to the multivariate relations but tended towards underestimation.

7.7.4 Combined Models; and PCLake and BASIS as Management Tools

Predictions of biodiversity may be made by combining PCLake and BASIS. The output of PCLake can be used as input in a BASIS application. In this way, biodiversity can be estimated from variables such as lake hydrology, nutrient loading, dimensions (lake depth, size) and sediment characteristics. The input of the PCLake results has to be carried out manually. Figure 7.6 shows that the results for a case study resembled quite well those of the other methods.

The results of the model applications show that it is possible to completely cover the 'chain' from human management in the catchment, via lake ecosystem functioning to biodiversity output. Considering the uncertainty of the

PCLake input data as obtained for the BIOMAN lakes, the results match reasonably well. Yet, it must be realised that the uncertainties in each of the submodels are also combined. Especially PCLake needs reliable input data on nutrient loading, retention time and soil type. Use of the (combined) models can be a cost-effective way of assessing biodiversity in shallow lakes and provides a method of coping with 'missing data'.

7.8 Synthesis: Policy Implications of the Results

The patterns of association observed for biodiversity in the BIOMAN lakes strongly indicate that biodiversity tends to be rather unrelated between different taxa and trophic levels. This has important consequences for policy makers and lake managers.

A first practical consequence for conservation policy is that identification of suitable indicators for biodiversity is complicated by the lack of association among taxa: the use of one single variable, be it the diversity in one easy-to-study group or an environmental variable, has very limited predictive power for the diversity of different groups of organisms. Thus the search for one simple index reflecting overall ecosystem diversity is of little relevance. Rather, useful indicators may only be found for one or a restricted subset of organism groups that tend to be associated with a similar gradient. To assess richness at the ecosystem level, several of these indicators have to be combined and weighed according to the final aim of the assessment. Therefore, the policy maker or lake manager first has to define priorities, including, for instance, groups deserving special attention, the importance of rarity, etc.

Second, given the low association in diversity between groups, future biodiversity studies would gain from the inclusion of taxonomic or functional groups that are often ignored and were not considered in BIOMAN either, such as the microbial community in sediments, groundwater and the periphyton.

Third, the multidimensionality of diversity also has important implications for the definition of management objectives: management measures that may increase the diversity in one organism group may be ineffective or even be contra-productive for the diversity in other groups. The lake manager should therefore be aware that choices have to be made and that probably no management technique exists that leads to an enhancement of diversity at all trophic levels and for all taxonomic groups. One potential management implication is that a good biodiversity strategy should involve variation in the remediation and management of shallow lakes in a region, so as to increase regional diversity. Lake management should thus also take a regional perspective through optimizing regional diversity by a differential management of local shallow lakes. Although clear-water lakes are generally more species-rich than turbid lakes for many taxa (amphibians and birds; e.g. Hanson and Butler 1994; Hargeby et al. 1994; macrophytes, zooplankton, macro-invertebrates; this study), the existence of both turbid and clear-water lakes within a region will enhance regional biodiversity for specific organism groups, such as planktonic algae and ciliates. Whether this is desirable from an ecological point of view will also depend on the history and the natural reference state of the lakes in a region. Studies such as BIOMAN may contribute to the knowledge needed to define regional management objectives, for instance for implementation of the European Water Framework Directive. However, other considerations than just maximizing biodiversity must also be considered, such as recreational value, drinking water issues, the natural reference state of lakes in a region, etc., as stated in the Introduction.

Acknowledgements This study was financially supported by EU project EVK2-CT-1999-000046 (BIOMAN). We thank the many lake managers and administrations that allowed us access to the study lakes, and the numerous people that contributed to the field work. Mr. M.J. 't Hart is thanked for his contribution to the programming of the models. S.D. and K.M. are postdoctoral researchers with the National Fund for Scientific Research, Flanders (FWO-Vlaanderen).

References

Aamodt A, Plaze E (1994) Case-based reasoning: foundational issues, methodological variations, and system approaches. AI Commun 7:39–59

Aldenberg T, Janse JH, Kramer PRG (1995) Fitting the dynamic lake model PCLake to a multi-lake survey through Bayesian statistics. Ecol Model 78:83–99

Chase JM, Leibold MA (2002) Spatial scale dictates the productivity–biodiversity relationship. Nature 416:427–430

Chorus I, Bartram J (eds) (1999) Toxic cyanobacteria in water. Spon, London

Declerck S, Vandekerkhove J, Johansson L, Muylaert K, Conde-Porcuna JM, Van der Gucht K, Martinez CP, Lauridsen T, Schwenk K, Zwart G, Rommens W, Lopez-Ramos J, Jeppesen E, Vyverman W, Brendonck L, De Meester L (2005) Multi-group biodiversity in shallow lakes along gradients of phosphorus and water plant cover. Ecology 86:1905–1914

Dodson S (1992) Predicting crustacean zooplankton species richness. Limnol Oceanogr 37:848–856

Dodson SI, Arnott SE, Cottingham KL (2000) The relationship in lake communities between primary productivity and species richness. Ecology 81:2662–2679

Graham MH (2003) Confronting multicollinearity in ecological multiple regression. Ecology 84:2809–2815

Hanson MA, Butler MG (1994) Responses to food web manipulation in a shallow waterfowl lake. Hydrobiologia 280:457–466

Hansson LA, Annadotter H, Bergman E, Hamrin SF, Jeppesen E, Kairesalo T, Luokkanen E, Nilsson PA, Søndergaard M, Strand J (1998) Biomanipulation as an application of food-chain theory: constraints, synthesis, and recommendations for temperate lakes. Ecosystems 1:558–574

Hargeby A, Andersson G, Blindow I, Johansson S (1994) Trophic web structure in a shallow eutrophic lake during a dominance shift from phytoplankton to submerged macrophytes. Hydrobiologia 280:83–90

Jackson A, McInnes A, Falconer I, Runnegar M (1984) Clinical and pathological changes in sheep experimentally poisoned by the blue-green algae *Microcystis aeruginosa*. Vet Pathol 21:102–113

Janse JH (1997) A model of nutrient dynamics in shallow lakes in relation to multiple stable states. Hydrobiologia 342/343:1–8

Janse JH (2005) Model studies on the eutrophication of shallow lakes and ditches. PhD thesis, Wageningen University, Wageningen

Janse JH, Aldenberg T, Kramer PRG (1992) A mathematical model of the phosphorus cycle in Lake Loosdrecht and simulation of additional measures. Hydrobiologia 233:119–136

Jeppesen E, Søndergaard Ma, Søndergaard Mo., Christoffersen K (eds) (1997) The structuring role of submerged water plants in lakes. (Ecological Studies 131) Springer, Berlin Heidelberg New York

Jeppesen E, Jensen JP, Søndergaard M, Lauridsen T, Landkildehus F (2000) Trophic structure, species richness and biodiversity in Danish lakes: changes along a phosphorus gradient. Freshwater Biol 45:201–213

Jochimsen EM, Carmichael WW, An J, Cardo D, Cookson ST, Holmes CEM, Antunes MBC, Melo-Filho DA, Lyra TM, Barreto V, Azevedo SMFO, Jarvis WR (1998) Liver failure and death following exposure to microcystin toxins at a hemodialysis center in Brazil. N Engl J Med 36:373–378

Leibold MA (1999) Biodiversity and nutrient enrichment in pond plankton communities. Evol Ecol Res 1:73–95

Mittelbach GG, Steiner CF, Scheiner SM, Gross KL, Reynolds HL, Waide RB, Willig MR, Dodson SI, Gough L (2001) What is the observed relationship between species richness and productivity? Ecology 82:2381–2396

Mooij W (2002) OSIRIS ver. 3.01. Netherlands Institute of Ecology, Centre for Limnology, Nieuwersluis

Moss B (1998) Ecology of fresh waters, man and medium. Blackwell Scientific, London

Scheffer M (1998) Ecology of shallow lakes. Chapman & Hall, London

Scheffer M, Hosper SH, Meijer M-L, Moss B, Jeppesen E (1993) Alternative equilibria in shallow lakes. Trends Ecol Evol 8:275–279

Vandekerkhove J, Declerck S, Vanhove JM, Jeppesen E, Conde-Porcuna JM, Brendonck L, De Meester L (2004a) Use of ephippial morphology to assess anomopod richness: potentials and pitfalls. J Limnol 63[Suppl]:75–84

Vandekerkhove J, Niessen B, Declerck S, Jeppesen E, Conde-Porcuna JM, Brendonck L, De Meester L (2004b) Hatching rate and hatching success of isolated versus non-isolated resting eggs. Hydrobiologia 526:235–241

Vandekerkhove J, Declerck S, Jeppesen E, Conde-Porcuna JM, Brendonck L, De Meester L (2005a) Dormant propagule banks integrate spatio-temporal heterogeneity in cladoceran communities. Oecologia 142:109–116

Vandekerkhove J, Declerck S, Brendonck, L, Conde-Porcuna JM, Jeppesen E, De Meester L (2005b) Hatching of cladoceran resting eggs: temperature and photoperiod. Freshwater Biol 50:96–104

Van Nes E (2002) Best analogous situations information system. User's guide for BASIS version 2.0. Wageningen University, Wageningen

Zwart G, Hannen EJ van, Kamst-van Agterveld MP, Van der Gucht K, Lindstrom ES, Van Wichelen J., Lauridsen T, Crumps BC, Han SK, Declerck S (2003) Rapid screening for freshwater bacterial groups by using reverse line blot hybridization. Appl Environ Microbiol 69:5875–5883

8 River Basin Management to Conserve Wetlands and Water Resources

JAMIE PITTOCK, BERNHARD LEHNER, and LI LIFENG

8.1 Introduction

This chapter is based on the premise that the precipitous decline in freshwater wetlands and species can only be arrested through conservation and sustainable management at a large scale, based on water (usually river) basins. A number of approaches to large-scale freshwater wetlands conservation are presented and assessed to draw conclusions on future conservation priorities.

The Millennium Ecosystem Assessment (MEA) concludes (MEA 2005, p. 44): "*Freshwater ecosystems tend to have the highest proportion of species threatened with extinction.*" While information is patchy and inconsistent, it is estimated that, since 1900, 50 % of the world's wetlands have been lost (Finlayson and Davidson 1999). The World Wide Fund for Nature (WWF) suggests that there has been a 50 % decline in the freshwater species population index in the 30 years since 1970, compared to 30 % for forest and marine species between 1970 and 2000 (Loh and Wackernagel 2004; Loh et al. 2005; Fig. 8.1). Does this mean that those of us in government, academia and non-government organisations charged with conservation have failed freshwater biodiversity far more than for other biomes?

The world's governments have agreed to establish "*comprehensive, effectively managed and ecologically representative national and regional systems of protected areas that collectively, inter alia through a global system, collectively contribute to achieving … the 2010 target to significantly reduce the current rate of biodiversity loss*" (CBD 2004a). As freshwater habitats are often a focus for human settlement and can be heavily used and modified, establishing freshwater protected area systems requires extensive negotiations and partnerships with local people. Extensive use of International Union for the Conservation of Nature (IUCN) protected area categories V and VI for less strict protection and sustainable resource use is often required to meet the needs of both local peo-

Ecological Studies, Vol. 191
R. Bobbink, B. Beltman, J.T.A. Verhoeven, and D.F. Whigham (Eds.)
Wetlands: Functioning, Biodiversity Conservation, and Restoration

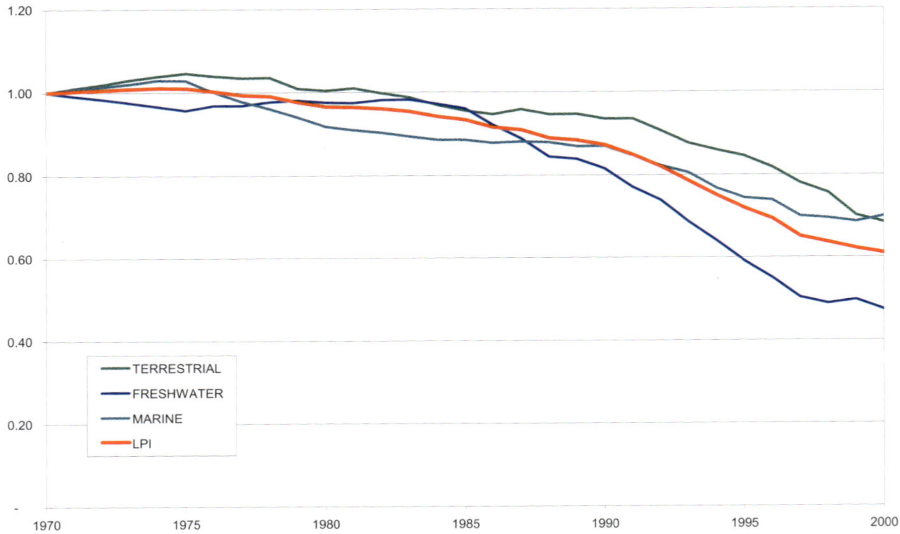

Fig. 8.1 Indices showing changes in the abundance of 323 freshwater species, indicating a decline of around 50 % for freshwater species from 1970 to 2000. Temperate and tropical species abundances are shown separately, plus the combined freshwater species population index. The combination of all these species is shown as the Living Planet Index (LPI). From: Loh and Wackemagel (2004)

ples and conservation. Further, the ecological health of wetlands depends on sustaining fragile ecological processes, such as water flows that can originate far from the site concerned. These flows are easily disrupted, for example by dams, which can reduce the volume, timing and quality of water essential for the survival of downstream species and habitats. Hence, wetland conservation advocates need to influence use of natural resources within entire river catchments to successfully conserve freshwater biodiversity.

This is a formidable task. Climate change threatens to significantly alter stream flow. Likely changes vary regionally and include: loss of glaciers and the base flows they sustain, changes in runoff and stream flow, and more frequent extreme flood and drought events (Arnell and Chunzhen 2001). Adaptations proposed by governments and others for greater hydroelectricity generation and increased water storage would further exacerbate the impact on freshwater ecosystems. Clean development mechanism funding for hydroelectric dams is one example of the impact of climate change policies.

There are already more than 45 000 large dams around the world with another 1,500 planned or under construction (WCD 2000), including some of the most biodiverse rivers like the Yangtze, Mekong, Amazon and Orinoco Rivers (WWF 2004). Already 59 % of the world's large river systems are fragmented by dams (Nilsson et al. 2005). Currently, humans appropriate 54 % of accessible runoff (WWAP 2003). Global water diversions are estimated to increase by 30–85% over the next 50 years (MEA 2005) to meet growing demands for more food for a growing population and more water-intensive food such as red meat for affluent populations (SIWI and IWMI 2004). Some rivers no longer reach the sea and many remaining waters are polluted and afflicted by invasive species (MEA 2005).

The world's governments are committed to address water management problems, but their agreements may exacerbate the decline in wetland habitats. There is unacceptable water-related poverty: 1.1×10^{12} people are without a safe water supply, 2.4×10^{12} are without adequate sanitation, and over 2.0×10^{6} people are killed each year from water bourne diseases (WWAP 2003). The 2002 World Summit on Sustainable Development (UN 2002) added targets to the UN's 2015 Millennium Development Goals (UNGA 2000) to: complete integrated water resource management plans, extend water supplies, sanitation and energy to the poor, and expand agricultural production that currently consumes 70% of the water diverted world wide (WWAP 2003). Coupled with growing demand for cleaner electricity to minimise greenhouse gas emissions, the pressure to build more dams and consume freshwater resources is growing.

WWF assessed 30 developing countries's World Bank Poverty Reduction Strategy Papers (PRSPs) and their African Caribbean Pacific–European Union Country Support Papers (CSPs; Pittock 2006). Of the 16 countries with PRSPs, only two (13%) prioritise water supply and sanitation, whereas eight (50%) record water as an issue without making it a clear priority. The results are worse for CSPs, where only four of the 26 countries (15%) with CSPs include water and sanitation as priorities as opposed to 22 (85%) who do not. In ten PRSPs that were assessed in more detail, agriculture is considered a driver of growth: irrigation development is favored in seven and construction of hydropower facilities is envisaged in five, whereas integrated water resources management does not feature strongly in any.

Thus, wetland conservation advocates have a huge challenge to influence their governments' national poverty reduction, sustainable development and integrated water resources management plans to deliver services for people while conserving freshwater biodiversity.

How then can the international community and wetland conservation advocates and managers enhance freshwater biodiversity conservation? The authors believe that societies by their land and water use activities deliberately or indirectly choose the extent of the natural wetlands they retain; and these choices are strongly influenced by the degree to which different stakeholders in society discuss and agree on water and wetland management objectives. In this chapter, it is argued that wetland advocates and managers need to apply large-scale planning and strategic interventions at the river basin, national and regional scale to conserve the ecological processes, particularly the hydrological regime, needed to conserve wetlands to the degree that the societies concerned choose.

8.2 Systematically Prioritising Wetland Conservation: Freshwater Ecoregion Conservation

Given limited resources of governments and non-government agencies, in WWF's view, a critical first step is to delineate conservation units or ecoregions (usually based on water basins) and prioritise those for intervention, based on species diversity and representation. Then, within each ecoregion there is a need to prioritise the ecological processes and geographic units at a finer scale (landscapes or watersheds) for conservation. The rapid pace of approval of new developments impacting on river basins, such as the 11 large dams planned and under construction in the Mekong River basin (WWF 2004), requires conservation advocates to propose measures to minimise impacts even where there is limited environmental data. WWF and partners are beginning to develop and apply GIS-based rapid assessment tools to identify and prioritise watersheds and floodplain landscapes within river basins for conservation interventions. These methods are briefly described in this section.

8.2.1 Freshwater Ecoregions

WWF and several other organisations have developed and adopted an approach to large-scale conservation planning referred to as ecoregion conservation (ERC). An ecoregion is a large unit of land or water containing a geographically distinct assemblage of species, natural communities and environmental conditions (Dinerstein et al. 1995). Freshwater ecoregions have been delineated, drawing primarily on distribution data for freshwater fish (Fig. 8.2). The boundaries of an ecoregion encompass an area within which important ecological and evolutionary processes most strongly interact (Olson and Dinerstein 2002). Large rivers and other wetlands and their contributing watersheds often fit this definition. A subset of the world's freshwater ecoregions, tagged the 'Global 200', have been prioritised for conservation interventions, based on species diversity and representation (Olson and Dinerstein 2002).

8.2.2 Planning Conservation of Freshwater Ecoregions

WWF has published a guide for practitioners to plan interventions for the conservation of freshwater ecoregions (Abell et al. 2002). As a key element, ERC advocates generation of a vision for what an ecoregion should look like in 50 years if its biodiversity is to be maintained. A conservation blueprint should be designed around the protection both of sufficient habitat, even for

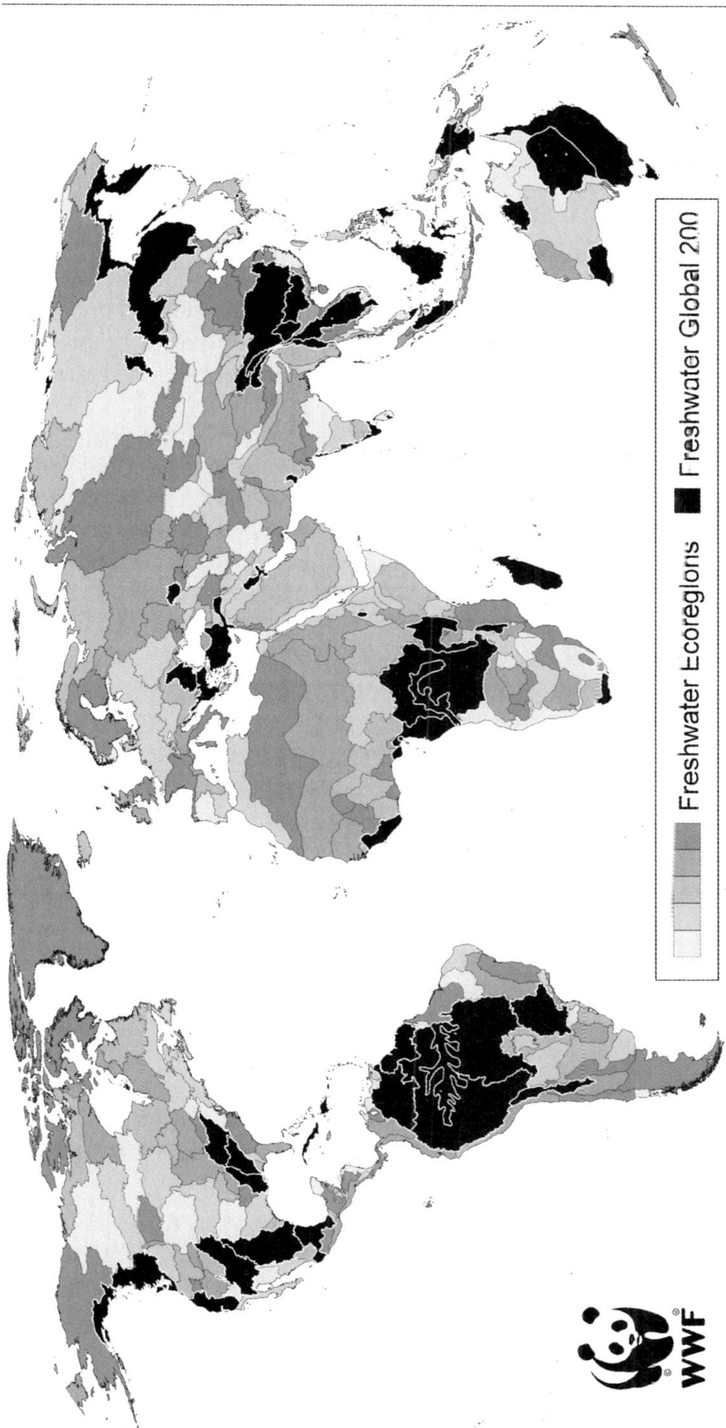

Fig. 8.2 'Global 200' freshwater ecoregions prioritised for conservation by WWF, based on species diversity and representation. From: Olson and Dinerstein (2002), WWF (2006)

the most wide-ranging and sensitive species, and of the physical processes that create and maintain those habitats. To address these requirements, a suite of priority areas is identified based on species representation, biological distinctiveness and conservation status. For the physical processes, however, the nature of freshwater systems requires that we go beyond these aquatic priority areas. A vision for a freshwater system must take into account the importance of lateral, longitudinal and even vertical connectivity, examine threats originating upland, upstream and even downstream, incorporate strategies for protecting hydrologic processes operating over large scales, and consider the implementation of land-based conservation strategies in the larger basin. Special attention should be paid to the identification and analysis of environmental flow requirements, critical areas for runoff generation and the effects of flow alterations on downstream wetland areas.

WWF, working with partners, has undertaken ERC in a number of freshwater systems, including the Amazon, Congo, Niger, Yangtze and lower Mekong Rivers. Figure 8.3 illustrates the biological vision developed for the Niger River basin that is now the basis of the Niger Basin Initiative involving the Niger Basin Authority, WWF, Wetlands International, IUCN and the Niger-

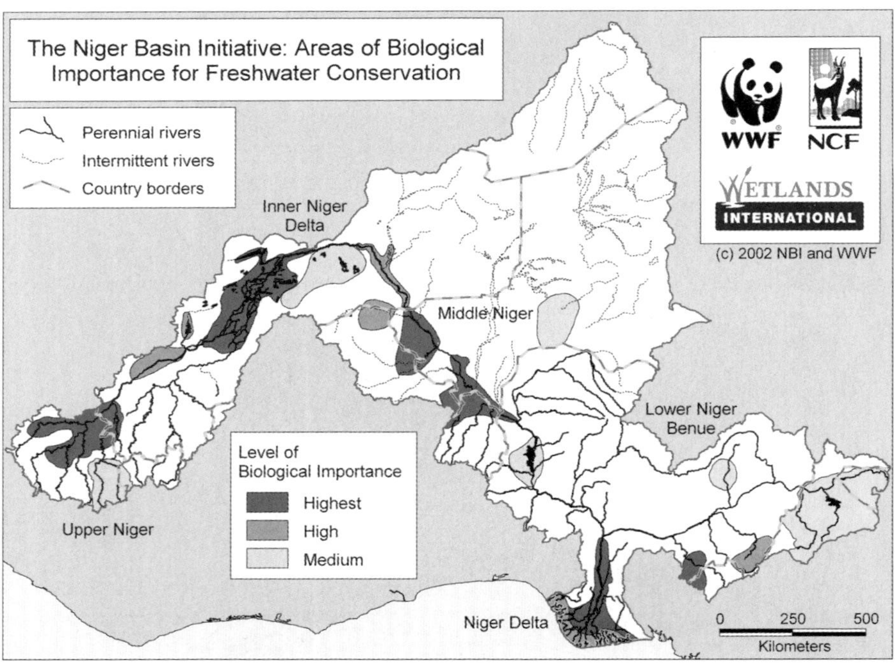

Fig. 8.3 Niger River basin biodiversity vision, showing priority landscapes and watersheds identified within the basin for intervention to conserve the biodiversity and ecological processes. From: Wetlands International (2002)

ian Conservation Foundation. The many lessons we have derived from this work include the critical need to integrate the expertise of hydrologists with that of biologists, the importance of starting with catchments rather than small 'hotspots' and the value of integrating freshwater strategies with parallel efforts in adjacent terrestrial and marine systems. This process has proved valuable in engaging, communicating and developing ownership for the conservation vision by government and non-government stakeholders and co-ordinating their activities.

WWF's next steps in developing this methodology involves estimating future threats like climate change, incorporating that information into conservation strategies and conducting research to begin to identify thresholds in land use that translate into threats to aquatic biodiversity.

8.2.3 Rapid Assessment of Watersheds and Landscapes in Data-Short River Basins

In river basins like the Amazon and Mekong, developments such as dams and deforestation for agricultural expansion require urgent decisions by conservation advocates as to which areas require protection, those areas that can be developed and the mitigation measures to be applied.

Consequently, WWF and partners are developing and applying GIS-based rapid assessment tools to identify and prioritise watersheds and floodplain

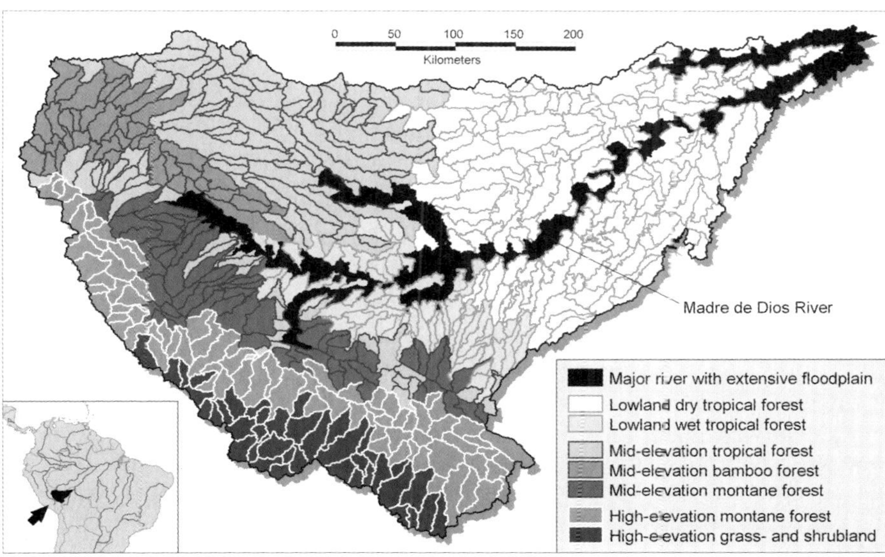

Fig. 8.4 Classification of sub-watersheds in the southwest Amazon basin, based on physical characteristics and remote sensing information. From: Abell et al. (2004)

landscapes within river basins for conservation interventions. For sub-watersheds, the approach used combines data for vegetation, soils, geology, flows and elevation to delineate similar watershed types as a surrogate for freshwater biological diversity (Fig. 8.4). This can be used to identify areas with a greater diversity of habitats to prioritise for conservation, to select examples of each type for conservation, or to enable trade-offs for the conservation of one watershed in return for the development of another similar type of watershed.

8.3 Using Treaties to Conserve Wetlands and River Basin

Intergovernmental agreements and treaties are essential for the conservation of wetlands by promoting better methods and standards for application by each country, and in particular of the 263 river basins shared by more than one country. While there are a great number of river basin specific agreements (UNEP 2002), unfortunately at this time these mostly concern national borders and water infrastructure development and thus make little or no contribution to conserving wetlands.

The effectiveness of three major multilateral agreements for the conservation of wetlands is discussed here. The Ramsar Convention on Wetlands and Convention on Biological Diversity (CBD) are international treaties while the European Union's Water Framework Directive (WFD) is sub-regional, binding legislation. The governments that are party to each of these agreements have committed to conserving freshwater biodiversity within a framework of integrated river basin management.

8.3.1 The Ramsar Convention on Wetlands

The Convention on Wetlands (www.ramsar.org) is often misunderstood as only a waterbird conservation treaty, when in fact its 146 contracting parties (nations) have committed themselves to: 'wise use' of all wetlands (including rivers) on their territory, conservation of wetland sites of international importance, and international co-operation (Ramsar Convention 1971). Arguably this Convention is the most global of all treaties that focus on freshwater management.

Ramsar was one of the first treaties to adopt what is now called the 'ecosystem approach'. In particular the Convention recognises the fundamental ecological functions of wetlands as regulators of water regimes and as habitats supporting a characteristic flora and fauna.

The Convention is now the world's largest protected area system, covering 1458 sites (just over 1 % of the world's registered protected areas; Chape et al. 2003) and 1.254×10^6 km^2 (over 6 % of the world's 18.8×10^6 km^2 area of parks; Chape et al. 2003) on its Register of Wetlands of International Importance (as at 8 August 2005).

In 1999, the Convention adopted guidelines for integrating wetlands and river basin management (Ramsar Convention 1999). In 2002, Ramsar's focus on stream flow management was strengthened by the adoption of resolutions on "*guidelines for the allocation and management of water for maintaining the ecological function of wetlands*" (Ramsar Convention 2002a) and application of the report of the World Commission on Dams (Ramsar Convention 2002b). The *Memorandum of Cooperation and Third Joint Work Plan with the Convention on Biological Diversity* (Ramsar 2002 c) gives Ramsar's decisions even greater stature, for example, in the revision of the CBD's Inland Water Biological Diversity Programme of Work (CBD 2004b). Thus, the Ramsar Conventions's role in water management is growing.

The key question is whether these commitments are translated into effective action by member governments. The triennial reports of member governments (Ramsar Convention 2002d) on their implementation efforts do indicate progress. The fact that 119 of 125 member governments in 2002 submitted reports suggests that most members consider the treaty processes seriously, yet the scale of positive self-responses on most key obligations is variable. For example, some of the responses were:

- 31 % said they had implemented integrated management for river basins and/or coastal zones.
- 35 % have national wetland policies or similar instruments.
- 45 % said they had reviewed national institutions related to wetlands to ensure that resources are available to implement the Convention.
- 63 % said they require an environmental impact assessment for any action that can potentially affect any wetland.

This treaty has many benefits for wetlands conservation since it: (a) creates moral pressure for member governments to establish and manage wetland protected areas, (b) sets standards, provides guidance and facilitates collaboration on best practice, (c) has a triennial global reporting and monitoring system and (d) encourages participation of NGOs, local and indigenous peoples. As Ramsar sites can be established over any land tenure, it is also a tool for establishing protected areas in collaboration with indigenous and private land holders. The emerging trend of neighbouring countries collaborating to implement the Convention regionally (see Section 8.5) is particularly promising.

Ramsar's effectiveness is only as good as the quality of legal protection and management afforded to Ramsar sites by different member governments and their commitment to fully implement their treaty obligations. While Ramsar may be one of the best implemented of the world's environment treaties, wetland conservation advocates can do more to help national governments fulfill their obligations by adopting a 'strategic framework for the Ramsar list' and a national wetlands strategy, as they have previously promised at the Ramsar Conferences of Parties.

8.3.2 Convention on Biological Diversity

The CBD (www.biodiv.org) has been slow in building momentum for conservation of freshwater habitats, with the adoption of key freshwater measures at its February 2004 Conference. The governments adopted a Programme of Work on Protected Areas (CBD 2004a) that:

- declared that protected areas systems are essential to achieving the Plan of Implementation of the World Summit on Sustainable Development and the Millennium Development Goals, as well as the 2010 target to significantly reduce the current rate of biodiversity loss;
- requested the Global Environment Facility to work with other donors, to address the long-term financial sustainability of protected areas to help to secure sufficient resources by 2008;
- decided to establish by 2006 time-bound and measurable protected area targets to establish and strengthen national and regional systems of protected areas, complete national protected area system gap analyses by 2006 and declare the necessary additional reserves by 2009.

The adopted definition of 'terrestrial' includes inland water ecosystems and so ensures they are explicitly included in these targets, and a provision in the Programme calls on the countries to rectify the under-representation of inland water ecosystems in their protected area systems. However, the Plan does not call for the adequate allocation and reservation of water for protected areas.

A Programme of Work for Inland Water Biological Diversity with a wide set of goals and activities was adopted without targets. Integrated river basin management was adopted as a primary tool along with a provision calling on Parties to facilitate minimum water allocations to maintain function and integrity of freshwater ecosystems. It is too soon to judge the effectiveness of implementation of these recent decisions. The collaboration with the Ramsar Convention (Ramsar 2002 c), amounting to endorsement of Ramsar as the CBD's wetlands implementing agency, is a promising step in efficiency, coordination and combining resources for wetlands biodiversity conservation.

8.3.3 European Union's Water Framework Directive

The European Union (EU) in 2000 adopted a new, overarching water law that will apply to all rivers in all EU member countries, plus accession states and some neighbouring countries; that is, over 29 countries to date. *The Directive of the European Parliament and of the Council 2000/60/EC Establishing a Framework for Community Action in the Field of Water Policy* – known as the *EU Water Framework Directive* – entered into force on 22 December 2000 (EU 2000). The Directive establishes a series of procedural and environmental obligations that each country must meet for each river. Eight 'minimum compliance deadlines' have been set for each member government to operationalise the Directive, starting with transposing it into national laws by the end of 2003, to public consultation on basin management plans for each river by the end of 2006, among other tasks, through to the achievement of environmental objectives by the end of 2015.

The Directive requires national governments to:

- establish river basin districts and a basin authority in collaboration with all basin governments;
- identify and agree key water management issues;
- adopt river basin management plans to achieve ecologically 'good status' of rivers;
- identify and protect water bodies used for drinking water abstraction;
- introduce water-pricing policies that provide adequate incentives for efficient use of water;
- control all pollutant emissions and discharges into surface waters;
- introduce specific controls for certain higher-risk pollutants.

The Directive obliges member states "to encourage the active involvement of all interested parties in the implementation of this Directive" and is backed up by EU rules on access to data and provisions for the public to take legal action for the Directive's enforcement. The Directive works on the principle of subsidarity, that is, empowering local governments to take locally appropriate action to meet common European standards. The Directive is enforced with tough fines levied by the EU on offending national governments. However, as at November 2003 the majority of the 23 European countries surveyed had yet to improve their water management, despite the Directive (WWF 2003).

This European example has a number of lessons for governments and other stakeholders at the national and provincial scales in managing rivers basins that cover many jurisdictions. It is a good example of setting common standards at the basin or EU scale and devolving responsibility for achieving them to more local levels of government who can develop locally accepted solutions.

The three examples of multilateral agreements discussed above highlight the potential of treaties to raise standards for conservation of wetlands at a river basin and national scale and to strengthen strategic action by governments. However, these measures were agreed relatively recently and a lot of action is required by all stakeholders for their effective implementation. The current gap between these recent government resolutions and the state of wetlands conservation also shows how much more work is required.

8.4 Poverty Reduction Through Wetlands Conservation

WWF's experience is that the conservation of biodiversity is not the primary motivation for most governments or stakeholders in managing wetlands, especially where local people live in poverty (Jones et al. 2004). Further, poverty reduction is currently the primary focus of international co-operation for sustainable development (e.g., UNGA 2000; UN 2002) and channels extensive resources that may diminish rather than improve wetland conservation unless directed strategically. This section discusses conservation and development in the Yangtze River basin as an example of how poverty reduction, improved river basin management and better wetlands conservation can be mutually reinforcing.

8.4.1 Yangtze River and Dongting Lake

The Yangtze is the world's third-longest river at 6300 km. The basin covers 20 % of China and contains 37 % of its freshwater resources, 39 % of its freshwater fish species and 53 % of its hydroelectricity production. One-third of China's people – 427×10^6 in 2002 – live within the basin and enjoy an economic growth rate of 15 % p.a., compared with the national rate of 9 % p.a. (State Statistical Bureau 2003). Notable or endangered species living in the basin range from the Dawn redwood tree (*Metasequoia glyptostroboides*) and the Siberian white crane (*Grus leucogeranus*) to Pere David's deer (*Elaphurus davidianus*) and the Yangtze river dolphin (*Lipotes vexillifer*; Mackinnon 1998).

The Yangtze was once connected to a huge floodplain with thousands of lakes but, due to reclamation since the time of the Song Dynasty (AD 960–1279), the lakes and wetlands in this floodplain have declined by 30 % or 13 000 km^2. Between 1940 and 1990, 800 lakes disappeared and hundreds of lakes were disconnected from the river (Shen 2001). Between 1949 and 1987, there were 1700 km^2 of wetlands reclaimed from Dongting Lake alone (Dou and Jiang 1999; Zhou et al. 2002). Clearing and logging from 1950 to 1990 reduced natural forest coverage from 30 % to 10 % of the basin and resulted in the sedimentation of wetlands. This significantly reduced the water storage capacity of Dongting Lake and increased the flood risk for the 10×10^6 local residents. Extensive pollution and eutrophication of wetlands has also diminished habitat quality (Dou and Jiang 1999; Zhou et al. 2002).

The floods of 1998 in the Yangtze River basin killed 1562 people and resulted in an estimated U.S.$$25 \times 10^{12}$ in damages (Ministry of Water Resources 1999). After the Yangtze River flood in 1998, the Chinese Government adopted a '32-character policy' (CCCP China and State Council 1998) to: stop logging in the upper reaches of the Yangtze, convert cultivated slopes into forest and grassland, breach some polders in flood emergencies, convert cultivated land to lakes to increase flood storage capacity, resettle people threatened by floods to newly established towns, reinforce the key embankments, dredge the river course to increase discharge, and provide people with jobs instead of relief subsidies. WWF supported this policy and started its Yangtze Programme in the Dongting Lake area to restore floodplain wetlands at four selected sites, two of which are described below in more detail, including the author's (L. Lifeng) unpublished data.

8.4.2 Xipanshanzhou Polder

Xipanshanzhou Polder covered 110 ha of Dongting Lake and was home to 580 people. The polder was built in 1972 and was breached by floods in 1996, 1998 and 1999. From 1999, WWF helped this village to develop alternative livelihoods that achieved significant economic, environmental and social successes. Households in Xipanshanzhou had a poverty-line income of CNY 2000 p.a. in 1999 before the project started. By 2004 income per household resulting from the project rose to CNY 3196 p.a. (CNY 1.0 = ca. U.S.$ 0.12 in 2003; Schuyt 2005). This change was brought about by a farming pattern change: from farming only rice to mixed farming (that included breeding fish, pigs, livestock, poultry, planting and producing high-value organic foods), developing ecotourism and sustainable fishing. It also involved subsidising households to install biogas cooking systems to free the women from the drudgery of collecting firewood, to mitigate the environmental impact of wood removal, to provide a healthier home environment and to produce organic fertiliser. Further, income continued to increase for participating households after external funding stopped in 2001.

There have been significant environmental benefits, with six rare fish species, ten aquatic plant species and ten bird species returning to the restored wetland. At the same time, 12 fish species of commercial value are increasing and the total number of birds in the winter season had tripled by 2003. The villagers reported having more leisure time, as they did not need to work to maintain the dykes to resist flooding their former rice polder. By contrast, the village now needs the floods to provide more freshwater and nutrients for their fish. The residents also reported more off-farm employment in nearby towns and cities, which contributed about 38 % to their total income by 2004. Because the demonstration site showed successful wetlands restoration and became a model for a flood adapted economy and for organic food production, the site has attracted many visitors for ecotourism, research and study tours.

8.4.3 Quinshan Polder and Lessons Learned

Qingshan Polder is an area of 11 km^2 that was built in the 1970s and was home to 5820 people. The polder was abandoned after it was breached in the 1998 flood and the area was included in the West Dongting Lake Nature Reserve. To resolve conflict, the WWF project helped the local fishers to form a 'co-management association' and establish an agreement overseen by a 'commission of co-management' for the Nature Reserve. In return for legal access to the Reserve, the fishers adopted a more sustainable fishing regime that minimised impacts on over wintering migratory birds. The fishers now support the Nature Reserve by contributing RMB 200 000 in lake utilisation fees to the Nature Reserve each year for management.

This Dongting Lake work demonstrates that poverty reduction is key to achieving the support of local people and other stakeholders for the conservation and restoration of key wetlands. These Yangtze demonstration projects are small but can be scaled-up, based on an ecoregion vision to the river basin scale to sustain the ecological processes of wetlands.

In mid-2002, the Yangtze River flooded Dongting Lake six times but was much less damaging than in previous years, attributed in part to 540 km^2 of restored polders that now safely hold floodwater. In November 2004, the China Council's Integrated River Basin Management Taskforce, involving the Yangtze River Resources Commission and WWF, among others, presented proposals to the Chinese State Council for more sustainable river basin management in China, based in part on this field experience.

This 'Living Yangtze' project highlights how river basin management measures can conserve wetlands to benefit people and nature. Working with communities at a sub-catchment or local scale is essential to demonstrate solutions. Learning the lessons from the field work and replicating it at the basin, provincial and national scales may begin to secure the conservation of ecological processes and wetlands at a river basin scale.

8.5 Conservation and Wise Use of Wetlands: a Regional Partnership Approach

The conservation of wetlands on a large scale may be accelerated through regional collaborations "*as they can build upon bio-geographic commonalities, shared wetland systems and wetland-dependent species, and solidly established common social and cultural links*" (Ramsar 2002e). This section discusses some of these initiatives within the framework of the Ramsar Convention.

8.5.1 Ramsar Convention and Regional Initiatives

International co-operation is one of the 'three pillars' of the Ramsar Convention under Article 5. This obligation of contracting parties was elaborated by the *Guidelines for International Cooperation under the Ramsar Convention* (Resolution VII.19) that provides the appropriate framework for promoting international collaboration among Parties and other partners, enhanced by Resolution VIII.30: *Regional initiatives for the further implementation of the Convention* (Ramsar Convention 2002e). Further discussion is scheduled at the Ninth Conference of the Parties in November 2005.

Phillips (2004), a former Deputy Secretary-General of the Ramsar Secretariat, says "*... the Ramsar Convention, largely through its Secretariat, has been an active player in seeking out and formalising working partnerships over the past decade. A primary motive for this has been to add-value to the efforts of the partners in each case, so that precious resources available to each can achieve more on the ground. There is strength in numbers. A second motive has been to 'bring Ramsar in from the cold' and mainstream it as part of the global efforts to promote the sustainable development and water use agenda. Thirdly, Ramsar has recognised that partnerships bring technical skills, and sometimes resources to help with implementation at the global, regional or national levels*".

The Ramsar Convention has a large number of global partnerships and initiatives but this discussion focuses on regional initiatives. The Convention's regional (continental or smaller-scale) partnerships (with the date of the establishment of the partnership) fall into three categories:

- Regional natural resources and environmental treaties, often focused on regional seas, namely: Cartagena Convention for the Protection and Development of the Marine Environment in the Wider Caribbean Region (2001), Mediterranean Action Plan of the Barcelona Convention for the Protection of the Marine Environment and the Coastal Region of the Mediterranean (2001) and the South Pacific Regional Environmental Programme (2002);
- River basin management organisations, namely: International Commission for the Protection of the Danube River (2000), Lake Chad Basin Commission (2002) and Niger Basin Authority (2002);
- Specialist regional wetlands initiatives: EUROSITE, the network of European natural heritage management bodies (1999) and MedWet – the Mediterranean Wetlands Initiative (since 1991).

In 2002, the Convention decided to formally endorse regional initiatives and provide seed funds for their establishment: a total of around CHF 80 000 (CHF 1.0 = ca. U.S.$ 0.79) p.a. in each triennium, starting with MedWet (Ramsar Convention 2002e). This sparked great interest in developing new initiatives in regions as diverse as Africa, the Andes, Carpathian Mountains, Baltic Sea, Black Sea and the Great Asian Mountains. While this interest is welcome, it also creates a dilemma, since the small budget of the Convention Secretariat limits the number of initiatives it can fund. This section briefly reviews the oldest successful regional initiative, MedWet, and one of the promising emerging initiatives, for the Great Asian Mountains, to identify some of the key elements for success.

Guidelines for the development of and support for regional initiatives to implement the Convention were adopted in 2002 (Ramsar Convention 2002e) that include: (a) involving all key stakeholders from the start, including regional governments and their ministries responsible for the environment and water issues, intergovernmental bodies, NGOs, academia, and economic actors, (b) having clearly defined terms of reference, (c) having strong scientific and technical backing, (d) aligning the strategic and operational targets to the Strategic Plan of the Convention, (e) securing political and financial support from governments and other partners for planned activities and projects, (f) becoming financially self-sufficient, (g) involving all the stakeholders in advisory mechanisms and (h) co-ordinating with the Ramsar Convention.

8.5.2 MedWet – The Mediterranean Wetlands Initiative

Under MedWet, the Ramsar Parties surrounding the Mediterranean have established a mechanism for dialogue, information and expertise exchange and assistance with mobilising support funds (Ramsar Convention 2004a). Concern was first clearly expressed at the destruction of Mediterranean wetlands at the MAR Conference held in the Camargue, France, in 1962, which eventually led to the establishment of the Convention on Wetlands at Ramsar, in Iran in 1971. The Grado Symposium in 1991 adopted as its goal: "*To stop and reverse the loss and degradation of Mediterranean wetlands*". From this grew MedWet, a partnership between the European Commission, the Ramsar Secretariat, the governments of France, Greece, Italy, Portugal and Spain, and several non-government organisations (NGOs) – Wetlands International, WWF International, the Station Biologique de la Tour du Valat and the Greek Biotope/Wetland Centre. The first MedWet project (1993–1996) produced and tested a number of tools of relevance to conservation and wise use of wetlands throughout the Mediterranean.

At the culmination of the first phase of MedWet, the partners drew up a Mediterranean Wetland Strategy for 1996–2006 to facilitate regional implementation of the Ramsar Strategic Plan, covering the whole Mediterranean. The goal of the strategy is to stop and reverse the loss and degradation of Mediterranean wetlands as a contribution to the conservation of biodiversity and to sustainable development in the region. The Strategy is intended to advance conservation and wise use of Mediterranean wetlands by:

- promoting the implementation of policies, including National Wetland Policies;

- advancing conservation of the biological diversity of Mediterranean wetlands through sustainable management, restoration and rehabilitation;
- increasing awareness of the ecological, cultural, recreational, scientific and social values among decision-makers, local communities and visitors;
- enhancing the capacity of institutions to conserve wetlands;
- making effective use of the MedWet tools and methodologies;
- guiding the effective use of funding;
- establishing a mechanism for consultation and co-operation between institutions.

MedWet has been working for a compatible inventory of all Mediterranean wetlands by the year 2010. Also through MedWet, collaborative conservation activities have been initiated involving countries sharing wetlands: (a) the Prespa Park comprising Ramsar sites in Albania, Greece and the FYR of Macedonia, (b) the Dojran Lake, shared between Greece and the FYR of Macedonia and (c) the Neretva River/Hutovo Blato shared between Bosnia and Herzegovina and Croatia.

The success of MedWet illustrates a number of lessons for developing regional initiatives (Kouvelis, personal communication):

- the partnership of scientific, government and non-government organisations is essential to pool and benefit from the different but complementary strengths of each participant;
- a staged approach to developing the initiative was required starting with a programme of demonstrable scientific and technical actions before engaging governments at the policy level;
- start with a simple governance structure until a more complex one is affordable;
- a common vision and co-ordination can help better target existing funds and raise new funding;
- it takes a long time, with MedWet being fully embraced by the Ramsar Convention in 2002 (Ramsar Convention 2002e), 11 years after its establishment;
- the willingness of participating governments to fund the initiative is essential to ensure there is core capacity, a factor that may be harder to replicate in other, less wealthy regions.

8.5.3 Great Asian Mountains Wetlands

Highland wetlands in the Great Asian Mountains – the Himalaya–Hindu Kush–Allay region and Mekong River basin – are of great importance, providing water resources and other environmental services to 250×10^6 people upstream and approximately 3×10^{12} people downstream in Asia. Highland wetlands in Asia have unique cultural and ecological values, yet they are deteriorating rapidly, due to both global climate change and increasing human activities that increasingly threaten people's livelihood and regional sustainable development.

To advance the conservation and wise use of highland wetlands, the Ramsar Secretariat, International Centre for Integrated Mountain Development (ICIMOD) and concerned national governments initiated a regional partnership in the Great Asian Mountains (Jian, personal communication; Ramsar Convention 2004b). The participating governments include: China, Pakistan, India, Kyrgyzstan, Thailand, Tajikistan, Nepal, Bangladesh, Bhutan and the Lao PDR. An initial series of workshops has been held in Urumuqi (2002), Kathmandu (2003), Sanya (2004) and Evian (2004).

Consensus was reached amongst representatives from national governments, regional and international agencies at the 2004 workshops on the objectives of this initiative (Ramsar Convention 2004b) for a regional and open partnership between country members that includes:

- inventory and collection of wetlands information;
- regional co-operation and sharing of knowledge;
- the development of wetlands policies;
- designation and management of transboundary Ramsar sites;
- enhancing capacity to manage wetlands;
- community education and public awareness to develop a common vision to increase awareness of the values of ecosystems;
- community-driven participatory management;
- valuation of wetland ecosystem services as a tool for informing decision-making;
- equitable management policies that support livelihoods both upstream and downstream.

Participants have agreed to the following actions:

- analysing key gaps in ongoing projects and initiatives in the regions;
- developing a regional strategy for delivery of a common vision, simplifying co-ordination and governance structures and identifying priorities for project development for initiating implementation;
- drafting a 'Himalayan Initiative Framework Agreement' for endorsement by national governments in 2005;
- identifying donors interested in supporting work on the priority issues;
- seeking funding for a 'Water Resources Management and Wise Use of Highland Lakes in Great Asian Mountains' to: (a) generate a wetlands inventory knowledge base, (b) propose suitable technical options for management and (c) form a platform for policy development and advocacy;
- establishing a regional wetland centre in the Great Asian Mountains region to form an institutional base to assist the regional governments in: (a) improving regional co-operation, (b) harmonising the data collection and reporting and (c) training and capacity building.

It has taken three years to reach the point of preparing a common strategy and there will be difficulties in attracting sufficient funding to maintain this initiative. Early progress has been made by a number of participating countries that have better documented wetland sites and designated some under the Ramsar Convention, for example, in China. These early stages in the Great Asian Mountains initiative show the potential for regional collaboration to greatly improve wetlands conservation.

8.6 Target-Driven Wetland Conservation: Lessons from WWF's Program FY99–FY04

WWF – the non-government conservation organisation – has to choose how and where to invest globally to achieve the most conservation with limited resources. This section outlines the 'target-driven' approach adopted by WWF and, in particular, the lessons learnt from its work to achieve its FY02–FY04 set of programme milestones, based largely on the author's (J. Pittock) unpublished data.

As a complex global network working in over 100 countries, WWF decided to focus its work in order to better link field and policy work on six priority themes, including conservation of freshwater biodiversity and habitats. For its priority themes, WWF set ten-year conservation targets it would like the world to achieve, plus a sequential series of three year milestones it sets itself. The targets and milestones were set to ensure WWF's work was focused on the most important interventions it can make and was sufficiently ambitious to make a significant difference commensurate with the scale of the challenge for conservation. The targets were also intended to aid communication and help instill a culture of accountability, monitoring and evaluation.

In 2001, WWF adopted the following freshwater targets and milestones, which are reviewed here. In this discussion, WWF has counted as a success only outputs and outcomes external to WWF and usually implemented by governments and others. These results are counted where WWF was a catalyst through: provision of funds, participation in an implementing consortium, or through advocacy where, in its judgment, the measures WWF promoted were largely adopted. Usually, other organisations have also played a role in these successes.

8.6.1 Target 1: Protect and Sustainably Manage 250 × 106 ha of Freshwater Ecosystems Worldwide

WWF set itself the milestone of instigating 55×10^6 ha of new freshwater protected areas between July 1999 and June 2004. WWF's support and advocacy significantly increased the designation by governments of new freshwater protected areas globally, especially as Ramsar sites with over 49 × 10^6 ha of new reserves being established (Fig. 8.5) and 6.5 × 10^6 ha more awaiting processing of designation documents. Small grants from WWF of up to CHF 40 000 (CHF 1.0 = ca. U.S.$ 0.79) to developing country conservation agencies have been a key catalyst. WWF supported designation of wetland reserves in 17 countries with total funding of CHF 626 000 over five years, resulting in the establishment so far of an additional 124 wetland reserves. Appropriate and public recognition of governments who have taken action to establish these reserves was another key factor in generating political support for new protected areas.

If the area of 'inland waters' habitats is 10.3 × 10^6 km^2 and 12 % is in protected areas (MEA 2005) and WWF has catalysed the designation of 49 × 10^6 ha of various types of freshwater protected areas since 1999, the WWF instigated reserves are around 4.76 % of global freshwater habitat and about 39.6 % of all freshwater protected areas. WWF has instigated since 1999 more than 28 % of the total area of all Ramsar sites that have been designated since the Convention was established in 1971.

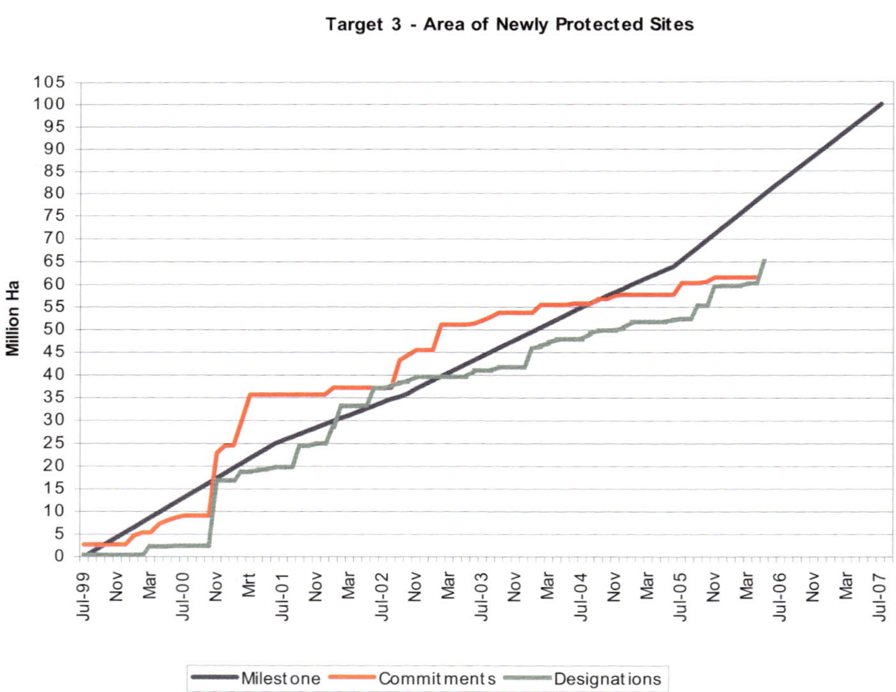

Fig. 8.5 Designation of freshwater protected areas initiated by WWF between July 1999 and December 2004 (WWF, unpublished data)

The Ramsar Convention has adopted a 250 × 10^6 ha target for the world freshwater protected areas (Ramsar Convention 2002f: Operational Objective 10.1.1) and the Convention on Biological Diversity is has adopted a 275 × 10^6 ha target. Consequently, WWF has adopted a 45 × 10^6 ha milestone for additional freshwater protected areas to be designated in 2004–2007.

WWF's milestone for best practice management of five high-profile wetland protected areas in retrospect appears insufficiently ambitious and unsystematic. WWF easily achieved its milestone, helping to improve management of 51 sites. In response, WWF has adopted a milestone of better management for 30 × 10^6 ha of freshwater protected areas between 2004–2007 and is working through the Ramsar Convention to produce a management guide based on Ramsar guidelines and a quantifiable 'tracking tool' to measure progress in improving the management of protected areas.

8.6.2 Target 2: Restore and Conserve Ecological Processes in More Than 50 River or Lake Basins

WWF considered that conserving rivers through sustainable integrated river basin management initiatives was required to significantly advance freshwater conservation. WWF's approach to planning large-scale conservation of the world's freshwater biome with freshwater ecoregion conservation is outlined earlier in this chapter. Its milestone of securing at least ten agreements that put management of rivers on a sustainable footing was achieved with 16 successes. A major highlight was the adoption of a rigorous EU Water Framework Directive, resulting in part from WWF advocacy, which will promote good river basin management in over 29 European countries. While WWF welcomed this progress, in its experience it takes at least seven years for such river basin initiatives to become sustainable. In the three years to 2007, WWF increased its ambition to cover 25 river basins globally.

Integrated River Basin Management

Integrated river basin management (IRBM) is the process of co-ordinating conservation, management and development of water, land and related resources across sectors within a given river basin, in order to maximise the economic and social benefits derived from water resources in an equitable manner while preserving and, where necessary, restoring freshwater ecosystems (Jones et al. 2004).

In 2004, WWF assessed 14 projects from its river basin programme portfolio to identify common lessons learnt (Jones et al. 2004). Though all projects have produced substantial outputs, none of them presents a complete, functioning IRBM process. Instead, each case study demonstrates the use of one or more particular approaches, tools or processes intended to promote and catalyse wider IRBM schemes within the respective river basin. The common global lessons identified are:

1. Long-term investment is needed by governments, NGOs and any other stakeholder promoting integrated river basin management, with initiatives taking at least seven years to produce substantial outputs.
2. River basin management requires a holistic and strategic approach, based on a vision and agreement between stakeholders on the values to be conserved and for sustainable livelihoods for the people of the basin.
3. Biodiversity conservation may have to take a back seat as a motivating factor for the community for river basin management in place of more immediate needs of local peoples for secure livelihoods in terms of such elements as water and food supplies and reduced risk of natural disasters and pollution.
4. It is important to work at different geo-political levels simultaneously.
5. Effective partnership building is an essential ingredient of IRBM and enables organisations to accomplish far more than by working alone.
6. Be ready to seize unexpected events such as flood or pollution incidents to generate government and community support for more sustainable river management measures.
7. Sustained efforts are needed to raise public awareness and to gain the support of local communities.
8. River basin conservation must build on a strong information and science base.
9. River basin management must be established as a political priority of top government officials if the required inter-agency and inter-government collaboration is to occur.
10. Formal protected area designations may be vital to underpin river basin management by promoting icon habitats or species whose ecological health acts as a benchmark for restoration and maintenance of required hydrological regimes.
11. The conservation community can catalyse and demonstrate but effective and sustained implementation of river basin scale solutions depends on governments, the corporate sector, communities and individuals accepting and committing to the principles of IRBM.

In essence, the assessment suggests that community support for sustainable river management requires a long term process of engaging all key stakeholders to develop a common vision and implementation mechanisms based on sound information that meets the most important needs of stakeholders. This requires conservation organisations to build socio-economic partnerships, capacities and expertise.

Dams and Other Water Infrastructure

Mitigating the threat posed by poor water infrastructure development was the focus of the WWF Dams Initiative that sought to stop or reorient ten dam developments. This was achieved with 13 major dam proposals stopped or operations reoriented to reduce environmental damage by establishing environmental flows. Key successes included contributing to abandonment of the Spanish Government's proposed National Hydrological Plan for inter-basin transfers, blocking Australia's Nathan Dam proposal in court, persuading the Polish Government to cut funding for the proposed Nieszawa Dam and gaining agreement in Zambia to restore environmental flows to the Kafue Flats.

As an example (Schelle et al. 2006), the 6500 km^2 Kafue Flats in Zambia was one of the world's richest wildlife habitats. The Itezhi-tezhi Dam was constructed in 1978 upstream of the Flats to store water for the Kafue Gorge hydroelectricity dam that is downstream of the Flats and which supplies 50 % of Zambia's power. The loss of wet season flooding greatly diminished the wildlife of the 50 000 ha Blue Lagoon and 60 000 ha Lochinvar National Parks. The populations of Kafue lechwe (*Kobus leche kafuensis*), a threatened antelope species, have declined by 60 %. In May 2004, an agreement negotiated between WWF, the Zambian Government and its Electricity Supply Company was signed to restore a more natural flooding regime to benefit both the environmental health of the Flats and local people's livelihoods. Power generation is maintained by rescheduling water releases from the upstream dam. In collaboration with the local community and commercial sugar farmers, a new 50 000 ha Mwanachingwala Conservation Area will also be established.

While WWF was successful in these instances, it has concluded that this is insufficient to reduce the threat to freshwater habitats from water infrastructure, with over 45 000 large dams in place and 1500 large dams currently planned or under construction (WCD 2000). Consequently, WWF's emphasis in the three years to 2007 will be to work with the programmes of key national governments for agriculture, energy and poverty reduction, to reduce their water infrastructure impact on river ecosystems in at least 20 instances. For example, work is required on national energy strategies to help the governments concerned choose the extent to which they rely on hydroelectricity that will damage freshwater ecosystems, versus fossil fuels that exacerbate climate change, versus nuclear electricity with its waste and security problems, versus renewable energies.

8.6.3 Target 3: Best Practices in Water management are Adopted in Key Water-Using Sectors

This work aimed to stabilise the diversion of water for human use, by focusing on the irrigation sector (which uses 70 % of diverted water) and manufacturing (which diverts 20 % globally; WWAP 2003). WWF's work to achieve partnerships with three multinational companies to adopt better practices in sustainable water management was unsuccessful, despite some promising small-scale projects, and may have been unrealistic. Many of the targeted companies were undergoing considerable upheaval during the corporate governance crisis. A number of major companies were willing but their head offices appeared unable or unwilling to gain the support of its subsidiaries. Consequently, this work has been reduced in the three years to 2007 and merged with work on irrigated commodities.

Another milestone sought to establish market-based mechanisms that would promote more sustainable, water-efficient production of three major irrigated crops. While WWF made sound progress, the milestone was not achieved. WWF assessed irrigated agriculture in key freshwater ecoregions (De Nooy 2003) and identified cotton and sugar as the most damaging crops where market-based measures may most easily promote better practices. WWF started work to identify these 'better management practices' with the International Finance Corporation and to co-ordinate manufacturers and retailers who want to sell more environmentally sustainable products. However, the milestone was unrealistic and WWF should have learnt from its experience in establishing market mechanisms for timber and fish where 'Stewardship Councils' took at least seven years each to establish effective market based measures.

Subsequently, WWF altered this target to require: "water use policies and practices adopted by at least 100 governments, three food and fibre industry sectors and five international processes, contribute to conservation by 2010". In the three years to 2007 WWF will seek to:

- *establish five market-based or policy frameworks for producing agricultural commodities that promote better management practices for water use and effluent reduction by 2007;*
- *support ten countries to develop and enact sustainable water use programmes that promote poverty alleviation and biodiversity conservation by 2007, in order to devote more resources in developing countries to improve poverty reduction strategies.*

This 'target-driven' approach to WWF's freshwater conservation work brought considerable focus, improved communications and accountability, and delivered some large-scale conservation outputs. Some milestones WWF adopted were unrealistic, but the failure to achieve them enabled WWF to prepare a more practical programme for the three years to 2007. Importantly, in a global network, it helped WWF build and link internal and external technical capacity in key areas of work, such as with freshwater conservation and agriculture.

8.7 Conclusion

As environmental scientists, advocates and managers we have an obligation to lead efforts to conserve biodiversity, but for freshwater species and habitats we are failing thus far. Increasing global demand for water, irrigated agricultural products and hydroelectricity are exacerbating the threats to freshwater biodiversity. Not only does climate change threaten to change the hydrology of freshwater habitats, but the adaptations proposed by governments and others for greater hydroelectricity generation and increased water storage would further exacerbate the impact on freshwater ecosystems.

This chapter has briefly examined large-scale mechanisms for wetlands conservation based on: ecoregion prioritisation and vision setting, integrated river basin management, poverty reduction, multilateral treaties, regional collaboration between countries and target-driven work by a non-government organisation. These methods show some promise but also highlight the complexity of this task and long term investments required to establish sustainable freshwater conservation initiatives. The paper also highlights the need for wetland conservation advocates to build partnerships and capacities to meet the social and economic needs of local communities to sustain wetlands conservation.

This paper contends that large-scale planning and strategic interventions at the river basin, national and regional scale are needed to conserve the ecological processes – the hydrological regime in particular – required to sustain wetlands to the degree that the societies concerned choose. Perhaps our greatest challenge is to reach out and engage the many sectors of our societies that need to be part of these large-scale solutions and help them choose for wetlands conservation.

Acknowledgements The authors appreciate additional comments received from Robin Abell, WWF, Guangchun Lei, Ramsar Secretariat and Spyros Kouvelis, MedWet. This paper includes material based on the presentations of Dr. Gordana Beltram (Former Chairperson, Ramsar Convention on Wetlands Standing Committee and Ministry of Environment, Slovenia) and Prof. Jian Liu (formerly with ICIMOD, Nepal) and participants in the session "River basin management to conserve wetlands and water resources" on 26 July 2004 at the INTECOL International Wetlands Conference in Utrecht, the Netherlands.

References

Abell R, Thieme M, Dinerstein E, Olson D (2002) Freshwater ecoregions. (A sourcebook for conducting biological assessments and developing biodiversity visions for ecoregion conservation, volume II). World Wildlife Fund, Washington, D.C.

Abell R, Thieme M, Lehner B, Powell, G, Palminteri S, Riveros J-C, Hamilton S (2004) Southwest Amazon freshwater–terrestrial integration. (Final report to Moore Foundation) World Wildlife Fund, Washington, D.C.

Arnell N, Chunzhen L (2001) Hydrology and water resources. In: McCarthy JJ, Canziani OF, Leary NA, Dokken DJ, White KS (eds) Climate change 2001: impacts, adaptation and vulnerability. (Contribution of working group II to the third assessment report of the intergovernmental panel on climate change) Cambridge University Press, Cambridge, pp 191–233.

CBD (2004a) Decision VII/28 Protected areas. Articles 8(a) to (e). Secretariat of the Convention on Biological Diversity, Montreal

CBD (2004b) Decision VII/4 Biological diversity of inland waters. Secretariat of the Convention on Biological Diversity, Montreal

CCCP of China and State Council (1998) Opinions of the central committee of communist party of China and state council on reconstruction after Yangtze flood, river and lake management, and water infrastructure construction (in Chinese). (Document no. 15) Central Committee of the Communist Party of China and State Council, Beijing

Chape S, Blyth S, Fish L, Fox P (eds) (2003) 2003 United Nations list of protected areas. UNEP/WCMC/IUCN, Gland

De Nooy R (2003) Water use for agriculture in priority river basins. WWF, Gland

Dinerstein E, Olson DM, Graham DJ, Webster AL, Primm SA, Bookbinder MP, Ledec G (1995) A conservation assessment of the terrestrial ecoregions of Latin America and the Caribbean. World Bank, Washington

Dou H, Jiang J (1999) Dongting lake (in Chinese). China S&T University, Hefei

EU (2000) Directive 2000/60/EC of the European Parliament and of the Council, establishing a framework for community action in the field of water policy. European Commission, Brussels

Finlayson CM, Davidson NC (1999) Global review of wetland resources and priorities for wetland inventory: summary report. In: Finlayson CM, Spiers AG (eds) Global review of wetland resources and priorities for wetland inventory. (Supervising Scientist Report 144) Supervising Scientist, Canberra

Jones T, Phillips B, Williams C, Pittock J, Davis T (eds) (2004) Conserving rivers. Lessons from WWF's work for integrated river basin management. WWF, Gland

Loh J, Wackemagel M (eds) (2004) Living planet report 2004. WWF, Gland

Loh J, Green RE, Ricketts T, Lamoreux J, Jenkins M, Kapos V, Randers J (2005) The living planet index: using species population time series to track trends in biodiversity. Philos Trans R Soc Lond 360:289–295

Mackinnon J, Meng S, Zhang P, Jia Z, Zhu X, Mei W (1998) A biodiversity review of China (in Chinese). WWF International, Hong Kong

MEA (2005) Ecosystems and human well-being: synthesis. Island Press, Washington, D.C.

Ministry of Water Resources (1999) China's 98 big flooding (in Chinese). China's water infrastructure and hydropower publisher, Beijing

Nilsson C, Reidy CA, Dynesius M, Revenga C (2005) Fragmentation and flow regulation of the world's large river systems. Science 308:405–408

Olson DM, Dinerstein E (2002) The global 200: priority ecoregions for global conservation. Ann M Bot Gard 89:199–224

Phillips W (2004) The efficiency and effectiveness of the Ramsar Convention on Wetlands, Canberra. (unpublished manuscript)

Pittock (2006) Poverty reduction, sustainable water and river management and multilateral development banks. Rhetoric but little action? Proc Int River Symp 7 (in press)

Ramsar Convention (1971) Convention text and amendments. Ramsar Convention Secretariat, Gland

Ramsar Convention (1999) Resolution VII.18 on river basin management. Ramsar Convention Secretariat, Gland

Ramsar Convention (2002a) Resolution VIII.1: guidelines for the allocation and management of water for maintaining the ecological functions of wetlands. Ramsar Convention Secretariat, Gland

Ramsar Convention (2002b) Resolution VIII.2: the report of the world commission on dams (WCD) and its relevance to the Ramsar Convention. Ramsar Convention Secretariat, Gland

Ramsar Convention (2002c) Joint work plan between Ramsar and the convention on biological diversity. Ramsar Convention Secretariat, Gland

Ramsar Convention (2002d) Analysis of national reports for Ramsar CoP8. Ramsar Convention Secretariat, Gland

Ramsar Convention (2002e) Resolution VIII.30: regional initiatives for the further implementation of the Convention. Ramsar Convention Secretariat, Gland

Ramsar Convention (2002f) The Ramsar strategic plan 2003–2008. Ramsar Convention Secretariat, Gland

Ramsar Convention (2004a) The Mediterranean wetlands initiative – mediterranean wetlands strategy 1996–2006. Ramsar Convention Secretariat, Gland

Ramsar Convention (2004b) Report of the 2004 Evian encounter on the Himalayas and the Mekong. Ramsar Convention Secretariat, Gland

Schelle P, Collier U, Pittock J (2006) Restoring the Kafue Flats. Proc Int River Symp 8 (in press)

Schuyt K (2005) Freshwater and poverty reduction: serving people, saving nature. WWF International, Gland

Shen G (2001) China's eco-construction and water resource conservation and utilization (in Chinese). China's water infrastructure and hydropower publisher, Beijing

SIWI, IWMI (2004) Water – more nutrition per drop. Stockholm International Water Institute, Stockholm

State Statistical Bureau (2003) China statistical year book (in Chinese). China's State Statistical Bureau, Beijing

UN (2002) Report of the world summit on sustainable development. Johannesburg, South Africa, 26 August to 4 September 2002. United Nations, New York

UNEP (2002) Atlas of international freshwater agreements. United Nations Environment Programme, Nairobi

UNGA (2000) United Nations millennium declaration. Resolution adopted by the General Assembly 55/2. United Nations General Assembly, New York

WCD (2000) Dams and development, a new framework for decision-making. (The report of the World Commission on Dams) Earthscan, London

Wetlands International (2002) Special issue on the Niger Basin initiative (a partnership of Wetlands International, World Wildlife Fund, and Nigerian Conservation Foundation). Fadama (Newsletter of the Wetlands International Africa Office, Dakar–Yoff) No. 5

WWAP (2003) Water for people water for life. (The United Nations world water development report) United Nations Educational, Scientific and Cultural Organization/ Berghahan Books, Barcelona

WWF (2003) Too many European countries failing to manage their water resources. (Media release on 13 November 2003) WWF, Gland

WWF (2004) Rivers at risk, dams and the future of freshwater ecosystems. WWF, Gland

WWF (2006) Freshwater ecoregions of the world. WWF, Washington, DC

Zhou H, Wang Y, Yu X (2002) Wetlands restoration and sustainable development in the central Yangtze area (in Chinese). Economic Science Press, Beijing

9 Aspects of Adaptive Management of Coastal Wetlands: Case Studies of Processes, Conservation, Restoration, Impacts and Assessment

PAT E.R. DALE, MIKE B. DALE, JULIE ANOROV, JON KNIGHT, MARC C. MINNO, BERNIE POWELL, RICHARD C. RAYNIE, and JENNEKE M. VISSER

9.1 Introduction

Coastal wetlands are dynamic and include the freshwater–intertidal interface. In many parts of the world such wetlands are under pressure from increasing human populations and from predicted sea-level rise. Their complexity and the limited knowledge of processes operating in these systems combine to make them a management challenge. Adaptive management is advocated for complex ecosystem management (Hackney 2000; Meretsky et al. 2000; Thom 2000; National Research Council 2003). Adaptive management identifies management aims, makes an inventory/environmental assessment, plans management actions, implements these, assesses outcomes, and provides feedback to iterate the process (Holling 1978; Walters and Holling 1990). This allows for a dynamic management system that is responsive to change.

In the area of wetland management recent adaptive approaches are exemplified by Natuhara et al. (2004) for wild bird management, Bunch and Dudycha (2004) for a river system, Thom (2000) for restoration, and Quinn and Hanna (2003) for seasonal wetlands in California. There are many wetland habitats for which we currently have only rudimentary knowledge (Hackney 2000), emphasizing the need for good information as a prerequisite for effective management. The management framework must also provide a way to incorporate the best available science into management decisions and to use management outcomes as opportunities to improve scientific understanding and provide feedback to the decision system. Figure 9.1 shows a model developed by Anorov (2004) based on the process–response model of Maltby et al. (1994) that forms a framework for the science that underlies an adaptive management system in the wetland context.

Ecological Studies, Vol. 191
R. Bobbink, B. Beltman, J.T.A. Verhoeven, and D.F. Whigham (Eds.)
Wetlands: Functioning, Biodiversity Conservation, and Restoration

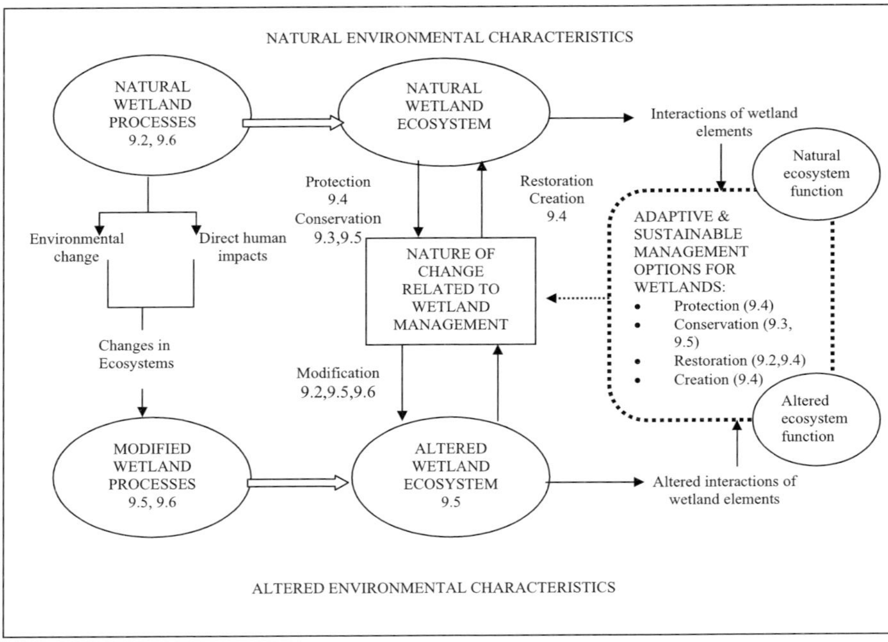

Fig. 9.1 A process model for wetlands (numbers refer to case studies in this chapter). Adapted from Maltby et al. (1994)

The model indicates that 'Natural Environmental Characteristics' provide the environmental context in which coastal wetland processes operate, including, for example, estuarine geology and geomorphology, hydrological regime, climate, vegetation, and soils. The'Natural Wetland Processes' lead to the development of a 'Natural Wetland Ecosystem'. The interactions of the natural wetland elements lead to a unique pattern of 'Natural Ecosystem Functioning'. Human actions, both direct and indirect, have the potential to alter the natural wetland processes and this results in altered ecosystems. Wetland management actions can restore the altered ecosystem towards a more natural level or may alter wetlands to provide for specific needs, such as pest management. The model is particularly suited to natural systems that are under pressure (such as coastal wetlands). It can lead to a greater understanding of the response of a coastal ecosystem to both anthropogenic and ongoing environmental changes; and sustainable management options may include wetland creation, restoration or maintenance of remnant wetlands.

This chapter provides case studies from the United States and Australia that exemplify the role science can play in informing the adaptive management process. The first case study (Section 9.2) demonstrates the application of a wide range of tools that can assist the understanding of long-term and recent processes, of both natural and anthropomorphic origin. The next case

Table 9.1 Elements of adaptive management and the case studies, with emphasis on the key issues and activities for each

Section/case study	Aim	Inventory	Plan	Action	Monitor/assess	Feedback
9.2/Logan, Qld	Understand process to minimize impacts of land use	Soils/sediment profiles, climate, vegetation, and land use	Plan research	Collect data	Analyse data, identify monitoring priority – soil acidification	Process model with feedback
9.3/Florida	Conserve wetlands, understand process	Vegetation, soils, water table depth, rainfall	Plan research	Collect data	Analyse data, identify key elements – water quality	Reference for future monitoring, recommendations
9.4/Louisiana	Restore lost wetlands (four approaches)	Sites identified and assessed	Plan new restoration projects and improve existing projects	Implement new projects and change existing projects	Assess outcomes	Recommendations
9.5/Coomera, Qld	Minimize impact of habitat modification on wetlands	14 years data: vegetation (and other environmental variables)	Plan research	Collect data (14 years)	Multivariate analysis, identify impacts, processes	Information to managers and permitting bodies
9.6/Coombabah, Qld, remote sensing[a] (details in Table 9.4)	Minimize impact and mosquito problem	Identify and classify mosquito breeding habitats	Plan hydrologic modification	Modify	Monitor impacts	Provide feedback to managers

[a] Remote sensing is a generic tool that has been used in all the case studies reported here.

study (Section 9.3) shows how monitoring information can be used to inform management decisions to conserve wetlands. The use of adaptive management principles to review past management actions is demonstrated next, which focuses on aspects of restoration (Section 9.4). Section 9.5 shows how adaptive management has refined the understanding of the impact of management actions on wetland process and suggests an extension to the application of the results. The last case study (Section 9.6) builds on the previous case study and integrates remote sensing into each part of the adaptive management framework. Each case study is related to the process model in Fig. 9.1 and to the adaptive management framework, as indicated in Table 9.1. The chapter concludes with an overview of how each approach is significant for the adaptive management of coastal wetlands.

9.2 Diverse Tools to Identify Processes and Long-Term Changes

Coastal wetlands are dynamic and geomorphologically complex ecosystems that are highly susceptible to change due to natural and human factors. In order to sustainably manage a wetland, as is required by state and commonwealth legislation in Australia, the natural behaviour of these ecosystems that operate over both long and short time-frames needs to be identified and understood. This case study illustrates a variety of tools to explore and explain processes along the freshwater–tidal interface in a subtropical environment in southeast Queensland, Australia. It addresses all components of Fig 9.1; and its role in an adaptive management framework is to provide an inventory of historic and recent processes and assess the impacts or outcomes of management actions in order to inform future planning and management.

The study area, located within a small sub-tropical coastal catchment that drains into the Logan River and Moreton Bay in southeast Queensland, Australia, is known as the Carbrook Wetlands, and comprises four remnant plant communities: intertidal mangroves fringing the tidal Logan River, a supratidal saltmarsh, inland a non-tidal slightly elevated rise supporting swamp she-oak (*Casuarina glauca*), and further inland a lower lying, flood-prone tea tree community (*Melaleuca quinquinervia*). A detailed history of environmental change is preserved within the estuarine sedimentary record of the study area and is a valuable indicator of natural environmental change. More recently, human-induced changes within the study area were superimposed on the natural process of environmental change, resulting in the wetlands suffering varying degrees of disturbance since European settlement in the late 1820s. The most significant changes occurred during European settlement, when vast areas of coastal lowlands were cleared for timber, sheep and cattle grazing, and for agricultural purposes, particularly sugar cane production. A second period of change occurred from 1989 to 1995, when the Melaleuca community suffered dieback in response to hydrological modifications to the coastal creek for the development of a golf course.

A characteristic feature of the study area, and of almost all of Australia's coastal embayments and wetlands, is the presence of sediments and soils rich in iron sulfides (acid sulfate soils; ASS). Since the last glacial period, ASS formed as a result of estuaries infilling with marine and fluvial sediments. These sediments are a natural part of the terrestrial sulfur cycle in which dissolved, oxidized forms of sulfur from estuarine or marine waters are reduced to sulfides by anaerobic bacteria (Van Breeman 1982). The sulfides then combine with the iron from terrigenous sediments to form iron sulfides, the most common form being pyrite (FeS_2). The subtropical climate of southeast Queensland plays an important role in the geochemical behaviour of these soils. The variable rainfall, together with its duration, intensity, and seasonal distribution significantly influence the physical and chemical properties of the ASS that occupy the coastal wetland environments in southeast Queensland. The substrate is prone to sulfuric acidification if the water table is lowered either naturally through evapo-transpiration or artificially drained.

Approach: an Inventory of the Past to Explain the Present

The research examined the evolution and connectivity of four coastal wetland community types by integrating the results from three discrete studies that focused on:

- the geomorphic evolution and vegetation history, providing an understanding of how the wetlands evolved during the Holocene and rates of natural ecological change;
- the connection between estuarine deposits (ASS) and their inherent chemical attributes;
- the land-use history of the area, providing a context from which to assess the impact of disturbing ASS.

9.2.1 Geomorphic Evolution and Vegetation History

In order to examine the geomorphological evolution and vegetation history of the floodplain, undisturbed samples were obtained using continuous augering from a transect of six boreholes that extended from the *Melaleuca* backswamp to the banks of the Logan River, encompassing the entire study area. The soil was sampled for pollen and soil analysis, soil morphological data and ASS, pH tests were recorded in the field, and detailed stratigraphic logs were prepared. Radiocarbon dating was carried out on six intact shells extracted

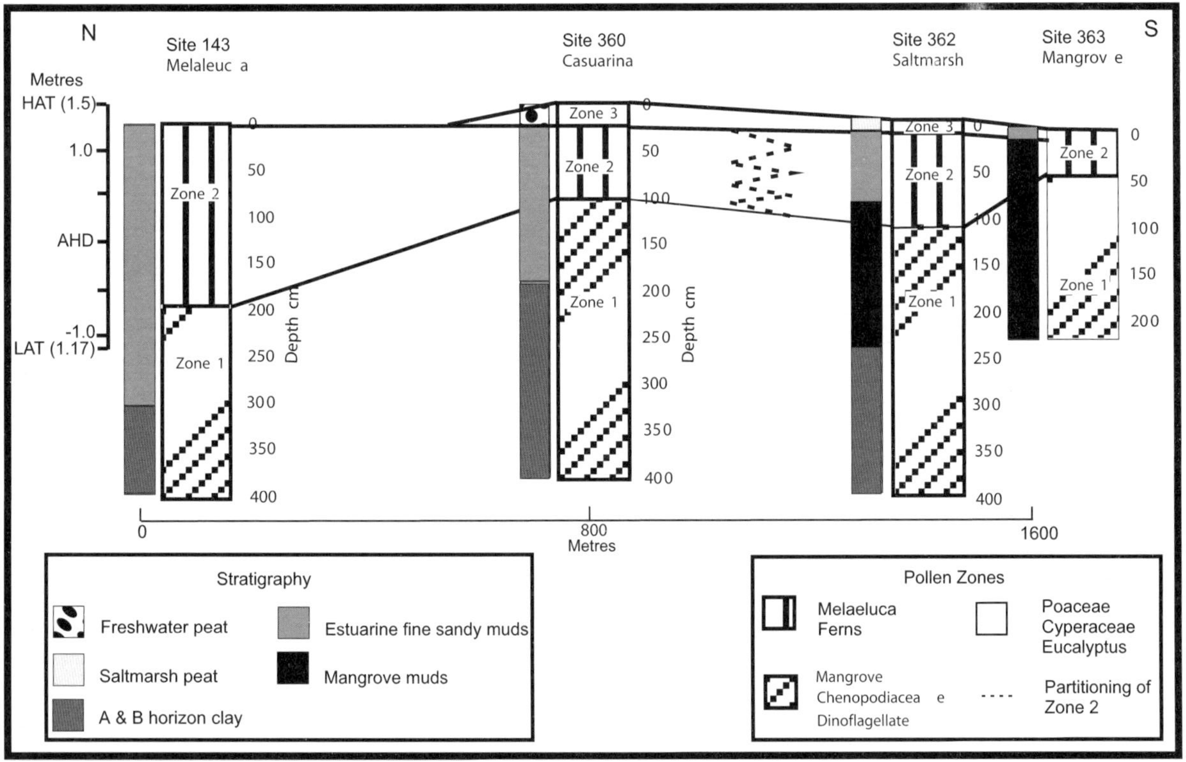

Fig. 9.2 Cross-section of study area, showing microfossil pollen zones and stratigraphy

from boreholes in the *Melaleuca* backswamp and in the *Casuarina* forest. The stratigraphy indicated that the floodplain evolved from an infilling estuary since the peak of the Holocene transgression 6500 BP. Pollen records from the four representative wetland communities (previously mentioned) were examined and compared to the stratigraphy to reconstruct the evolution of Holocene coastal wetland vegetation during the marine transgression and subsequent shoreline progradation (Fig. 9.2). The pollen spectra of the lower Zone 1 generally reflected the estuarine depositional environment that existed throughout the study area during the mid to late Holocene. The partitioning of Zone 2, between the *Casuarina* and saltmarsh sites, was influenced by the lateral extension of down-catchment freshwater deposits and by subtidal deposits from the Logan River.

The mid–late Holocene vegetation history was controlled by natural damming associated with meandering and levee building by the creek that influenced freshwater input, drainage, and salinity. When the stratigraphy and the pollen zones of the *Melaleuca* and *Casuarina* sites are compared, there is a 100 cm lag between the stratigraphic and pollen zone boundaries. Based on an extrapolation of radiocarbon dates from further down the profile at one site, the results suggest that it took 800 years during the late Holocene for a significant change to occur from estuarine- to freshwater-dominated vegetation at both sites. Thus, pollen analysis, when used in combination with stratigraphic modelling, provided an important point of reference for rates of natural ecological change in response to evolutionary changes to the physical environment. More recent changes related to human occupation are considered next.

9.2.2 Human Modification of Carbrook Wetlands

In order to examine possible explanations for ecosystem changes in the Carbrook Wetlands, information drawn from field observations during 2000–2003 and a field investigation (2002) were combined with evidence obtained from historical records, using a series of aerial photographs from 1934–1995, analysing historical records, maps, survey plans, and photos. Based on the analysis of aerial photos, the development of a golf course in 1989 led to the incursion of saline water into the freshwater wetlands. By 1990 the local creek was shortened by more than half its original length. Field observations during 2000–2003 noted that dead Melaleucas were unstable and haphazardly falling to the ground. The change in hydrological conditions from freshwater to saline conditions resulted in the progressive decline in health and destruction of the *Melaleuca* wetlands.

The purpose of the geochemical field investigation was to determine the status of the underlying ASS layers of each of the wetland plant communities (previously mentioned). Soil samples for upper and lower soil horizons were subjected to X-ray diffraction (XRD) for mineral composition. Redox status and soil pore water quality were also measured and the data fed into the MINEQL (ver. 4.2) model (Westall et al. 1976) for an assessment of geochemical equilibrium. In particular, the focus was on examining the conditions conducive to iron sulfide formation or breakdown and possible impacts for coastal wetland vegetation and receiving waters.

Soil pore-water results (Table 9.2) show that all the soils had relatively high levels of soluble iron, except for the deep subsoil (at 1 m) at the mangrove site. However, for all samples extracted from the mangrove and saltmarsh sites, soluble aluminum levels were low, pH conditions weakly acid to neutral, and $Cl:SO_4$ ratios greater than 7.2, indicating that minor iron sulfide oxidation (possibly acid volatile sulfides) may be occurring, but that seawater was buffering any acidity produced. Inland, soluble aluminium levels were extremely high in the shallow aerated subsoils under *Casuarina* and *Melaleuca* communities. This, together with the strongly acid pH conditions and $Cl:SO_4$ ratios of 2.2 to 5.5 in all samples, is consistent evidence of significant and accelerated oxidation of iron sulfides in the *Casuarina* and *Melaleuca* wetlands, caused by artificial landscape drainage. At Carbrook, these changes have impacted differently on the various wetland communities present. Even though the surface elevation between wetland soil sampling sites varied by no more than 0.25 m, the depths to watertable at time of sampling were much more variable (0.5–3.0 m).

In similar low-lying coastal wetlands elsewhere in east coast Australia, channel-shortening and the excavation of drainage networks for the rapid removal of floodwaters led to a reduced hydroperiod and accelerated oxida-

Table 9.2 Selected pore water data at Carbrook site. *btl* Below testing level

Source	Cl (mg/l)	SO_4 (mg/l)	$Cl:SO_4$	Fe (mg/l)	Al (mg/l)	pH
Seawater	19 500	2700	7.2	<1	<1	8.2
Mangroves						
Shallow	20 000	2650	7.5	0.19–14.0	<0.16	7.4
Deep	19 000	2600	7.3	0.5	btl	7.1
Saltmarsh						
Shallow	20 500	2075	9.8	0.09–58.0	btl	6.3
Deep	20 000	2300	8.6	35	btl	6.4
Casuarina						
Shallow	4800	1100	4.4	220–440	61–811	4.2
Deep	6475	1800	3.6	440	0.1	4.1
Melaleuca						
Shallow	4400	800	5.5	0.38–14.0	22–31	4.6
Deep	4375	2000	2.2	150	0.1	6.0

tion of iron sulfides. After a long dry period, oxidation by-products (iron, sulfuric acid, aluminum) are released into drains and receiving waters (White et al. 1996; Johnston et al. 2003). The legacy of over-drainage has been the change in wetland vegetation, salinization, acid scalding, soil acidity, and poor water quality linked to fish kills and loss of aquatic habitat. A common method to ameliorate ASS conditions is to re-instate and mimic a more natural hydrologic regime whilst modifying current land practices. This can be achieved by maintaining higher water levels in drains in order to reduce ground water gradients to the drains, and/or to contain water over ASS landscapes in order to reduce the wetting–drying cycle.

9.2.3 Conclusion

Recognition of the nexus between the geomorphic evolution of an area, the distribution of existing and potential acidity (ASS), and the interaction of climate, hydrology, and artificial drainage are critical for the effective management and sustainability of coastal ecosystems. This research provides a greater understanding of the response of a coastal ecosystem to both anthropogenic and ongoing environmental changes – a key consideration when devising sustainable management strategies. In order to devise ecologically sustainable management options for highly vulnerable coastal ecosystems such as the Carbrook Wetlands, the adaptive management approach must rely on sound scientific understanding of the processes operating in the national coastal wetland ecosystem. This will help to assess the consequences of various management options so that ecosystem values can be maintained.

The next section explores the monitoring needed to conserve wetlands affected by environmental change.

9.3 Managing for Conservation: Monitoring Ecological Changes in Coastal Wetlands in Northeast Florida, USA

Considering the model presented in Fig. 9.1, natural wetland processes may be modified through environmental change, as well as by human activities, resulting in altered wetland ecosystems. In an adaptive management framework, assessing change is a monitoring activity but can also be used to revise the inventory and provide feedback to future planning and actions.

Environmental change in natural areas of Florida happens frequently when the incidence of fire is reduced, causing woody plants to increase and herbaceous plants to decrease. Human-induced impacts to natural systems also occur at many scales and magnitudes from sea-level rise due to global warming (Williams et al. 1999) to land development, pollution, invasion by exotic species, and hydrologic change caused by drainage canals or groundwater withdrawals for public supply and agriculture. Monitoring not only provides empirical measures of ecosystem change, but also supplies environmental data to support the development and calibration of models that can be used to better manage conservation lands. Adaptive management of conservation lands is the best approach to maintaining natural areas and their biodiversity.

9.3.1 Approach

Stokes Landing Conservation Area (SLCA) is a 111-ha preserve located just north of the City of St. Augustine in St. Johns County, Florida. This natural area lies along the west bank of an Tolomato River estuary and has a variety of upland natural communities (scrub, pine flatwoods, hammocks) and wetlands (salt and freshwater marshes, swamps, bogs). In order to better understand the processes affecting SLCA and how to manage the natural resources, five wetland and five upland sites were selected for monitoring. At each monitoring site soils and vegetation were characterized over four years and water levels in shallow wells and rainfall were collected weekly, providing a baseline record of the system.

Sites were selected based upon plant community type, access, and security from vandalism. Water table characteristics were monitored in 3-m wells similar to those described by Miner and Simon (1997). Depth to water from ground surface was recorded weekly from January 1995 through December 1999. A bucket-type soil auger was used to sample the soil at each site. Color, texture, and hydrologic indicators (Hurt et al. 1998) were noted for each horizon encountered. Soils were classified to series (Soil Conservation Service 1999). Plant species richness was determined by delineating a plot (20 × 20 m) adjacent to each well. The plant species growing within each plot were recorded. A rain gauge was installed next to well 9 in the saltmarsh and rainfall amounts were recorded weekly.

9.3.2 Environmental Characteristics

Rainfall at SLCA varied greatly from month to month and between years. Typically, January through May were the driest months in northern Florida. During this study, 1997 was a very wet year (ca. 153 cm of rain) and 1999 was an extremely dry year (ca. 105 cm of rain).

Comparing wetland and upland soil characteristics, the wetland sites generally had higher organic content (muck) at the surface (A horizon). Most sites had sandy soils with well developed elluvial zones (E horizons) where iron, aluminum, and other materials had been leached lower, and spodic horizons (Bh) where these materials and organic compounds accumulated, indicating a fluctuating water table (Fig. 9.3).

Sites with the most extreme environmental conditions had the fewest number of plant species (Table 9.3). Only two species of plants were found in the saltmarsh. The freshwater marsh site (well 12) had the highest plant diversity (41 species) of any of the sites monitored. Wetland sites were relatively rich in graminoids (grasses, sedges) and herbaceous plants. For the upland communities, the driest site, oak scrub (well 16) had lowest diversity, with only 19 species of vascular plants, and the mesic hammock at well 10 had the

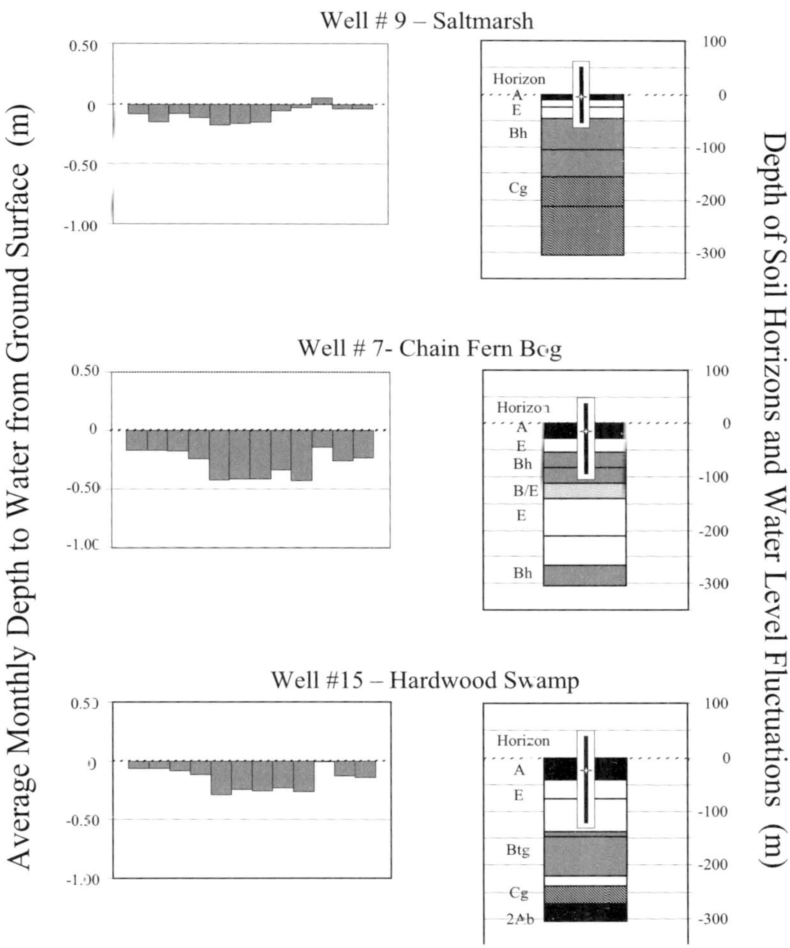

Fig. 9.3 Seasonal variation in the water table (*left*) and range of water table variation (*black bars*) and means (*stars*) at three wetland monitoring sites within Stokes Landing Conservation Area, St. Johns County, Florida

Table 9.3 Plant species richness at monitoring sites in the Stokes Landing Conservation Area, St. Johns County, Florida

Well/ wetland site number	Natural community	Number of species Trees	Shrubs	Vines	Herbs	Grami-noids	Total
Wetland sites							
7	Chain fern bog	3	2	1	3	2	11
9	Saltmarsh	0	0	0	1	1	2
12	Freshwater marsh	2	0	5	21	13	41
14	Chain fern bog	3	1	0	11	11	26
15	Hardwood swamp	8	6	3	3	0	20
Upland sites							
8	Xeric hammock	4	11	4	3	0	22
10	Mesic hammock	8	3	6	10	8	35
11	Mesic hammock	7	3	5	6	1	22
13	Pine flatwoods	5	11	3	4	0	23
16	Oak scrub	4	8	3	2	2	19

greatest, with 35 species. Trees, shrubs, and vines were much more diverse in the upland study sites compared to the wetlands.

At all of the monitoring sites, the water table was lowest at the end of the dry season and beginning of the wet season during May, June, and July. Generally the water table was much lower in the upland sites than in the wetlands. Of the upland sites, the mesic hammocks had the highest water tables and the oak scrub had the lowest. Average water levels were about the same for the xeric hammock and pine flatwoods. At the upland sites, average water levels were also generally near the Bh or spodic horizon and ranged from 56 cm to 113 cm below ground. Figure 9.3 shows examples for the wetland sites. Water levels were highest at the saltmarsh site (well 9) and lowest in the hardwood swamp (well 15). Average water levels were near the surface in the saltmarsh and ranged from 16 cm to 24 cm below ground at the other wetland sites.

These data provide a reference condition against which future monitoring data can be compared. Within the period of this initial study (1995–2000), the vegetation at wetland site 7 changed from a Virginia chain fern (*Woodwardia virginica*)-dominated understory to a black rush (*Juncus roemerianus*)-dominated system. The fern, a salt-intolerant species, as well as mature cabbage palms (*Sabal palmetto*), pines, and hardwood trees growing in or at the edge of the wetland died, while the black rush, a salt-tolerant species, survived. The evidence suggests that saltwater intrusion into the bog from the nearby saltmarsh, perhaps due to an extreme event, played a role in causing this shift in plant community. It also makes clear that water quality should be monitored in coastal wetlands. Re-sampling the vegetation every few years at each of the well sites will document changes in the plant communities due to land management activities or changes in environmental conditions. Long-term monitoring of groundwater levels will help to detect hydrologic impacts from the growing urban areas bordering the area.

9.3.3 Conclusion

This research shows how preserved wetland sites are affected by processes occurring outside the boundaries of the preserved area such as lowering of the regional water table. It provides a base from which to assess management activities. This assessment is a vital component of adaptive management.

Monitoring the effects of management activities is the focus of the next two sections.

9.4 Managing for Restoration: a Multi-Scale Adaptive Approach in Restoring Coastal Wetlands in Louisiana, USA

Coastal wetland loss was not recognized as a problem in Louisiana until the late 1970s (Craig et al. 1979; Gagliano et al. 1981). This wetland loss is related to direct human impacts (Fig. 9.1) as well as the natural dynamics based on the deltaic geology of the area. To address the coastal wetlands loss, the Coastal Wetlands Planning, Protection and Restoration Act (CWPPRA) was signed into United States law in 1990. Since then, the State of Louisiana and 6 six Federal agencies (US Environmental Protection Agency, National Marine Fisheries Service, Natural Resources Conservation Service, US Army Corps of Engineers, US Fish and Wildlife Service, US Geological Survey/ National Wetlands Research Center) have jointly developed a series of annual priority project lists to target Louisiana's wetland loss. After several years of planning and establishing the restoration program, construction was completed on the first project in April 1994. As of January 2004, a total of 71 CWPPRA projects have been constructed in Louisiana, projected to restore and/or protect 34 710 ha (CWPPRA 2004). As more projects were constructed, it became apparent that an opportunity existed to improve the program by learning from the existing projects through the application of a 'learning by doing' adaptive management approach (Walters and Holling 1990). The objectives for this review effort were: (1) improve the linkages among planning, engineering and results monitoring, (2) document changes made to the project in the different phases of project development and implementation (i.e., document the adaptiveness of the program), (3) recommend any changes that could improve the project (the assessment and feedback components of adaptive management, and (4) learn from implemented projects so that future projects could be improved.

9.4.1 Approach

The review effort involved members of six federal agencies, four universities, and the State of Louisiana representing the CWPPRA Environmental Work Group, Engineering Work Group, Academic Advisory Group, Monitoring Work Group, and Technical Advisory Group. Projects for review were selected based on length of time since construction, length of monitoring data record, number of variables monitored, availability of project information, and similarity to other projects proposed for review. It became evident during this process that most projects had not been constructed long enough to have a sufficient amount of data with which to conduct a thorough review that assessed the effectiveness of the projects. A total of 15 projects were reviewed by an interdisciplinary team of biologists and engineers from state and federal agencies as well as university scientists. The reviewed projects represent the following wetland restoration techniques: (1) wetland creation, (2) hydrologic restoration, (3) shoreline protection, and (4) freshwater re-introduction. At the smallest scale, project-specific engineering, operational, and biological response aspects were reviewed to answer the basic questions of: (1) is the project working physically as designed, (2) is the wetland response as planned, and (3) if not, why not, and can the project be made to work better? At an intermediate scale, projects were grouped by restoration technique. At the largest, programmatic, scale recommendations were developed for the improvement of design, implementation, operation, and evaluation processes.

9.4.2 Lessons Learned

Lessons learned and recommendations for improving project implementation were presented in project review reports (available from http://dnr.louisiana.gov/crm/coastres/adaptive/adaptive.asp). Below, we present the major findings for four restoration techniques as well as recommendations for improvement of the overall CWPPRA program.

Wetland Creation

This restoration technique attempts to create wetlands in areas that have changed from wetland to water. It generally consists of dredging sediments from adjacent lake or bay bottoms and depositing this material to achieve wetland elevation after compaction and dewatering. The reviewed projects show that it is very difficult to attain, or even determine, the correct elevation for a healthy wetland. This is partially due to the past difficulty in measuring actual elevations within the Louisiana coastal zone. Wetland creation requires very accurate elevation measurements in Louisiana's micro-tidal climate (average tidal range <30 cm). Based on the uncertainties of sediment stackability, available quantities, and subsidence rates in Louisiana, estimates of sediment consolidation and compaction rates sometimes resulted in final elevations higher than anticipated. Other problems encountered during construction included containment and de-watering issues confounded by vandalism. Closer construction and post-construction inspections were recommended and are being implemented. Because of subsidence and hence high rates of relative sea-level rise in Louisiana, it was often assumed that it is prudent to err on the high side when creating marsh with dredge material (i.e., overfill with the expectation that the material will eventually settle and compact to the correct elevation some time in the future). However, a more sustainable, although admittedly more costly, alternative may be to achieve the optimal elevation sooner and include a maintenance component in the project after several years. By achieving the correct marsh elevation earlier in the project life, natural, self-sustaining accretion processes (including mineral sedimentation and plant belowground and aboveground productivity) will be initiated and the created marsh will be providing some of the critical wetland functions at an earlier date than if overfilled. As the marsh settles over time, a subsequent, pre-planned maintenance component consisting of marsh nourishment with fluid-dredged material could be utilized to fine-tune the elevation and provide an infusion of sediments and nutrients that would stimulate productivity and enhance the overall natural sustainability of the project, so that the created area does not revert to the degraded habitat that existed prior to project construction.

Hydrologic Restoration

This restoration technique attempts to restore a more natural hydrologic regime in areas that have human-induced altered hydrology. The technique generally consists of creating a closed hydrologic unit by connecting natural and artificial hydrologic barriers and controlling water exchange through the use of different water-control structures. Two of the reviewed projects will likely not have the desired restoration effect without modification. For one, this is due to the fact that the construction, operation, and maintenance are under control of the landowner and some features were not completed. The water control structures funded by CWPPRA were a subset of a larger group of structures proposed by the landowner. Some of these proposed structures were not completed. In addition, the operations of the water-control structures are the landowners' responsibility and are not always consistent with the objectives of the CWPPRA project. Documentation of operational changes is not always available. For the second project, the construction of one of the designed features has been deferred to another project that is still in the design phase. The third project has structures that were built as designed, but one of the key structures was vandalized and has been inoperable for most of the project life. It was also discovered that the operational plan for this project was based on an erroneous marsh level, but this has been corrected. Preliminary monitoring data suggest that the project seems to have the desired effect of reducing the rate of land loss.

Two of the three hydrologic restoration projects reviewed had issues that hindered their effectiveness (e.g., landowner control of operations, deferred construction of a project feature). However, the three projects were constructed early in the CWPPRA program; and improvements in the planning, design, and construction process minimized these problems in recent projects. The use of hydrodynamic models for the design and operation of water-control structures, combined with geotechnical investigation of the substrate supporting these structures is strongly recommended for future hydrologic restoration projects and has been incorporated into the planning and engineering and design process of projects not yet constructed.

Shoreline Protection

This restoration technique uses engineered structures to reduce wave energy and protect the adjacent wetland shoreline. Two reviewed projects used techniques (gabion foreshore dike, PVC wall) that were developed for poor soil conditions. Both these techniques have promising results, with material accreting behind the structures and the reversal of shoreline erosion. The other five projects used foreshore rock dikes, which all appear to reduce and even reverse shoreline erosion.

Freshwater Re-Introduction

This restoration technique uses diversion structures to re-introduce Mississippi River water into coastal marshes in an attempt to restore the pre-flood protection levee connection between the river and the adjacent wetlands. The projects reviewed here are not CWPPRA projects, but were included since this restoration technique is highly recommended for future CWPPRA projects. Operating diversion structures is a challenge in Louisiana as they affect many different resource user groups along the estuarine gradient. This results in conservative use of the structures, which limits evaluation of the potential benefits. However, in spite of the conservative operations, all reviewed projects seemed to have a positive effect on vegetation composition and biomass. Louisiana Department of Wildlife and Fisheries (LDWF) sampling in one of the project areas indicates that white shrimp (*Panaeus setiferus*), redfish (*Sciaenops ocellatus*), and speckled seatrout (*Cynoscion nebulosus*) catches, as well as alligator (*Alligator mississippiensis*), waterfowl, and muskrat (*Ondatra zibethica*) counts are all greater since the re-introduction of river water. Monitoring results show that the two projects that use siphons to divert water have minimal impacts on water levels in the project area and that concerns about erosion from the flow are not substantiated by observations.

9.4.3 Program Recommendations

The three main components for an effective adaptive management plan for a restoration project are: (1) a clear goal statement, (2) a conceptual model, and (3) a decision framework (Thom 2000). All of these should be developed during the CWPPRA planning process. Project goals should be identified early in the process and should not change unless the intent of the project changes. Currently, there is a perceived success/failure decision associated with goals and objectives. The goals and objectives are deliberately vague to prevent such a decision. This can be improved by clearly defined effectiveness criteria and targets. Evaluation of the project would document the project's ability to move towards the target (for an example of this approach, see Zedler and Callaway 2000). If a project is moving away from the target, the cause can be determined and the operation and maintenance plans can be adjusted. Causes of wetland loss were often based on landowner and local observation, not on data collection and scientific review. If causes of wetland loss are not accurate, the likelihood of failure to reduce land loss is increased. A good conceptual model of the controlling physical factors and the resulting system structure and function is necessary for a successful restoration project (Thom 2000). If the project is not moving towards its structural goals, the project manager can go back to the conceptual model to understand what is going wrong and recommend corrections to the operation and maintenance plan of the project (Thom 2000). The process of writing down the controlling factors along with the desired system structure and function is a useful exercise for providing the basis for the design of the project (Thom 2000). Zedler (1996) used flow charts to show how project results fit into the decision process.

Design changes, whether before or during construction, need to be coordinated with monitoring, operation, and maintenance to ensure that the project achieves its anticipated outcomes. This could be accomplished by having a project management team, consisting of the personnel involved in all aspects of the project and two academic advisors. This team should improve the communication among the biologists and engineers involved with the project. In addition, such a team can provide the institutional memory, which is sometimes lost with changes in personnel and associated changes in project managers.

Projects should be planned, monitored, and evaluated in the context of their place in the surrounding environment. Interdisciplinary support from the academic community is recommended for the evaluation of project effectiveness and cumulative effects. The Coast wide Reference Monitoring System (Steyer et al. 2003) recently approved for implementation by the CWPPRA Task Force could provide this context.

The review of CWPPRA projects resulted in multiple recommendations for individual project and program improvements (details for all of which are available from http://dnr.louisiana.gov/crm/coastres/adaptive/adaptive.asp). This is part of the adaptive management feedback loop. The CWPPRA program continues to adapt and improve; and these improvements are resulting in more efficient and beneficial projects. This ability for management programs to adapt based on current information is an important component for any adaptive management program.

9.4.4 Conclusion

This section stresses the need for planning, review, and monitoring; and it notes the absence of long-term monitoring as a constraint on project evaluation in dynamic landscapes.

The next example reports the results of a relatively long-term monitoring project that used a multivariate method to assess impacts of management in a situation where the priority management objective is to reduce disease vector mosquitoes in wetlands near human settlements.

9.5 Managing the Environment to Reduce Insect Pests: a Multivariate Approach to Assess Impacts of Disturbance on Saltmarsh Processes in Subtropical Australia

Not all environmental values of intertidal saltmarshes are beneficial to humans. To conserve and restore wetlands are commendable objectives, but wetlands may also be the source of mosquito-borne diseases such as malaria, West Nile virus in the United States, and Ross River virus in Australia. That usually generates a public demand to manage the insects; and this in turn may obviate restoration efforts and may be in conflict with conservation efforts. This section explores an innovative multivariate method applied to long-term data to identify processes and to assess the effects of marsh modification. The assessment links directly back to management aims and site inventory and thus advances the adaptive process. In terms of Fig. 9.1, it is exploring the change from a natural to an altered wetland ecosystem with an emphasis on process, assessing impacts of disturbance on the process, and informing future planning and management.

The study area is in southeast Queensland, Australia, close to a major tourist destination, and with a high rate of population growth. The disease vector mosquito population is managed by modifying the environment by an Australian innovation known as runnelling (Hulsman et al. 1989). This alters the marsh hydrology, allowing increased connection between marsh and tidal source that in turn allows access to predators and may flush larvae out into the estuaries or reduce the oviposition of adult mosquitoes. The aim of the research reported here was to assess the impacts of runnelling on the vegetation over 14 years, using a minimal message length (MML) clustering approach as described in Wallace and Dowe (2000). The results are used to inform an integrated adaptive management process, as described in Sipe and Dale (2003).

9.5.1 Approach

This case study is part of a more comprehensive study into the impacts of runnelling, using field and remotely sensed data. As vegetation reflects environmental characteristics and is relatively easy to observe, it is relevant for managers who are often resource-limited. Here we identified system states as reflected by the vegetation, using multivariate clustering models. First, we identified states, or classes, based on the density and size of the two species: *Sporobolus virginicus* and *Sarcocornia quinqueflora*. Changes of state were shown using a transition matrix and this indicated that the states or classes identified formed cycles within the system (Dale and Dale 2002). This confirms and illustrates the interactions shown in Fig. 9.1. There were three cycles with weak linkages (Fig. 9.4). One was from tall dense *Sporobolus virginicus* to bare ground, another was from tall dense *Sporobolus virginicus* to a mix with *Sarcocornia quinqueflora*, and the third was a cycle with *Sarcocornia quinqueflora* dominant in all states. Next we asked whether the disturbance affected the underlying processes of the system or whether there was only a shift in state (this is important in the process model shown in Fig. 9.1). This was modelled with a continuous hidden Markov model, allowing for multiple processes. The result was that a single model provided the best fit to the data. Thus only one process was operating and this was also unrelated to runnelling (Dale et al. 2002) and we concluded that runnelling had no significant impact on the saltmarsh vegetation processes. This was also supported by other research at the site (e.g., Dale et al. 1996; Chapman et al. 1998; Breitfuss and Dale 2004; Jones et al. 2004).

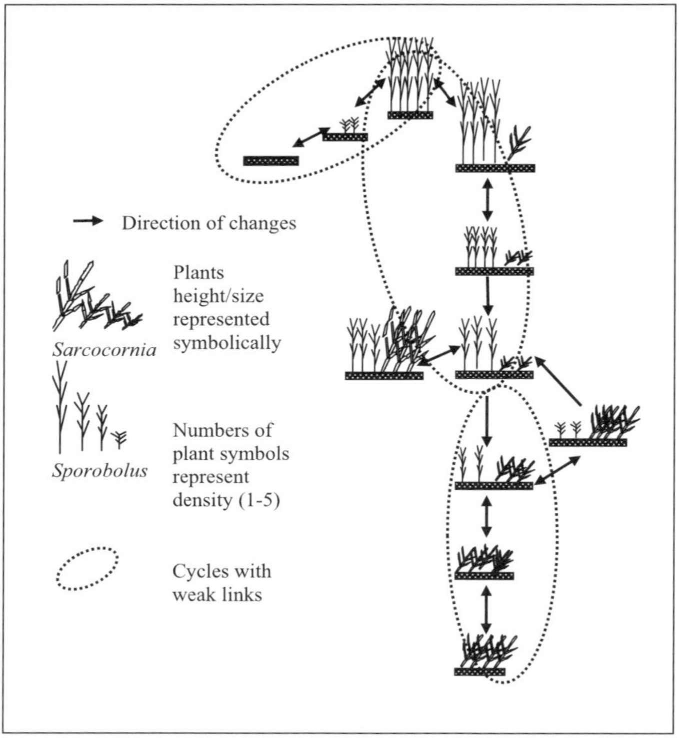

Fig. 9.4 Model of changes and their cycles in the saltmarsh. Adapted from Dale and Dale (2002)

9.5.2 Conclusion

Not only does the research inform mosquito management (an intended consequence) but it also has the potential to inform marsh management for restoration or creation purposes, by indicating which classes may optimize success. That is, a restoration project, for example, may be well advised to restore the most dynamic assemblage (*Sporobolus* and some *Sarcocornia*) to maximize saltmarsh sustainability.

The runnelling management concept is to be extended to mangrove forests, which may also be mosquito problem areas. Since hydrology is the driver for managing the saltmarsh mosquito, a critical issue is how to identify mosquito breeding sites under the mangrove canopy and how water moves between them. This is addressed in the next section.

9.6 Use of Remote Sensing to Monitor Hydrologic Processes in Mangrove Forests and to Integrate Across the Adaptive Management Framework

The last case study exemplifies the potential of remote sensing to inform all stages of an adaptive management strategy. It has been used in all the case studies reported here. In terms of the process model in Fig. 9.1, it mainly addresses the process (hydrologic) part of the system, whether natural or modified, but uses the environmental characteristics (vegetation, substrate) to identify the processes.

The study area is at the tidal Lake Coombabah, in southeast Queensland, Australia. The study site is a mangrove basin forest characterized by tidal flooding and summer rainfall. The basin forest has a complex hydrology that produces ideal conditions for mosquito breeding, often resulting in a significant human health risk and requiring mosquito management.

Extension of the runnelling concept to mangrove basin forests has potential for mosquito management. However this requires detailed information describing the forest substrate structure and complex hydrologic function such as water distribution and hydrologic connectivity. Methods for extracting information about extent of flooding beneath forest canopies do not usually describe substrate structure nor provide information about hydrologic processes such as water connectivity and depth. Responding to these information requires the current research to use thermal imagery in order to provide information about the substrate structure of the basin forest, including its hydrology.

9.6.1 Approach

In order to satisfy a range of scientific, planning, and legislative requirements, an adaptive management framework was proposed as a suitable tool to guide the research. This was because an adaptive management framework offers a coherent method for integrating initial research and evaluation with subsequent modifications and re-evaluations as the research develops towards its goal – in our case an operational mosquito control strategy.

An adaptive management framework for minimizing the health risk of mosquito-borne disease needs to consider a number of factors in developing a response strategy. As indicated in the examples given in other sections, the adaptive management framework identifies management aims, makes an inventory/environmental assessment, plans management actions, implements these, assesses outcomes, and provides feedback to iterate the process. Using this approach, an adaptive management framework for the research is presented in Table 9.4, where each management stage and corresponding strategy is outlined for this case study.

Table 9.4 An adaptive management framework for the control of mosquito breeding in a mangrove basin forest

Adaptive management framework stage	Application overview	Application strategy
1. Identify management aims	Minimal impact hydrologic intervention of mangrove basin forest	Aim: to achieve control of mosquito-breeding (oviposition and larval production) by modification to hydrologic processes leading to enhanced tidal flushing of potential mosquito-breeding habitats
2. Make inventory/initial assessment	Define mosquito-breeding habitat requirements; identify and describe related hydrology of mangrove basin forest	Identification of potential mosquito breeding habitats; remote sensing information for developing maps of forest structure depicting: a. potential mosquito breeding habitat b. substrate structure identifying basin connectivity
3. Management action plan	Plan hydrologic modification to establish/improve tidal flushing of mangrove basin forest	Plan for placement of channels increasing tidal flushing between mosquito-breeding areas and tide source
4. Implementation of action plan	Undertake hydrologic modification as planned	Construction of channels increasing magnitude and frequency of tidal flushing in target areas
5. Assess outcome	Evaluation of effectiveness of modification for required outcome	Has hydrologic modification produced an effective mosquito oviposition and larval production reduction result; is ongoing monitoring required?
6. Provide feedback	Iterative decision strategy: ongoing evaluation, modification response or alternative strategy response	Is longer-term environmental monitoring required; does modification require alteration; is an alternate strategy required, e.g., rotational impoundment management

Remote sensing is an integral tool for providing the necessary information for problem solving within an adaptive management framework approach (Table 9.4). Not only does the remote sensing activity (Table 9.4, Stage 2) provide the required hydro-environmental data, it is also needed to inform other adaptive management framework stages. In this particular research, at Stage 1, achieving management aims is feasible only when considered using remote sensing methods, as other data collection methods are either too expensive, impractical or potentially too damaging to the mangrove environment. Stages 3 and 4 depend on maps provided from Stage 2 to develop and undertake a plan of action. Evaluation in Stage 5 requires consideration of maps from Stage 2 and possibly revised remote sensing data, especially if the feedback in Stage 6 involves substantial modification or alternate planning.

9.6.2 Conclusions

The research reported here addresses a variety of dimensions of adaptive management in the context of wetlands. The overall aim of each piece of research is to contribute to the sustainable and adaptive management of the wetland systems. The main conclusion is that adaptive management needs to be informed by rigorous science, yet communicated in a way that can be easily interpreted and used by the end-user (the wetland manager).

Each case study approaches the use of scientific study to inform management action, using a multi-disciplinary or interdisciplinary approach. Reference to the process model of coastal wetlands (Fig. 9.1) and the first case study shows how a variety of approaches converge to help managers and others to understand the system and inform management actions. The geochemical analysis of acid sufate soils indicated the need also for in-depth assessment in order to identify those areas most vulnerable to acidification. The Louisiana case study is an example of applying adaptive management principles to a large wetland restoration effort. It used interdisciplinary teams consisting of the project managers, biologists, engineers, and coastal scientists, resulting in the direct transfer of information among the different disciplines.

The Florida example shows how the baseline inventory is used to assess the efficacy of management activities and to monitor the effect of urban development close to conservation areas. Feedback will improve environmental outcomes. The multivariate approach used in the saltmarsh research provides an example of how underlying processes can be discovered at a fine scale relevant to the system, providing an inventory of process, a method for monitoring and potential for feedback to the broader management system, including the potential to inform restoration. The final project illustrates, for all stages of the adaptive management framework, the use of remote sensing in the context of a difficult-to-survey dense forested wetland.

Some common themes emerge from these examples in the context of wetland management. Hydrology appears to be a driving variable in all the examples and may be impacted by various forms of development. The projects highlight the importance of monitoring water table levels and water quality, especially, though not exclusively, salinity; as saline intrusion may be an issue in some wetland environments. Because of the complexity of the system, all of these projects involve interdisciplinary teams, bringing together skills from a wide range of areas including pedology, geomorphology, palynology, hydrology, plant science, entomology, remote sensing, and, for all of them, aspects of management.

The final message is that the sustainable management of coastal wetlands relies on knowledge of ecosystem processes, so that the rates and direction of ecosystem change can be assessed, whether due to ongoing environmental changes or to direct human impacts. This facilitates the development of effective and sustainable management strategies to maintain, create or restore coastal wetlands. The adaptive management approach outlined by Walters and Holling (1990) is strongly recommended, to improve the use of both existing and new knowledge of ecosystem processes in order to inform wetland management actions.

References

Anorov JM (2004) Integrated study of coastal wetland characteristics and geomorphic processes in a south east Queensland catchment. PhD thesis, Griffith University, Nathan

Breitfuss MJ, Dale PER (2004) The endangered Illidge's ant blue butterfly, *Acrodipsas illidgei*, from an intertidal habitat managed for mosquito control. J Am Mosq Control Assoc 20:91–93

Bunch MJ, Dudycha DJ (2004) Linking conceptual and simulation models of the Cooum River: collaborative development of a GIS-based DSS for environmental management. Comput Environ Urban Syst 28:247–264

Chapman H, Dale PER, Kay BH (1998) A method for assessing the effects of runnelling on salt-marsh grapsid crab populations. J Am Mosq Assoc 14:61–68

Craig NJ, Turner RE, Day JW Jr (1979) Land loss in coastal Louisiana (USA). Environ Manage 3:133–144

CWPPRA (2004) Data available at: http://lacoast.gov/cwppra/

Dale PER, Dale MB (2002) Optimal classification to describe environmental change: pictures from the exposition. Commun Ecol 3:19–29

Dale PER, Chandica AL, Evans M (1996) Using image subtraction and classification to evaluate change in subtropical intertidal wetlands. Int J Remote Sensors 17:703–719

Dale MB, Dale PER, Li C, Biswas G (2002) Assessing impacts of small perturbations using a model-based approach. Ecol Model 156:185–199

Gagliano SM, Meyer-Arendt KJ, Wicker KM (1981) Land loss in the Mississippi River deltaic plain. T Gulf Coast Assoc Geol Soc 31:295–300

Hackney CT (2000) Restoration of coastal habitats: expectation and reality. Ecol Eng 15:165–170

Holling CS (ed) (1978) Adaptive environmental assessment and management. Wiley, Chichester

Hulsman K, Dale PER, Kay BH (1989) The runnelling method of habitat modification: an environment focussed tool for salt marsh management. J Am Mosq Control Assoc 5:226–234

Hurt GW, Whited PM, Pringle RF (eds) (1998) Field indicators of hydric soils in the United States, version 4.0. US Department of Agriculture Natural Resources Conservation Service, Fort Worth

Johnston, SG, Slavich, PG, Hirst P (2003) Alteration of groundwater and sediment geochemistry in a sulfidic backswamp due to *Melaleuca quinquenervia* encroachment. Aust J Soil Res 41:1343–1367

Jones J, Dale PER, Chandica AL, Breitfuss MJ (2004) Distribution changes of the grey mangrove *Avicennia marina* (Forsk.) using large scale aerial color infrared photographs: are the changes related to habitat modification for mosquito control? Estuar Coast Shelf Sci 61:45–54

Maltby E, Hogan DV, Immirzi CP, Tellam JH, Peijl MJ van der (1994) Building a new approach to the investigation and assessment of wetland ecosystem functioning. In: WJ Mitsch (ed) Global wetlands: old world and new. Elsevier Science, Amsterdam, pp 637–659

Meretsky VJ, Wegner DL, Stevens LE (2000) Balancing endangered species and ecosystems: A case study of adaptive management in Grand Canyon. Environ Manage 25:579–586

Miner JJ, Simon SD (1997) A simplified soil-zone monitoring well. Restor Manage Notes 15:156–160

Natuhara Y, Kitanos M, Goto K, Tsuchinaga T, Imai C, Tsuruho K, Takada H (2004) Creation and adaptive management of a wild bird habitat on reclaimed land in Osaka Port. Landscape Urban Plan. DOI 10.1016/j.landurbplan.2003.10.022

National Research Council (2003) Adaptive monitoring and assessment for the comprehensive Everglades restoration plan. National Academies Press, Washington, D.C.

Quinn NWT, Hanna WM (2003) A decision support system for adaptive real-time management of seasonal wetlands in California. Environ Model Software 18:503–511

Sipe NG, Dale PER (2003) A framework for developing integrated environmental management to minimise risk of mosquito-borne disease: environmental science and Ross River virus disease. Appl Environ Sci Public Health 1:139–144

Soil Conservation Service (1999) Keys to soil taxonomy. Pocahontas Press, Blacksburg

Steyer GD, Sasser CE, Visser JM, Swenson EM, Nyman JA, Raynie RC (2003) A proposed coast-wide reference monitoring system for evaluating wetland restoration trajectories in Louisiana. Environ Monit Assess 81:107–117

Thom RM (2000) Adaptive management of coastal ecosystem restoration projects. Ecol Eng 15:365–372

Van Breeman E (1982) Genesis, morphology and classification of acid sulfate soils in coastal plains. In: Acid sulfate weathering. Soil Science Society of America, Madison, pp 95–108

Wallace CS, Dowe DL (2000) MML clustering of multi-state, Poisson, von Mises circular and Gaussian distributions. Stat Comput 10:73–83

Walters CJ, Holling CS (1990) Large-scale management experiments, and learning by doing. Ecology 71:2060–2068

Westall JC, Zachary JL, Morel FMM (1976) MINEQL, a computer program for the calculation of chemical equilibrium composition of aqueous systems. (Technical Note 18, Department of Civil Engineering) Massachusetts Institute of Technology, Cambridge

White I, Sammut J, Wilson BP, Bowman GM (1996) In: Hunter HM, Eyles AG, Rayment GE (eds) Conference on downstream effects of land use, Rockhampton, April 1995. Queensland Department of Natural Resources, Rockhampton. pp 165–172

Williams K, Ewel KC, Stumpt RP, Putz FE, Workman TW (1999) Sea-level rise and coastal forest retreat on the west coast of Florida, USA. Ecology 80:2045–2063
Zedler JB (1996) Tidal wetland restoration: a scientific perspective and southern California focus. (California sea grant college system) University of California, LaJolla
Zedler JB, Callaway JC (2000) Evaluating the progress of engineered tidal wetlands. Ecol Eng 15:211–225

Section III:

Wetland Restoration and Creation

10 Contrasting Approaches to the Restoration of Diverse Vegetation in Herbaceous Wetlands

Aaron M. Boers, Christin B. Frieswyk, Jos T.A. Verhoeven, and Joy B. Zedler

10.1 Introduction

Undisturbed fens, bogs, and sedge meadows have high plant species diversity (Verhoeven and Bobbink 2001; Amon et al. 2002). High diversity is associated with a combination of features – a dependable water source, water with low nutrient concentrations, low nutrient influxes (which prevent any one species from developing high biomass), and a lack of dominance (allowing many oligotrophic species to co-exist). Fens and sedge meadows remain low in nutrients due to ample discharges of high-quality groundwater; the more species-rich fens are continuously leached by groundwater, while sedge meadows show a seasonal pattern (wet in spring, followed by a late-summer drawdown). Bogs, in contrast, are supplied primarily by rainwater, which is typically low in nutrients.

Diverse natural plant communities can shift to species-poor vegetation when either the water supply or the nutrient regimes change in response to human activities within the watershed (Day et al. 1988; Bedford and Godwin 2003) or when invasive species establish and subsequently outcompete native species (Galatowitsch et al. 1999; Kercher et al. 2004). Nutrient addition and altered hydroperiods can act alone or in combination to reduce wetland plant diversity (Wilcox and Meeker 1991; Verhoeven et al. 1993; Kercher and Zedler 2004). While we know that shifts in hydroperiod toward more flooding or less frequent inundation can shift diversity (Mack et al. 2000), such effects are less well known than the effects of nutrient addition. One reason is that changes in hydroperiod are difficult to separate from other impacts (such as nutrient/sediment influxes); another is that nutrient addition experiments are easier to conduct.

Wetlands experience increased nitrogen (N) and phosphorus (P) availability through both external and internal eutrophication. *External* eutrophica-

Ecological Studies, Vol. 191
R. Bobbink, B. Beltman, J.T.A. Verhoeven, and D.F. Whigham (Eds.)
Wetlands: Functioning, Biodiversity Conservation, and Restoration

tion occurs when nutrients are added from outside sources, such as runoff from agricultural fields and urban streets. Runoff carries N and soil particles with adsorbed nutrients, especially P. Atmospheric deposition is another external source of N (Berendse et al. 1993). *Internal* eutrophication occurs when previously unavailable nutrients become liberated from the wetland's soil or water. N bound-up in organic matter can become available to plants when water levels drop, soil becomes aerated, and decomposition rates increase (Burt et al. 2002). P can be liberated when naturally fluctuating water levels are stabilized by a dam, soils become anaerobic, and chemically bound P becomes soluble (Patrick and Khalid 1974; Craft 1997, Richardson and Vepraskas 2000). In addition, the supply of surface water with a chemical composition different from the "original" water in the wetland may lead to increased decomposition and associated mobilization of inorganic N and P (Beltman et al. 2000).

In combination, hydrologic alteration and eutrophication can facilitate the invasion of aggressive species, which in turn can greatly accelerate the loss of native plant diversity (Fig. 10.1; Huenneke et al. 1990; Burke and Grime 1996; Green and Galatowitsch 2002; Woo and Zedler 2002). In sites with indicators of hydrologic alteration, such as culverts and drainage ditches, Kercher et al. (2004) found more occurrences of invasive reed canary grass (*Phalaris arundinacea*) and fewer native species. In a recent mesocosm experiment, Kercher and Zedler (2004) showed that increased flooding killed native species and opened the canopy, so that reed canary grass could establish. Then, eutrophication increased the invader's growth and rate of spread. They also found strong synergisms among nutrient additions (N+P at high levels) and altered hydroperiods (especially longer-duration flooding) – biomass of reed canary grass doubled with simultaneous flooding and eutrophication. Where stormwater inflows increased due to urbanization, wet prairie converted to reed canary grass; and where the invasive became dominant, the number of native

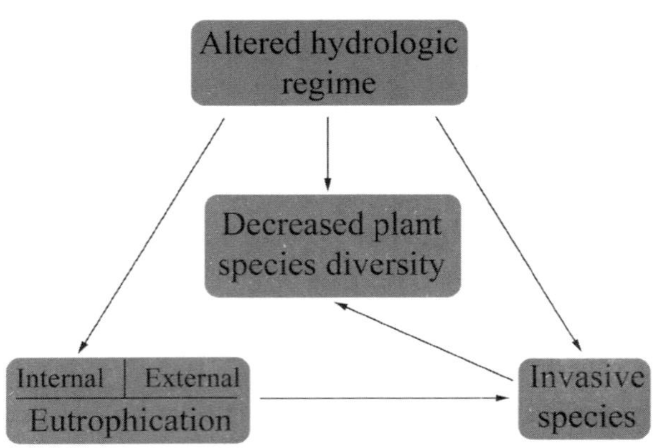

Fig. 10.1 Relationships among factors that decrease plant species diversity in herbaceous wetlands. *Arrows* indicate a positive interaction

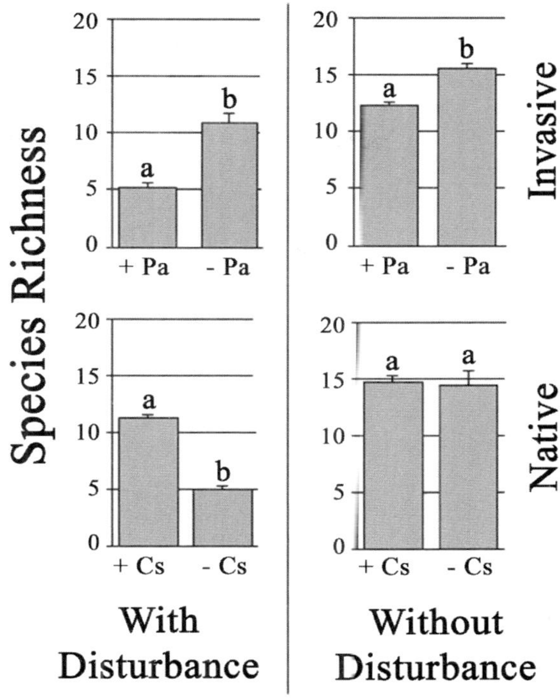

Fig. 10.2 Impact of the presence or absence of the invasive reed canary grass, *Phalaris arundinacea* (*Pa*), and native tussock sedge, *Carex stricta* (*Cs*), on species richness of plots in wetlands with and without indicators of hydrological disturbance (e.g., drainage ditches, culverts, etc.). Note that *P. arundinacea* reduces species richness, while *C. stricta* increases species richness in disturbed sites. Modified from Kercher and Zedler (2004)

species and their floristic quality declined (Fig. 10.2; Kercher and Zedler 2004). Boers et al. (2006) found a similar decline in species richness and floristic quality with increasing abundance of the invasive hybrid cattail (*Typha* x *glauca*; Fig. 10.3).

The risk of losing plant diversity is high for wetlands, especially those that occur low in the landscape, where nutrients collect from entire watersheds (Ehrenfeld 1983; Verhoeven et al. 1993). With biodiversity loss, native ecosystems may lose the ability to carry out ecosystem services that depend on diverse vegetation. For example, monotypes of wetland submersed macrophytes remove fewer nutrients from water than three-species assemblages (Engelhardt and Ritchie 2001); and halophytes grown alone accumulate less nitrogen than six-species assemblages (Callaway et al. 2003). In addition, initial invasions of exotic species are often followed by other aliens, especially animals, in what Simberloff and Von Holle (1999) call an "invasional meltdown".

Given that diverse wetlands have value and that wetland diversity is increasingly threatened by human land use, restoration ecologists are challenged to improve methods of restoring species-rich wetlands. The restoration of species-rich wetlands is accomplished by discouraging the invasion and dominance of species (whether native or alien) that can suppress diver-

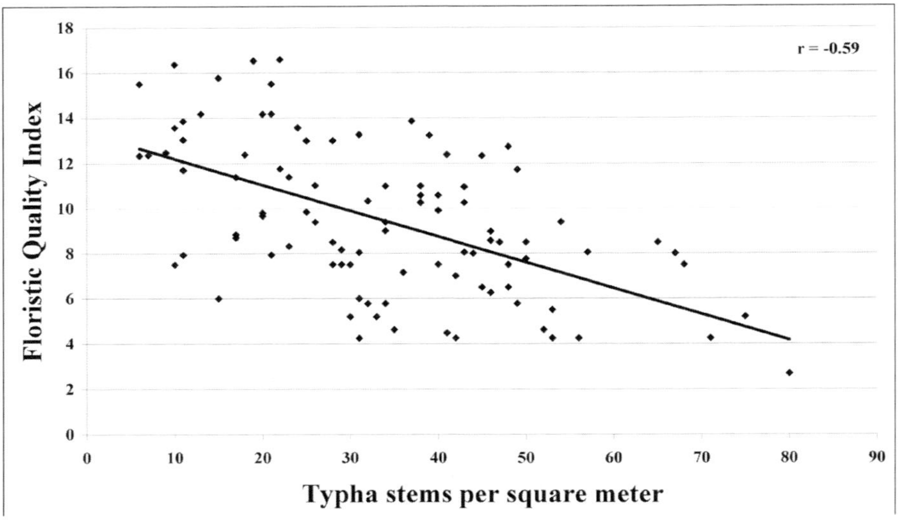

Fig. 10.3 Relationship between floristic quality and density of cattail (*Typha* spp). Data from created wetlands in Des Plaines, Ill. (Boers et al. 2006)

sity, while encouraging the growth and persistence of a diverse assemblage of native species. However, discouraging undesirable species and encouraging desirable species can often only be accomplished by addressing underlying factors, namely: hydrologic alteration, eutrophication, and propagule supply. Restorationists have many choices of restoration techniques; and desired outcomes for restoration sites vary. Here we compare efforts to develop diverse plant communities in temperate herbaceous wetlands drawing on our experiences in Wisconsin (USA) and in The Netherlands. We illustrate a range of approaches in two areas with ample wetlands and lost diversity, but with different restoration contexts (goals, approaches, techniques).

10.2 Restoration Contexts

Wisconsin is a medium-size state with two Great Lakes coastlines; The Netherlands is about one-sixth as large and has a strong oceanic influence that moderates its climate. Even though The Netherlands occurs further north (center = 52°30' N, 5°45' E), its winters are milder than in Wisconsin (center = 44°26' N, 89°46' W). The Netherlands is Europe's most densely populated country, with 389 people km^{-2}. Wisconsin has about 1/10th the density, at 37 people km^{-2-} (www.geography.about.com). Both places are fairly flat, making large areas available for agriculture. Although both regions have large-scale agricultural activities, the intensity of land use for dairy cattle and animal husbandry is much higher in The Netherlands. Wisconsin and The Netherlands are both relatively wet, with over 700 mm of precipitation per year; both have long, convoluted coastlines; and both have major rivers. Water control structures are much more prevalent in The Netherlands, where water is pumped out of large polders to allow agriculture. However, many Wisconsin rivers have multiple dams and many remnant wetlands have drainage ditches that artificially lower water levels.

Wetlands in Wisconsin are mostly privately owned and large areas have been drained for agriculture. In total, 46 % of the wetland area in the 1780s had been lost by the 1980s (Dahl 1990). The federal government has several programs that encourage landowners to volunteer former wetlands for restoration. Under such programs, farmers can agree to stop cultivating and be paid to allow drained lands to develop saturated soil or accumulate standing water. Wetland hydrologic conditions are commonly restored by excavating and breaking drain tiles and filling ditches. In some cases, new wetlands are created by digging ponds in upland areas. In both restoration and creation sites, the resulting wetland vegetation is often dominated by invasive plants.

A large part of the current land area of The Netherlands was once covered with wetlands – i.e., river floodplains, salt marshes, fens, and even bogs in the western coastal plain and the deltas of the rivers Rhine, Meuse, and Scheldt (Wolff 1993). The sandy hills to the east were locally covered by bogs, moist heathlands, and brook valleys. Since the eleventh century A.D., wetlands have been progressively drained, leveed, reclaimed for agriculture and severely modified. The current area of wetland nature reserves is less than 10 % of the historic area and many wetlands occur in a totally reconstructed landscape setting of polders and farmland. Nature reserves are mostly owned by the state or by private nature conservation organizations and they are often managed through grazing or mowing. Threats to the biodiversity in these reserves include loss of species through their small size and isolated location in fragmented landscapes, eutrophication, and desiccation when neighboring areas are drained. The use of polluted surface water to prevent desiccation creates another problem, as internal eutrophication may result from the increased inputs of sulfate and chloride (Beltman et al. 2000). Restoration of degraded wetlands is an activity of growing importance for the nature conservation agencies in The Netherlands. Enlargement of small reserves is seen as an important priority. The Dutch government adopted a national Nature Policy Plan in 1990, envisaging the building of an ecological network of larger nature reserves connected as much as possible by corridors. The realization of this Ecological Main Structure of The Netherlands involves the acquisition of land from farmers or other private landowners and therefore is a long process. The government estimated the total cost and adopted a long-term implementation plan. In recent years, the poor financial situation of the country has postponed the spending of finances for this purpose. Substantial restoration projects have been launched in Dutch wetlands. Bog restoration involves hydrological measures to increase water levels. Fens are restored by clear-cutting and mowing to restore early-successional stages. Floodplains that are still in contact with their river channels are being restored by abandoning agricultural activities, removing artificial water-level control structures, digging side-channels, and employing grazing by large herbivores (Baptist et al. 2004). Restoration of biodiversity of waterfowl, vegetation, and macrofauna is the most important target.

10.2.1 Restoration Targets

It is important to define a target plant community type and essential abiotic conditions prior to implementing a restoration strategy. In some cases, the target is high biodiversity; in other cases the target relates to functional attributes such as water quality improvement. Support of recreational uses, such as hunting, is also a common goal. The range of possible restoration targets may be limited by site conditions. For example, sites that have nutrient-rich inflowing water would not likely sustain high levels of biodiversity. For such wetlands, a more appropriate target may be a wetland with high water quality improvement or flood water storage capabilities, and as a result, would likely have low biodiversity. Restorations that are likely to support high levels of biodiversity would have low to moderate nutrient levels and naturally fluctuating water levels. In order to achieve a diverse community of native species, initial seeding may be required and long-term monitoring and management are vital. In The Netherlands, plant community associations are well known and the goals of restoration projects are often a predefined community type, such as calciphilic vegetation (van Duren et al. 1997) or mesotrophic grassland (Jansen et al. 2001) where the target is a given assemblage of species.

10.3 Discouraging Undesirable Species

The set of species considered undesirable for a restoration site depends on the target of the restoration project. Wetlands being restored for biodiversity might have plantings of rarer species, but not aggressive dominants that could outcompete shorter, slower-growing plants. Such sites need to be nutrient-poor, because there are strong feedbacks between nutrients and diversity. For wetlands designed to improve water quality, the desirable species would be opportunistic, fast-growing plants that grow best under high-nutrient conditions, accumulate nutrients in aboveground tissues, and can potentially be harvested to move the nutrient crop off-site. Aggressive species would be undesirable in the former and desirable in the latter.

In some cases, the goal of a restoration project is to restore a specific plant community type; under these circumstances any species not found in the target community would be undesirable. A project designed to return a site to a known pre-disturbance state or to approximate a specific reference site might be judged off-target if it attracts other species, even if they are all native. Our focus is restoration of biodiversity, therefore undesirable species would be those that tend to suppress diversity.

10.3.1 Undesirable Invaders

Invasive plants are defined as species or strains that rapidly increase their spatial distribution by expanding into native plant communities (Richardson et al. 2000). Such species seem to be of less concern in The Netherlands than in Wisconsin, where several invasions are occurring at rates that are measurable in short time-frames. For example, the area dominated by invasive cattail (*Typha angustifolia*, *T.* x *glauca*) tripled in eight years in a Wisconsin Great Lakes coastal wetland (Fig. 10.4). The wetland invasive plants of greatest con-

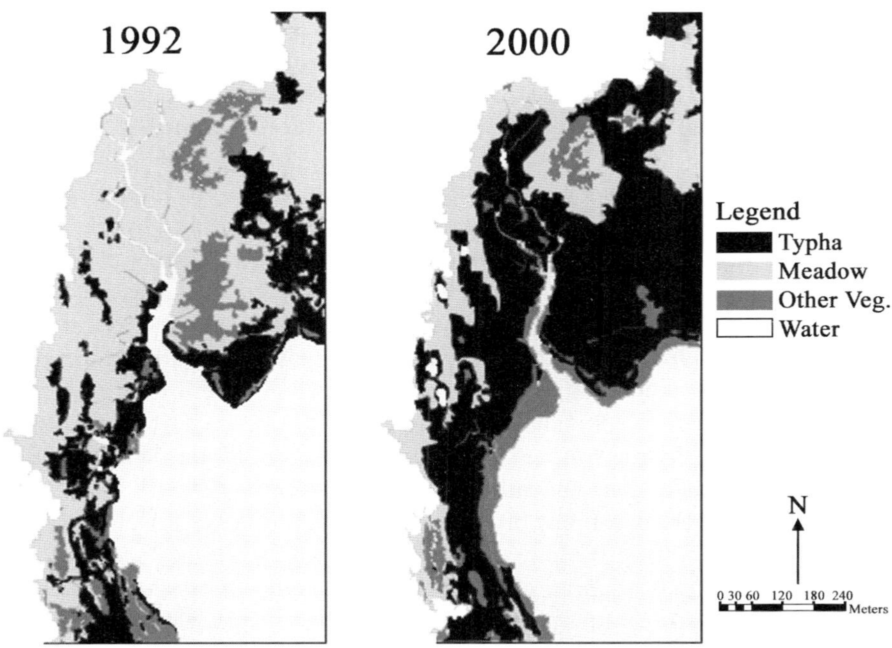

Fig. 10.4 Expansion of invasive cattail (*Typha angustifolia*, *T.* x *glauca*) in Peter's Marsh, near Green Bay, Wis., as interpreted from 0.5-m resolution black and white air photos and ground truth data by L. Ladwig and C. Frieswyk (University of Wisconsin–Madison Botany, unpublished data). This coastal wetland occurs in a watershed that is highly impacted by agricultural and urban land uses

Table 10.1 Invasive plants of greatest concern in herbaceous wetlands of Wisconsin and The Netherlands

Latin name	Common name	Plant type
Wisconsin invasives		
Phalaris arundinacea	Reed canary grass	Grass
Typha angustifolia	Narrow leaf cattail	Graminoid
Typha x *glauca*	Hybrid cattail	Graminoid
Phragmites australis	Common reed	Graminoid
Lythrum salicaria	Purple loosestrife	Forb
Frangula alnus	Glossy buckthorn	Shrub
The Netherlands invasives		
Molinia caerulea	Purple moorgrass	Grass
Calamagrostis epigejos	Wood small-reed	Grass

cern in Wisconsin (Table 10.1) are reed canary grass, hybrid cattail, narrow leaf cattail, and purple loosestrife. In addition, common reed and glossy buckthorn have more recently invaded Wisconsin and are becoming problem invaders. Each of these species has the ability to spread rapidly and dominate large areas and is thought to have exotic ancestry. Purple loosestrife, narrow leaf cattail, and glossy buckthorn are introduced species; populations of reed canary grass and common reed are a mix of native and introduced strains; and hybrid cattail has one native and one introduced parent (USDA NRCS 2004).

Purple loosestrife and glossy buckthorn spread by producing copious seeds. Reed canary grass, common reed, narrow leaf cattail, and hybrid cattail are able to reproduce vegetatively via rhizomes and to disperse as floating plant fragments. These plants possess abundant aerenchyma tissue, which allows gas transfer between the roots and the air. Gas exchange to the roots improves growth in flooded anoxic soils. These invasives are strong competitors for nutrients and light; and they are capable of forming large, dense, nearly monotypic stands that exclude native species (Galatowitsch et al. 1999). For example, reed canary grass dominates 40 000 ha of wetlands in southern Wisconsin alone (Bernthal and Willis 2004).

Invasive species cause many problems in The Netherlands as well, but most of these species were native to the Dutch flora and only recently became invasive because of their adaptations to growth in a nutrient-rich environment. Examples are purple moorgrass (*Molinia caerulea*) and wood small-reed (*Calamagrostis epigejos*), grass species which invade moist heathlands and dune slacks, respectively (Aerts and Berendse 1989; Koerselman 1992). Exotic species are of less concern in The Netherlands, where exotic plants have been moved across Africa, Asia, and Europe for many centuries. Introduced species probably established and spread widely before ecologists began to characterize wetland vegetation. In the central United States, many invasions have been documented in the past 100 years (Galatowitsch et al. 1999).

10.3.2 Controlling Invasives

Controlling invasive species is a very difficult task, because it requires limiting nutrient availability (Huenneke et al. 1990; Burke and Grime 1996; Green and Galatowitsch 2002). Reducing nutrient inputs might be the key factor controlling invasibility of an ecosystem (Sher and Hyatt 1999; Davis et al. 2000). Conversely, any restoration action that increases nutrient supply should give an advantage to invaders (Sher and Hyatt 1999).

The measures that have been employed to control invasive species in Wisconsin include burning, application of herbicides, removal of nutrient-rich sod, drowning, and manual removal (Fig. 10.5; Apfelbaum and Sams 1987). The Wisconsin Department of Natural Resources and The Nature Conservancy use fire to control woody invasive species such as buckthorn [*Rhamnus* spp; R. Hoffman (Wisconsin Department of Natural Resources), personal communication; H. Spaul (Wisconsin Chapter, The Nature Conservancy), personal communication]. The University of Wisconsin–Madison Arboretum controls woody invaders by applying Garlon herbicide to cut stumps [S. Glass (University of Wisconsin–Madison Arboretum), personal communication]; and Applied Ecological Services uses glyphosate herbicide to control reed canary grass [S. Lenhardt (Applied Ecological Services), personal communication]. Deep water can be used to drown reed canary grass [A. Kitchen (US Fish and Wildlife Service), personal communication], and cutting plants followed by flooding has been used to kill hybrid cattail (Boers et al. 2006). The Wisconsin Waterfowl Association has used sod removal to eliminate invasive species and their seed bank at restoration sites [J. Nania (Wisconsin Waterfowl Association), personal communication].

10.3.3 Minimizing Eutrophication

An increase in available nutrients in a wetland often leads to the replacement of a diverse native plant community by undesirable species (Day et al. 1988; Wilcox and Meeker 1991; Verhoeven et al. 1993). Under eutrophic conditions, wetlands tend to be dominated by a few aggressive species, often invasives. The two most common limiting nutrients in wetland systems are N and P. Research indicates that a large proportion of wetlands on organic soil are P-limited and marshes are often N-limited; also, some wetlands shift between N and P limitation (Verhoeven et al. 1993). Only experimental nutrient additions can verify which nutrient is limiting; and results can vary at the site or species level (Bedford et al. 1999). It is often difficult to control the N and P

Fig. 10.5 Common wetland restoration techniques. Controlled burns (**a**) and/or herbicide application (**b**) are used in areas dominated by reed canary grass to reduce the cover of the invasive species and allow for establishment of native species. **c** Following the removal of undesirable species, a mixture of native species are seeded to establish the desired plant community Photos: Michael Healy

that enters a site via external eutrophication. The major external nutrient sources include agricultural runoff, atmospheric deposition, and urban runoff. These sources can be quite distant from the affected wetland and depend on land use in the watershed. In some cases, site modifications can be done to limit external eutrophication, such as creating buffers around wetlands to minimize overland flow of nutrient-rich water, or building settling ponds upstream of the wetland to catch sediment before it enters.

Nutrients can also become available to plants through release from the soil, called internal eutrophication. Internal eutrophication is a potentially critical, but poorly understood source of excess nutrients. For example, chemically bound P can be released when water stands and the soil becomes anaerobic (Richardson and Vepraskas 2000). Although wetlands, by definition, have standing water or saturated soils for at least part of the growing season, artificially prolonging the hydroperiod can allow excess P to become mobile and available for plant uptake. Most wetland soils are acidic (Bedford et al. 1999; Amon et al. 2002) and typically contain P that is bound to iron (Fe) on soil particles and thus is unavailable for use by plants (Szilas et al. 1998). Under reducing conditions, the existing Fe^{3+} is reduced to Fe^{2+} and as a result P is released into solution and becomes available for use by plants (Patrick and Khalid 1974; Craft 1997). Beltman et al. (2000) showed that the availability of P may be strongly increased when surface water rich in sulfates and chlorides is supplied to groundwater-fed wetlands to prevent desiccation. Sulfates may be reduced to sulfides, which precipitate with iron. This accelerates decomposition and the release of iron-bound phosphates.

Lowering the water table can reduce the availability of P, but not necessarily all nutrients. De Groot and Fabre (1993) found that water level drawdown of wetlands increases the P adsorption capacity of the sediment, thereby reducing P availability. In an emergent marsh system, N was found to be limiting under flooded conditions, but not when the water table was below the soil surface (Neill 1990). This may indicate lower P availability, and therefore limitation, when soil is not inundated. In wetlands with highly organic soils, however, lowered water levels could lead to internal eutrophication, if soil aeration increases the rate of organic matter decomposition. Mineralization of organic matter releases N and P directly and increased decomposition also leads to acidification, which indirectly releases N and P from soil (Verhoeven et al. 1993). More research on internal eutrophication is required to understand its potential impacts on restoration. Much more of this research is being done in Europe than in the United States (e.g., Koerselman et al. 1993; Beltman et al. 2000).

10.3.4 Establishing Appropriate Hydrology

Many wetlands experience hydrologic regimes that are either more or less variable than historically. Where dams or dikes have been installed for flood control, some upstream wetlands have prolonged periods of inundation during the growing season (Hill et al. 1998), while some downstream wetlands have water levels that are unnaturally low, because flood peaks have been eliminated. Both are at risk of increased internal eutrophication. Stable water levels might liberate excess P in wetlands at abnormal times during the growing season (Craft 1997). As a result, native plant species might not be able to take up the excess nutrients (Woo and Zedler 2002). Any non-native species that are in the system and adapted to take up and use a nutrient pulse would then have an opportunity to expand (Sher and Hyatt 1999; Davis et al. 2000).

Increases in the amount of impervious surface in a watershed cause water levels to have greater fluctuations. Many native species cannot tolerate the rapid flooding and drying cycles caused by increased runoff, causing declines in biodiversity. The effects of increasing either the inundation or desiccation of wetlands need to be understood in order to limit undesirable species in both natural and restored wetlands. In wetland restorations where the target is a diverse community of native species, the hydrologic regime should be restored to natural conditions, free of artificial water-control structures. Water levels that mimic ones found in a nearby reference wetland are likely to be most favorable to the native plant community. Wetlands designed for water quality improvement or flood abatement would require hydrologic regimes to optimize their specific functions, which may be very different from natural conditions.

10.4 Encouraging Desirable Species

There are several approaches restorationists can take to restore native plant diversity to a site. Once the natural hydrologic and other abiotic conditions are restored, the desirable wetland plant species might reestablish and flourish through natural recruitment processes. However, many sites may lack sufficient natural recruitment and may require seeding or planting of desired species.

10.4.1 Site Modifications

Site conditions influence the diversity of a wetland. Where water and nutrients are in excess, sites can be modified to encourage a diversity of plants to establish. Because hydrology is the principal determinant of wetland structure and functioning (Mitsch and Gosselink 1993; NRC 2001), it makes sense that restoring the hydrologic conditions that were disturbed and then caused the wetland to degrade would be needed to recover native species. In Wisconsin, this involves ditch plugging or filling, stream realignment, dam removal, and/or breaking drain tiles (Thompson and Luthin 2004).

Once hydrologic conditions are restored, other techniques can also encourage diversity. Restoring lost topographic heterogeneity might help encourage both diversity and water-quality-improvement functions. This might come in the form of variable hydroperiods, topography or substrate. Middleton (2000) emphasizes the need for flood pulsing. Peach and Hall (Uni-

Fig. 10.6 Series of artificial tussocks created in a wetland restoration. The resulting increase in topographic heterogeneity produced a greater variety of sites for the establishment of plant species and supported greater species richness than flat areas

versity of Wisconsin–Madison Botany, unpublished data) examined the relationship between topographic heterogeneity and plant diversity in sedge meadows. In a restoration project in Madison, they built mud "bumps" and doubled the number of species relative to flat plots (Fig. 10.6).

In The Netherlands, site manipulation techniques often focus on the chemistry of a wetland. Techniques for defertilizing eutrophic sites include manipulating water levels to increase denitrification (Koerselman et al. 1993), sod removal (Olff and Bakker 1991; Beltman et al. 1996), and harvesting vegetation to remove N, P, and K (Olff and Bakker 1991; Koerselman et al. 1993). In sites where acidification is problematic, lime additions are used to raise the pH (van Duren et al. 1997).

10.4.2 Natural Recruitment

After site conditions have been restored, the appropriate wetland plant species might reestablish and flourish on their own (passive restoration). The restoration of thousands of prairie potholes in the United States Upper Midwest relies on "volunteer" recruitment from seed banks and dispersal, although the resulting species list is shorter than in natural potholes and more exotics are likely to be among the volunteers (Budelsky and Galatowitsch 2000). A wait-and-see strategy is being employed at large-scale restorations of riparian wetlands along the Illinois River [sites at Emiquon, Spunky Bottoms, and Hennepin; G. Sullivan (The Wetlands Initiative), personal communication]. Aquatic plants, including submerged and emergent species, respond well to rewetting, but wet prairie and sedge meadow species are less predictable (G. Sullivan, personal communication).

Natural recovery of vegetation has two components, seed bank and propagule source. Seed banks have been cited as the key to vegetation recovery in natural systems prone to disturbance. In Great Lakes coastal wetlands, where large-scale water level fluctuation continually displaces plant communities, species in the seed bank follow rising and falling water levels by recruiting when conditions are suitable (Keddy and Reznicek 1986). However, the longevity of plant seeds varies (Baskin and Baskin 1998). Without continual, or at least sporadic, seed rain, a species' seed bank can be depleted. The longer an area has been degraded and the more severe that degradation, the less likely it is to have a rich seed bank. For example, Wienhold and van der Valk (1989) found both species richness and seed density to decline with time since drainage in prairie potholes of the glaciated United States Midwest.

The status of the seed bank can be assessed before restoration to decide whether plant material needs to be introduced. In planning to restore Lower Greene Prairie in the University of Wisconsin–Madison Arboretum, we conducted a simple seed bank study and found few viable native seeds; we then tested seeding experimentally in an adaptive restoration approach.

The second component of passive vegetation recovery is propagule source. For restoration sites that are near a diverse source of seeds and other propagules, seeding provides little benefit and passive restoration can save costs (Kellogg and Bridgham 2002; Beukes and Cowling 2003). However, as natural areas become more fragmented, propagule sources become less common and natural immigration of native species slows (Kindscher and Tieszen 1998). The lack of a nearby native seed source can leave a restoration site more susceptible to invasion by invasive species, especially those that are wind- or water-dispersed.

10.4.3 Sowing Seed

The goal of many restoration projects is to restore the full range of biodiversity; and in Europe the evaluation of "success" often involves the number of red-listed species present (Grootjans et al. 2002). In Wisconsin, landowners are guided to sow seed in restoration sites (Thompson and Luthin 2004). Sowing seed is feasible for small restorations, as seed can be collected from nearby wetland remnants (with landowner permission) or purchased from nurseries that sell local genotypes.

Sowing of seed to facilitate restoration is less common in The Netherlands, where efforts focus on restoring the natural hydrology and chemistry, while relying on seed banks and dispersal to provide the appropriate vegetation (Nienhuis et al. 2002). Seed banks are very important in the restoration of species-rich fens in The Netherlands (van der Valk and Verhoeven 1988; Beltman et al. 1996), although van der Valk and Verhoeven (1988) suggested sowing or transplanting species that are missing from the seed bank. Roelofs et al. (2002) concentrated on restoring natural chemistry and improving seed dispersal to prevent local extinction in fragmented landscapes; Grootjans et al. (2002) relied on seed banks to provide historic vegetation after nutrients are reduced; and Eertman et al. (2002) simply monitored the natural restoration after tidal influence was reintroduced in Sieperda Marsh. The general feeling in The Netherlands is that sowing seed may hamper the development of the targeted vegetation type. The species that establish first in a restored site will retard further germination and inhibit establishment of other species. If seeding is chosen, it is important to supply seeds of a complete set of species characteristic of the plant community. The seed source is also a critical issue (van Diggelen et al. 2001). Commercially produced seeds will be genetically very different from seeds of wild varieties. Seeds collected in the field may also be from different ecotypes, especially if they were collected a long distance from the restoration site. Generally, nature managers who apply sowing begin by collecting seeds (or ideally, hay if there is a mowing regime) at nearby sites where the targeted vegetation type still occurs and distributing these at the restoration site.

The need to sow seed is rarely examined, but where it has been evaluated, the evidence is somewhat conflicting. Kellogg and Bridgham (2002) found more emergent species but fewer meadow species in wetland restorations that had been planted versus those in which only the hydrology had been restored. In contrast, Mulhouse and Galatowitsch (2003) found that, in 12-year-old prairie pothole restorations, submerged aquatics and emergent species had recruited well, but common wet meadow species were lacking, and the authors recommended planting. In Wisconsin, Wilcox (University of Wisconsin–Madison Botany, unpublished data) sowed seed of 33 species, but only 12 were abundant in sowed plots; unsowed plots had four species. Salt marsh plots near San Diego demonstrated the ability of three species to recruit readily from seed, while five did not (Lindig-Cisneros and Zedler 2002); of the three that produced copious seedlings, only one is dispersed widely (Morzaria-Luna, University of Wisconsin–Madison Botany, unpublished data). The five salt marsh species that virtually never recruit seedlings in nature need to be planted as seedlings for them to develop viable populations in restoration sites (O'Brien and Zedler 2006).

European advocates of seeding propose that planting is necessary in extreme situations, such as bare substrate (Muller et al. 1998) and when hydrology is irrevocably changed (Clevering and van Gulik 1997). In the United States, Budelsky and Galatowitsch (2000) and Mulhouse and Galatowitsch (2003) provide evidence that certain classes of species (sedges, other wet meadow perennials) are slow to recruit naturally, especially under competition from invasive species. In contrast, some experimental studies show no clear advantage to seeding in restoration projects (Kindscher and Tieszen 1998; Kellogg and Bridgham 2002; Beukes and Cowling 2003). Despite the controversy, the fact remains that species diversity is often very difficult to restore.

10.4.4 The Decision to Plant

As the science of restoration develops, it is increasingly apparent that every restoration is unique and that approaches must be evaluated on a project-by-project basis. The decision to seed/plant falls into this category, especially since planting is expensive and time-consuming. Three factors should be considered: (1) the likelihood that diverse, native vegetation will recover on its own, (2) the presence and potency of invasive species, and (3) the ability to continue site maintenance (Fig. 10.7).

If invasive species are present on the site or nearby, they will likely colonize the restoration site much more quickly than native species. When invasive species become dominant, native species are less able to establish. Seeding or planting diverse native species in such situations can jump-start the native vegetation, giving it an advantage over invasives. This technique was tested in the University of Wisconsin–Madison Arboretum: after repeated applications of herbicide to reduce cover by reed canary grass, Wilcox (University of Wisconsin–Madison Botany, unpublished data) sowed seed of 33 native species and immediately found high species richness in planted areas but not in unplanted areas. Several species that were not planted volunteered in low abundance, for an overall richness of 52 species. However, one year after seeding, reed canary grass reestablished its dominance of the site.

The practicality and possibility for site maintenance should inform the decision to seed or plant. Very large sites are costly to seed and time-consuming to maintain. Beukes and Cowling (2003) propose that seeding in very large areas or in those with unsuitable soil conditions may not be worth the cost. Further, Mulhouse and Galatowitsch (2003) suggest that, without site maintenance, at least in the beginning, the effort and money spent on planting can be wasted. Continuing site maintenance is often not budgeted in Wisconsin and other restoration plans in the United States (NRC 2001). If possible, even areas where continued site maintenance is not practical, an initial seeding of a cover crop of native species may decrease the invasibility of the site. While the goal

Fig. 10.7 Process for deciding whether or not to seed or plant a wetland restoration site

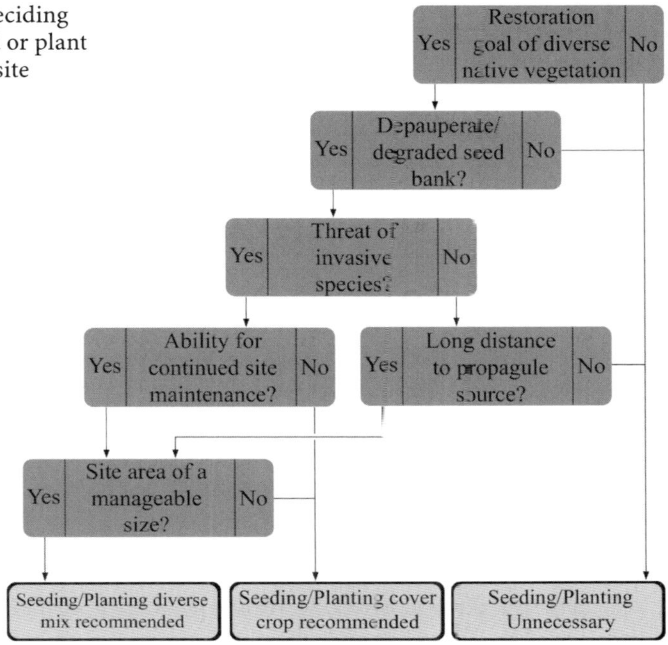

of most restorations is to be self-sustaining, invasive species often make this unlikely. Expectations differ in Europe, where plans for long-term site maintenance are common (Pfadehauer and Klötzli 1996).

10.4.5 Suitable Sources for Propagules

Ideally, one would plant species that were historically present in a site in order to facilitate a return to what was lost. If hydrologic and other environmental conditions have changed, however, historically present species might not survive (Muller et al. 1998). Experimental plantings in small plots can reduce some of the uncertainty in what will establish under novel conditions.

Seed and propagule source is an important consideration. The ecotype of an individual species often affects its establishment, such that locally adapted ecotypes are much more successful than those from commercial stocks (Willard and Levine 1990). Plant performance, especially competitive ability, can be affected by seed source, leading Gustafson et al. (2004) to suggest that using non-local seed sources is unacceptable in restoration. In addition to ecotype, the genetic diversity of a species is important. Not only is a genetically diverse population better able to adapt to specific site conditions, but there is also evidence that it reduces the rate of species richness decline and promotes a convergence of plant community composition (Booth and Grime

2003). Genetic diversity can also be promoted by using multiple local seed sources (Gustafson et al. 2002).

In Europe, there is particular concern about the rare occurrence of woody species in strongly developed areas. In restoration projects, it is common to plant tree and shrub species; and there is increasing awareness of the importance of using propagules or juveniles of local genotypes. In The Netherlands, an inventory has been made of sites where rare woody species with their historical genetic composition still occur. Specialized nurseries have now started breeding programs to propagate material from these sites; and this material is now strongly favored in current restoration projects involving shrub or tree species.

10.5 Emerging Principles

- *Many of the constraints on restoration are abiotic.* Hydrologic conditions and water chemistry are primary examples. A decline in groundwater levels at a restoration site can limit the ability to restore vegetation that depends on a consistent subterranean water supply (Beltman et al. 1996; Hunt et al. 1999). van Duren et al. (1997) found that application of lime did not decrease soil porewater acidity to the levels required for calciphilic plant species. Extreme events can shift the species composition of a site from a desired community to a less desired community. For example, a flood caused a sedge meadow restoration in Wisconsin to become dominated by hybrid cattail, which persists on the site and prevents restoration of the sedge meadow community (Ashworth 1997).
- *Changes in the biota can limit a site's potential to be restored.* The seed bank found at a site degrades over time. For example, former wetlands that had been converted to agriculture for many years will lack a rich seed bank (Wienhold and van der Valk 1989; Olff and Bakker 1991). Klötzli and Grootjans (2001) describe semi-stable states in restorations where an undesired species becomes dominant and creates conditions that favor itself. Such feedbacks are often the case with invasive species.
- *Wetland restoration can be more strategic.* Land managers become better equipped to prioritize sites for restoration as the science of restoration advances and tools become available to deal with abiotic and biotic constraints. When degradation is not reversible within a reasonable timescale, another site might have a higher priority for restoring diversity. When resources are available to work on sites with "irreversible" damage, the alternative goals will become clearer. In landscapes with many former wetlands and limited resources for restoration, it is strategic to develop management priorities first, then prioritize the sites where each goal can best be achieved.

- *It may not be possible to recover exactly what was lost.* In such cases, recovery of a general vegetation type and recovery of ecosystem functions might be more realistic goals (van Diggelen and Marrs 2003). Where degradation is irreversible, an appropriate goal would be the establishment of a diversity of native plants, even if the assemblages are novel to the site or the vegetation type. Project goals can be adjusted to suit the site's potential to be restored.
- *Projects can be designed to allow learning as restoration proceeds.* To date, our understanding of the underlying processes that influence ecosystem development is poor; and specific outcomes are not highly predictable. Klötzli and Grootjans (2001) note that, despite over 40 years of restoration experience in The Netherlands, unexpected and unexplainable developments still occur in restoration projects. If restoration projects were implemented as experiments, alternative approaches could be tested and results monitored; then, approaches that best achieve the goals could be used in subsequent phases of the project. While restoration proceeds, hypothesis-testing and scientific principles can reduce uncertainty and restoration efforts that are conducted as experiments can advance science.

References

Aerts R, Berendse F (1989) Above-ground nutrient turnover and net primary production of an evergreen and a deciduous species in a heathland ecosystem. J Ecol 77:343–356

Amon JP, Thompson CA, Carpenter QJ, Miner J (2002) Temperate zone fens of the glaciated Midwestern USA. Wetlands 22:301–317

Apfelbaum SI, Sams CE (1987) Ecology and control of reed canary grass. Nat Area J 7:69–74

Ashworth SM (1997) Comparison between restored and reference sedge meadow wetlands in south-central Wisconsin. Wetlands 17:518–527

Baptist MJ, Penning WE, Duel H, Smits AJM, Geerling GW, Van der Lee GEM, Van Alphen JSL (2004) Assessment of the effects of cyclic floodplain rejuvenation on flood levels and biodiversity along the Rhine river. River Res Appl 20:285–297

Baskin CC, Baskin JM (1998) Seeds: ecology, biogeography, and evolution of dormancy and germination. Academic, San Diego

Bedford BL, Godwin KS (2003) Fens of the United States: distribution, characteristics, and scientific connection versus legal isolation. Wetlands 23:608–629

Bedford BL, Walbridge MR, Aldous A (1999) Patterns in nutrient availability and plant diversity of temperate North American wetlands. Ecology 80: 2151–2169

Beltman B, Van Den Broek T, Bloemen S, Witsel C (1996) Effects of restoration measures on nutrient availability in a formerly nutrient-poor floating fen after acidification and eutrophication. Biol Conserv 78:271–277

Beltman BTG, Rouwenhorst MB, Van Kerkhoven T, Van der Krift, Verhoeven JTA (2000) Internal eutrophication in peat soils through competition between chloride and sulphate with phosphate for binding sites. Biogeochemistry 50:183–194

Berendse F, Aerts R, Bobbink R (1993) Atmospheric deposition and its impact on terrestrial ecosystems. In: Vos CC, Opdam P (eds) Landscape ecology of a stressed environment. Chapman & Hall, London, pp 104–121

Bernthal TW, Willis KG (2004) Using Landsat 7 imagery to map invasive reed canary grass (*Phalaris arundinacea*): a landscape level wetland monitoring methodology. (Wisconsin Department of Natural Resources Publication SS-992-2004) US Environmental Protection Agency Region V, Madison

Beukes PC, Cowling RM (2003) Evaluation of restoration techniques for the succulent Karoo, South Africa. Restor Ecol 11:308–316

Boers AM, Zedler JB, Veltman RLD (2006) *Typha* x *glauca* dominance and extended hydroperiod constrain restoration of wetland diversity. Ecol Eng (in press)

Booth RE, Grime JP (2003) Effects of genetic impoverishment on plant community diversity. J Ecol 19:721–730

Budelsky RA, Galatowitsch SM (2000) Effects of water regime and competition on the establishment of a native sedge in restored wetlands. J Appl Ecol 37:971–985

Burke MJW, Grime JP (1996) An experimental study of plant community invasibility. Ecology 77:776–790

Burt TP, Pinay G, Matheson FE, Haycock NE, Butturini A, Clement JC, Danielescu S, Dowrick DJ, Hefting MM, Hillbricht IA, Maitre V (2002) Water table fluctuations in the riparian zone: comparative results from a pan-European experiment. J Hydrol 265:129–148

Callaway JC, Sullivan G, Zedler JB (2003) Species-rich plantings increase biomass and nitrogen accumulation in a wetland restoration experiment. Ecol Appl 13:1626–1639

Clevering OA, Gulik WMG van (1997) Restoration of *Scirpus lacustris* and *Scirpus maritimus* stands in a former tidal area. Aquat Bot 55:229–246

Craft CB (1997) Dynamics of nitrogen and phosphorus retention during wetland ecosystem succession. Wetland Ecol Manage 4:177–187

Dahl TE (1990) Wetlands losses in the United States 1780's to 1980's. US Department of the Interior, Fish and Wildlife Service, Washington, D.C.

Davis MA, Grime JP, Thompson K (2000) Fluctuating resources in plant communities: a general theory of invasibility. J Ecol 88:528–534

Day RT, Keddy PA, McNeill J, Carleton T (1988) Fertility and disturbance gradients: a summary model for riverine marsh vegetation. Ecology 69:1044–1054

De Groot CJ, Fabre A (1993) The impact of desiccation of a freshwater marsh (Garcines Nord, Camargue, France) on sediment–water–vegetation interactions: part 3. The fractional composition and the phosphate adsorption characteristics of the sediment. Hydrobiologia 252:105–116

Diggelen R van, Marrs RH (2003) Restoring plant communities – introduction. Appl Veg Sci 6:106–110

Diggelen R van, Grootjans AP, Harris JA (2001) Ecological restoration: state of the art or state of the science? Restor Ecol 9:115-118

Duren IC van, Boeye D, Grootjans AP (1997) Nutrient limitations in an extant and drained poor fen: implications for restoration. Plant Ecol 133:91–100

Eertman RHM, Kornman BA, Stikvoort E, Verbeek H (2002) Restoration of the Sieperda tidal marsh in the Scheldt estuary, the Netherlands. Restor Ecol 10:438–449

Ehrenfeld JG (1983) The effects of changes in land-use on swamps of the New Jersey Pine Barrens. Biol Conserv 25:353–375

Engelhardt KAM, Ritchie ME (2001) Effects of macrophyte species richness on wetland ecosystem functioning and services. Nature 411:687–689

Galatowitsch SM, Anderson NO, Ascher PD (1999) Invasiveness in wetland plants in temperate North America. Wetlands 19:733–755

Green EK, Galatowitsch SM (2002) Effects of *Phalaris arundinacea* and nitrate-N addition on the establishment of wetland plant communities. J Appl Ecol 39:134–144

Grootjans AP, Bakker JP, Jansen AJM, Kemmers RH (2002) Restoration of brook valley meadows in the Netherlands. Hydrobiologia 478:149–170

Gustafson DJ, Gibson DJ, Nickrent DL (2002) Genetic diversity and competitive abilities of *Dalea purpurea* (Fabaceae) from remnant and restored grasslands. Int J Plant Sci 163:979–990

Gustafson DJ, Gibson DJ, Nickrent DL (2004) Competitive relationships of *Andropogon gerardii* (Big Bluestem) from remnant and restored native populations and select cultivated varieties. Funct Ecol 18:451–457

Hill NM, Keddy PA, Wisheu IC (1998) A hydrological model for predicting the effects of dams on the shoreline vegetation of lakes and reservoirs. Environ Manage 22:723–736

Huenneke LF, Hamburg SP, Koide R, Mooney HA, Vitousek PM (1990) Effects of soil resources on plant invasion and community structure in Californian serpentine grassland. Ecology 71:478–791

Hunt RJ, Walker JF, Krabbenhoft DP (1999) Characterizing hydrology and the importance of ground-water discharge in natural and constructed wetlands. Wetlands 19:458–472

Jansen AJM, Eysink FT, Maas C (2001) Hydrological processes in a Cirsio-Molinietum fen meadow: implications for restoration. Ecol Eng 17:3–20

Keddy PA, Reznicek AA (1986) Great Lakes vegetation dynamics: the role of fluctuating water levels and buried seeds. J Great Lakes Res 12:25–36

Kellogg CH, Bridgham SD (2002) Colonization during early succession of restored freshwater marshes. Can J Bot 80:176–185

Kercher SM, Zedler JB (2004) Multiple disturbances accelerate invasion of reed canary grass (*Phalaris arundinacea* L.) in a mesocosm study. Oecologia 138:455–464

Kercher SM, Carpenter QJ, Zedler JB (2004) Interrelationships of hydrologic disturbance, reed canary grass (*Phalaris arundinacea* L.) and native plants in Wisconsin wet meadows. Nat Area J 24:316–325

Kindscher K, Tieszen LL (1998) Floristic and soil organic matter changes after five and thirty-five years of natives tallgrass prairie restoration. Restor Ecol 6:181–196

Klötzli F, Grootjans ABP (2001) Restoration of natural and semi-natural wetland systems in central Europe: progress and predictability of developments. Restor Ecol 9:209–219

Koerselman W (1992) The nature of nutrient limitation in Dutch dune slacks. In: Carter RWG (ed) Coastal dunes. Geomorphology, ecology and management for conservation. Balkema, Rotterdam, pp 189–199

Koerselman W, Van Kerkhoven MB, Verhoeven JTA (1993) Release of inorganic N, P, and K in peat soils: effect of temperature, water chemistry, and water level. Biogeochemistry 20:63–81

Lindig-Cisneros R, Zedler JB (2002) Halophyte recruitment in a salt marsh restoration site. Estuaries 25:1175–1184

Mack RN, Simberloff D, Lonsdale WM, Evans H, Clout M, Bazzaz FA (2000) Biotic invasions: causes, epidemiology, global consequences, and control. Ecol Appl 10:689–710

Middleton BA (2000) Hydrochory, seed banks, and regeneration dynamics along the landscape boundaries of a forested wetland. Plant Ecol 146:169–184

Mitsch WJ, Gosselink JG (1993) Wetlands. Wiley, New York

Mulhouse JM, Galatowitsch SM (2003) Revegetation of prairie pothole wetlands in the mid-continental US: twelve years post re-flooding. Plant Ecol 169:143–159

Muller S, Dutoit T, Grevilliot F (1998) Restoration and rehabilitation of species-rich grassland ecosystems in France: a review. Restor Ecol 6:94–101

NRC (2001) Compensating for wetland loss under the clean air act. (National Research Council committee on wetland mitigation) National Academy, Washington, D.C.

Neill C (1990) Effects of nutrients and water levels on emergent macrophyte biomass in a prairie marsh. Can J Bot 68:1007–1014

Nienhuis PH, Bakker JP, Grootjans AP, Gulati RD, Jonge VN de (2002) The state of the art of aquatic and semi-aquatic ecological restoration projects in the Netherlands. Hydrobiologia 478:219–233

O'Brien E, Zedler JB (2006) Accelerating the restoration of vegetation in a southern California salt marsh. Wetlands Ecol Manage (in press)

Olff H, Bakker JP (1991) Long-term dynamics of standing crop and species composition after the cessation of fertilizer application to mown grassland. J Appl Ecol 28:1040–1052

Patrick WH, Khalid RA (1974) Phosphate release and sorption by soils and sediments: effect of aerobic and anaerobic conditions. Science 186:53–55

Pfadehauer J, Klötzli F (1996) Restoration experiments in middle European wet terrestrial ecosystems: an overview. Vegetatio 126:101–115

Richardson DM, Pysek P, Rejmánek M, Barbour MG, Panetta FD, West CJ (2000) Naturalization and invasion of alien plants: concepts and definitions. Divers Distrib 6:93–107

Richardson JL, Vepraskas MJ (2000) Wetland soils: genesis, hydrology, landscapes, and classification. CRC/Lewis, Boca Raton

Roelofs JGM, Brouwer E, Bobbink R (2002) Restoration of aquatic macrophyte vegetation in acidified and eutrophicated shallow soft water wetlands in the Netherlands. Hydrobiologia 478:171–180

Sher AA, Hyatt LA (1999) The disturbed resource–flux invasion matrix: a new framework for patterns of plant invasion. Biol Invasions 1:107–114

Simberloff D, Von Holle M (1999) Synergistic interactions of nonindigenous species: invasional meltdown? Biol Invasions 1:21–32

Szilas CP, Borggaard OK, Hansen HCB (1998) Potential iron and phosphate mobilization during flooding of soil material. Water Air Soil Pollut 106:97–109

Thompson AL, Luthin CS (2004) Wetland restoration handbook for wisconsin landowners, 2nd edn. Wisconsin Department of Natural Resources, Madison

USDA NRCS (2004) The PLANTS database, ver. 3.5. (available at http://plants.usda.gov) National Plant Data Center, Baton Rouge

Valk AG van der, Verhoeven JTA (1988) Potential role of seed banks and understory species in restoring quaking fens from floating forests. Vegetatio 75:3–13

Verhoeven JTA, Bobbink R (2001) Plant diversity of fen landscapes in The Netherlands. In: Gopal B (ed) Biodiversity in wetlands: assessment, function and conservation. Backhuys, Leiden, pp 65–87

Verhoeven JTA, Kemmers RH, Koerselman W (1993) Nutrient enrichment of freshwater wetlands. In: Vos CC (ed) Landscape ecology of a stressed environment. Chapman and Hall, London, pp 33–59

Wienhold CE, Valk AG van der (1989) The impact of duration of drainage on the seed banks of northern prairie wetlands. Can J Bot 67:1878–1884

Wilcox DA, Meeker JE (1991) Disturbance effects on aquatic vegetation in regulated and unregulated lakes in northern Minnesota. Can J Bot 69:1542–1551

Willard DE, Levine DA (1990) Regional analysis of fringe wetlands in the midwest: creation and restoration. In: Kusler JA, Kentula ME (eds) Wetland creation and restoration: the status of the science. Island Press, Covelo, pp 299–325

Wolff WJ (1993) Netherlands-Wetlands. Hydrobiologia 265:1–14

Woo I, Zedler JB (2002) Can nutrients alone shift a sedge meadow towards dominance by the invasive *Typha* x *glauca*? Wetlands 22:509–521

11 Fen Management and Research Perspectives: An Overview

Beth Middleton, Ab Grootjans, Kai Jensen,
Harry Olde Venterink, and Katalin Margóczi

11.1 Introduction

A fen has vegetation that is actively forming peat and is fed by ground- or surface water (Joosten and Clarke 2002). In Europe a "fen meadow" is a ground- or surface water-fed mown grassland that does not form peat, since it was formed after modest drainage of a fen or it developed on a predominantly moist soil (Grootjans and Van Diggelen 1995). Therefore, fens and fen meadows are considered to be different ecosystems by most European authors. Others do not make a distinction between fens and fen meadows because the species composition of both ecosystems may overlap considerably (Wheeler et al. 1995). In North America, fens dominated by tussock-forming sedges are referred to as "sedge meadows", which are often grazed. Since there is not yet scientific agreement on whether sedge meadows are fens or fen meadows, we will refer to them as fens in this chapter.

Fens are the most diverse temperate community types (Bedford and Godwin 2003). The biodiversity of fens is threatened by agricultural and urban development, fragmentation, hydrologic changes, shifts in grazing and mowing, structural changes in vegetation, and reproductive problems related to isolated populations. Species are declining in numbers in fens in Europe and North America. Estimates of species lost, for example, from the United Kingdom, are that 95–98 % of species-rich fens present before 1940 have been lost (García 1992). Similar losses are reported for France (Muller et al. 1998) and the Netherlands (Jansen et al. 2000).

The fragmentation of fens occurs via agricultural and urban development; and as these encroach, the wetlands become smaller and more isolated because of hydrological modification (Figs. 11.1, 11.2). In particular, any alteration in surface water flow from streams and rivers creates a barrier to seed dispersal (Galatowitsch and van der Valk 1996; Middleton 1999). Species loss

Ecological Studies, Vol. 191
R. Bobbink, B. Beltman, J.T.A. Verhoeven, and D.F. Whigham (Eds.)
Wetlands: Functioning, Biodiversity Conservation, and Restoration

Fig. 11.1 Mire distribution in Central Europe under natural conditions (Succow 1988; reproduced by permission of Fischer Verlag, Jena). Only peatlands exceeding 300 ha have been included. *Black* 3D bogs, *gray* 3D fens

occurs because of the lack of habitat (Bedford and Godwin 2003), but also because of reproductive problems associated with small population sizes (Ellstrand and Elam 1993; Fischer and Stöcklin 1997). While these problems might be alleviated by the reconnection of dispersal corridors (Taylor et al. 1993), higher amounts of dispersal between these disconnected wetlands could contribute to more opportunities for an invasion of exotic species in North America, where invasive species in wetlands are a major problem (Galatowitsch et al. 1999).

Water quality and sedimentation patterns can be altered by agricultural and urban development; and these changes contribute to shifts in species composition in fens. Sedimentation is on the increase in fens adjacent to cities in the Midwestern United States; and this encourages invasive species (*Phalaris arundinacea*, *Typha* x *glauca*) at the expense of *Carex stricta* (Werner and Zedler 2002). Urban water runoff increases the amounts of N and P in the surface water, and subsequently, the biomass of *T.* x *glauca* increases. *C. stricta* does not increase in biomass, so that fens with urban runoff become dominated by *T.* x *glauca* (Woo and Zedler 2002). The nutrient dynamics in fens changed with land use changes. Atmospheric nitrogen

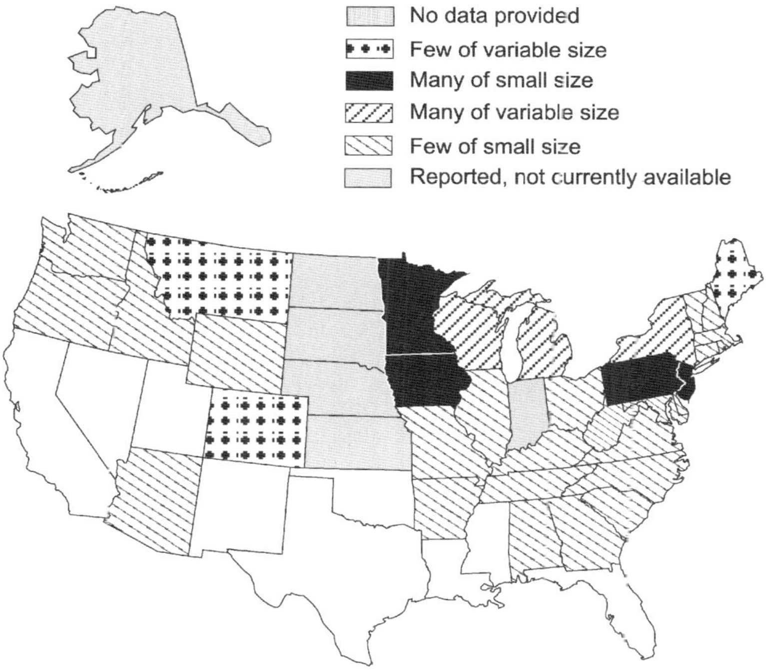

Fig. 11.2 Distribution of North American fens (Bedford and Godwin 2003; reproduced by permission of Society of Wetland Scientists)

deposition, in particular, increased 3- to 10-fold during the past decades in these wetlands (Olde Venterink et al. 2001, 2002). Eutrophication causes increases in the plant biomass of certain species and contributes to the encroachment of large herbaceous and woody species (Bobbink and Roelofs 1995).

11.2 Hydrological Systems of Fens

Fens with similar vegetation types exist in very different landscapes and soil conditions. Small sedge vegetation with many Red List species (*Caricion davallianae*, i.e., species of alkaline fens), for instance, occur in coastal dune slacks on sandy soils (Ranwell 1959), in brook valleys of the European lowlands on peat soils (Succow and Joosten 2001), but also in calcareous spring mires in alpine regions of Europe (Ellenberg 1986).

Hydrological systems can create similar habitats within a wide range of landscapes. Groundwater fed fens are "flow through" systems, in which nutri-

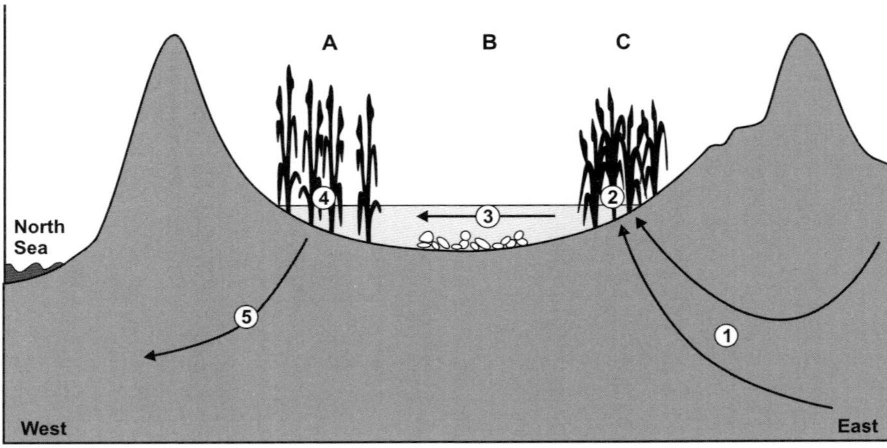

Fig. 11.3 Schematic presentation of a through-flow dune slack on the Dutch Wadden Sea island. *1* Incoming calcium and iron-rich groundwater, *2* exfiltration of groundwater in a vegetation zone with sedges, *3* surface flow and precipitation of iron and $CaCO_3$ in a zone where nutrient-poor pioneer vegetation persists, *4* infiltration of iron- and calcium-poor surface water in a *Phragmites australis* stand, *5* sulfate reduction during infiltration (after Adema 2002; reproduced by permission of Opulus Press)

ents become less available for plant growth due to geochemical changes along the flow paths of the groundwater (Fig. 11.3).

11.2.1 Large and Small Hydrological Systems

Much research has been done on the hydrological conditions of fen systems (Komer 1994; Wassen et al. 1996; Van Diggelen 1998; Jansen et al. 2000; Schot et al. 2004). Some authors present very detailed hydrological research in species-rich fens on mineral soils, a case in which most of the groundwater comes from a very small catchment area (Jansen et al. 2000), while others worked in very large mire systems, like the Biebrza in Poland (Wassen et al. 1996) and the Ob River in Russia (Loos and Schipper 2003; Fig. 11.4). Sometimes the surface area of a fen is rather small, but the catchment area supplying the fen with groundwater is very large. This is often the case with fens and spring mires that are supplied with artesian water, originating from large aquifers in which the water flow has been blocked due to shifting of geological strata (Van Andel and Arondson 2005). Fens are very vulnerable to erosion by groundwater flow. The digging of drainage ditches may result in the down-cutting of the ditch channel, which effectively lowers the local water table within the fen. The peat may begin to decompose and erode into the ditch. After the ditch is established, the degradation process may be very difficult to

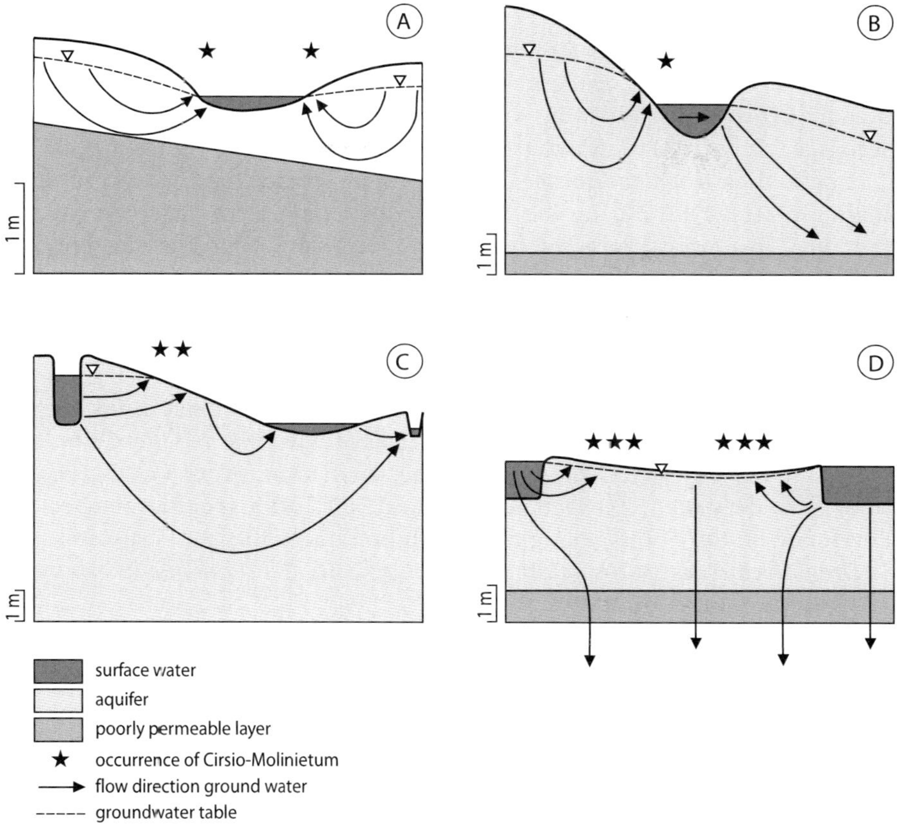

Fig. 11.4 The fen meadow community *Cirsio–Molinietum* (indicated by stars) occurs in the Netherlands under various hydrological conditions. **A** Heathland pool situated above shallow loam layer with groundwater supply from local sand ridges, **B** local (small) hydrological system with through flow of groundwater, **C** artificial hydrological system with lateral discharge of calcareous surface water from large canals, **D** surface water system with discharge of calcareous surface water from ditches (modified after Jansen et al. 2000; reproduced by permission of Opulus Press)

contain, and may result in the total destruction of the mire (Wolejko et al. 1994). Due to rapid peat mineralization, these systems become very eutrophic and tall herbs (such as *Urtica dioica*), shrubs, and trees move in quickly (Wolejko et al. 1994).

11.2.2 Natural Fens Can Be Very Stable

Fens are mires that receive water from the mineral surroundings and hence respond to changes in regional hydrology. Large fens, however, also raise their water level and that of their surroundings autogenously, leading to the expansion of fens on progressively higher positions in the landscape. Such rising water levels may also result in the gradual paludification of hitherto dry upland depressions and may even reverse water flow from groundwater recharge to groundwater discharge conditions (Succow and Joosten 2001).

Due to their thick and rather flexible peat layer, such fens became increasingly more resilient during the course of their long-term development and remained in a more or less stable situation for several centuries, despite considerable climatological and hydrologic changes. They only changed character when certain hydrological thresholds were crossed, the exact magnitude of these thresholds being dependent on the stage of mire development. Schult (2002) carried out detailed palynological research in the Trebel area and found that between 8000 and 7000 years BP, a terrestrialization mire developed in a calcareous lake, with species such as *Menyanthes trifoliata*, *Cladium mariscus*, and *Carex rostrata*. At this stage, small amounts of $CaCO_3$ (travertine) were deposited. Between 7000 BP and 5000 BP, a groundwater-fed fen developed on top of this young mire, with predominantly moss species, such as *Drepanocladus* spp, *Homalothecium nitens*, and *Meesia triquetra*. *Menyan-*

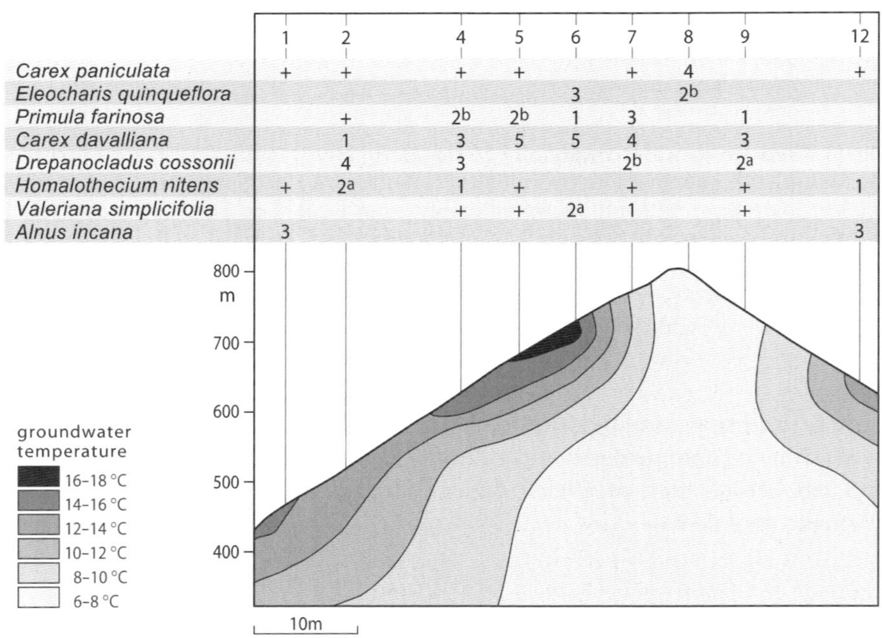

Fig. 11.5 Temperature profile and vegetation composition across a small Slovak spring mire near the High Tatra Mountains. The species coverage was estimated using a Braun–Blanquet cover/abundance scale (Grootjans, unpublished data)

thes trifoliata remained present throughout this period and small sedges such as *C. diandra* and *C. limosa* also occurred. Chalk deposition occurred until ca. 3000 BP and then stopped. This developmental sequence was remarkably stable; and similar results have been found in Western Germany (Schwaar 1980) and Poland (Wassen and Joosten 1996).

Nowadays such vegetation types are very rare and chalk deposition is almost no longer found in the North German lowlands. However, remnants of these vegetation types can still be found in the mountain areas of Bavaria (Germany) and Slovakia. Here we find relatively small calcareous spring mires, where travertine is still being formed in pools with *Drepanocladus* spp, *Menyanthes trifoliata*, and *Chara* spp. This travertine formation is a rather recent phenomenon in some wetlands and mainly occurs in oligotrophic (calcareous) pools where supersaturated groundwater discharges (Grootjans et al. 2005). During summer the surface water of these pools is warmed (Fig. 11.5). Outgassing occurs and $CaCO_3$ precipitates. Especially moss species and some water plants, such as *Chara* spp, can accelerate this travertine formation by using the high concentration of CO_2 in the pools as a carbon source. $CaCO_3$ precipitates on the leaves and is later added to the soil (Van Breemen and Buurman 2002). The added travertine prevents acidification and associated eutrophication in the soil and sustains a very high biodiversity in such fens. However, such small groundwater-fed systems are vulnerable to hydrological changes in the landscape and also to climatic change. These small groundwater-fed fens can easily shift from peat-forming systems toward eroding stages within a short period of time (Wolejko et al. 1994).

11.2.3 Hydrochemical Processes Stabilizing the Biodiversity of Fens

Formerly, hydrological systems that provided natural fens with a large supply of base-rich groundwater were able to stabilize nutrient poor fen vegetation for many centuries, without any management by man. However, during the past few centuries, almost all fens have been slightly drained and changed into low-productive meadows and pastures that cannot be maintained without management (Kotowski 2002). When large hydrological changes in the surrounding landscape occur, the through-flow of anaerobic groundwater should be re-established to restore the high biodiversity of fen species. A relatively slow groundwater flow in fen systems is essential for maintaining a high biodiversity. Slow-flowing groundwater prevents erosion and stimulates the precipitation of iron and $CaCO_3$ in the root zone, thus stabilizing nutrient cycling at a low level (Lamers et al. 2002; Olde Venterink et al. 2001, 2002). The study of historical descriptions and well preserved remnants of fen systems may prove very helpful in understanding why some systems can easily maintain a high biodiversity and why other cannot perform this function.

11.3 Eutrophication in Fens

Eutrophication is considered one of the main factors contributing to the loss of biodiversity in fens in Western Europe, because increased nutrients are related to increased productivity and the competitive exclusion of shorter species. Species-rich communities of low productivity are eventually transformed into communities of high productivity dominated by a few tall sedges, grasses or shrubs (Grime 1979). This increased dominance of tall herbaceous and woody species occurred both in Europe and North America, not only due to eutrophication, but also because of changes in rural land usage related to grazing (Middleton 2002a) and drainage (Wheeler et al. 1995; Grootjans et al. 1996).

11.3.1 Change in Management

After 1945, agriculture shifted away from small farms and moved toward larger mechanized ones that no longer utilized fens for pastures (Middleton 2002a) or hay-mowing (García 1992). After that, the short diverse vegetation shifted toward communities dominated by woody and tall herbaceous species. These changes in vegetation structure are directly attributable to the cessation of cattle grazing and mowing. Especially after moderate to heavy grazing, shrubs and trees grow and shade the rarer fen species (Middleton 2002a, b).

Strategies for dealing with these changes in community structure and related losses in biodiversity have taken different routes in Europe and North America. Managers in North America control the dominance of woody and tall species with fire, whereas European managers rely on mechanical mowing and hay removal (Schrauzer et al.1996). Fire temporarily increases the species richness, flowering and seed set of fen species in North America (Middleton 2002b). Because eutrophication is a contributing factor in the growth of these large fen species, fire may also be useful because it removes nutrients such as nitrogen from the system. Similarly, hay removal also removes nutrients from these systems, which eventually may lead to phosphorus limitation, and is therefore of value in maintaining the biodiversity of fens (Verhoeven et al. 1996; Tallowin and Smith 2001).

11.3.2 Change in Nutrient Budgets

The increase in productivity in fens is the result of increased availabilities of potentially growth-limiting nutrients such as nitrogen (N), phosphorus (P), and potassium (K; Verhoeven et al. 1996; Van Duren and Pegtel 2000). To

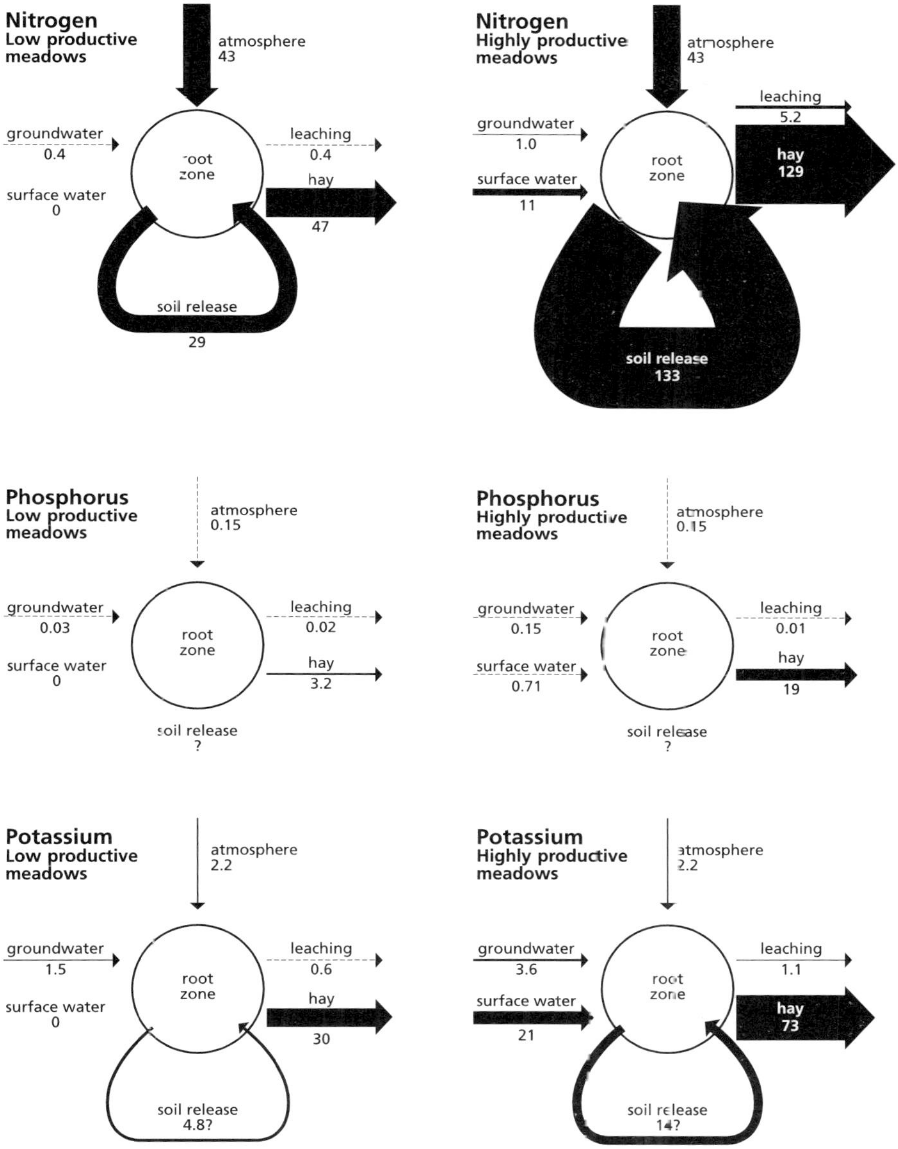

Fig. 11.6 Annual N, P, and K fluxes into, out of, and within the root zone of low-productive species-rich meadows and highly productive species-poor meadows. All fluxes are in kg N, P, or K per hectare per year (Olde Venterink et al. 2002; reproduced by permission of the Ecological Society of America)

understand the causes of nutrient enrichment and evaluate the effects of management, one should have insight into annual nutrient budgets, i.e., the rates of nutrient fluxes into and from external sources as well as net release rates of nutrients in the soil (Koerselman et al. 1990).

Olde Venterink et al. (2001, 2002) quantified N, P, and K budgets along productivity gradients in fens to evaluate the importance of atmospheric deposition, flooding, groundwater flow, leaching, and soil release rates. Differences in N availability along productivity gradients were caused by release rates in the soil (mineralization; Fig. 11.6). Besides N release in the soil, atmospheric N deposition made up large proportions of the annual N budgets. P and K availabilities along the productivity gradients were mainly influenced by soil processes, as indicated by soil extractable nutrient. The input and output fluxes of nutrients by groundwater flow were very low, while the input by flooding seemed only important for K (Fig. 11.6). However, it should be noted that the input of nutrients adsorbed to sediment was not included in the flooding assessment, which could be a very important input source of P in fens (Mitsch et al. 1979). The dominant role of soil processes in setting nutrient availabilities along the productivity gradients demonstrated that human alterations of site conditions, such as drainage, have been a major cause of nutrient enrichment for N, P, and K (Olde Venterink et al. 2002). Furthermore, considering that atmospheric N deposition has increased from 5–20 kg N ha^{-1} $year^{-1}$ (Erisman and Draaijers 1995) to an average 43 kg N ha^{-1} $year^{-1}$ in this part of Europe (Fig. 11.6), increased atmospheric N deposition also has been a major source of nutrient enrichment.

The long-term effects of annual hay-harvesting on nutrient limitation were also assessed by Olde Venterink et al. (2002). In fens with low productivity, N output by hay-harvesting accounted for N input from atmospheric deposition, whereas there was a net output of P and K (Fig. 11.7; Olde Ventervink et al. 2002). At highly productive sites, hay-harvesting resulted in net output of N, P, and K (Koerselman et al. 1990). When the annual nutrient output was expressed as percentages of the total soil pools, it was found that net output of K (1–20 % of soil K pool) was mostly much larger than that of P (generally 0.5–3.0 %) or N (0–3 %). Hence, in the long-term, hay-harvesting seems to induce K limitation, particularly when they are drained for a long time (Van Duren et al. 1981; Van Duren and Van Andel 1997).

11.3.3 Internal Eutrophication

More intensive agricultural use not only leads to lowered water tables but also to increased concentrations of NO_3^- and SO_4^{2-} in ground- and surface water. Effects of water table alterations receive much attention, but there is confusion about its consequences for nutrient kinetics. Desiccation often leads to eutrophication in peatlands (Grootjans et al. 1996). However, this is not always

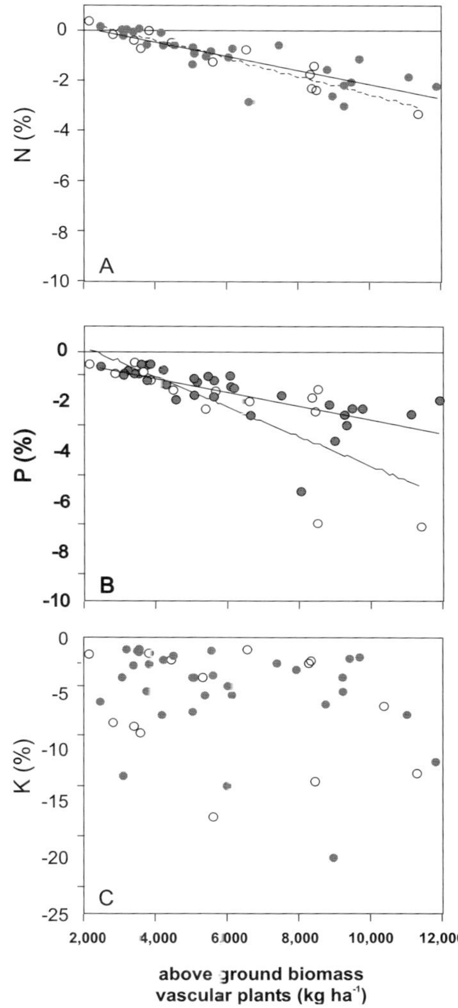

Fig. 11.7 Net annual output of N, P, and K by hay-harvesting as a percentage of the total N, P, and K pools in the soil (0–10 cm), in relation to the above-ground biomass of vascular plants in fens (*open circles*) and meadows (*filled circles*) (Olde Venterink et al. 2002; reproduced by permission of the Ecological Society of America)

true. Fens that are limited by P and K, in particular, usually respond to drainage with a decreased productivity. This had already been noticed by Russian researchers in the early 1930s during the communist period, when extensive mires were reclaimed for agriculture. They found a marked decrease in groundwater-fed fens after drainage (Ozushko 1934, cited in P'yavchenko 1976). Ozushko warned: "that money invested in drainage, is wasted unless the drained area is then used for the cultivation of agricultural crops". At that time, Russian agro-scientists already knew that P and K were limiting in many drained mires and not N.

Restoring desiccated fens is not simply a matter of increasing water levels. Recent research shows that rewetting with sulfate-rich surface water, in particular, may lead to increased eutrophication in fens. High SO_4^{2-} availability can induce PO_4^{3-} mobilization in wetland soils. This may occur when the former discharge of iron-rich groundwater is not restored. In such a case, the groundwater is replaced by sulfate-rich surface water. When the water stagnates, all oxygen is consumed and sulfate reduction is stimulated when the outside temperatures are sufficiently high. Sulfate reduction produces sulfides, which in the presence of iron form FeS_x (pyrite). PO_4^{3-} bound to iron hydroxides is then released to the pore water or the stagnating water body (Smolders et al. 1995; Lamers et al. 1998).

Although the active manipulation of the internal mobilization of nutrients has been practiced since medieval times by farmers trying to improve their crop yield by regular flooding with surface water, present biogeochemical research in peatlands often overlooks the alterations in nutrient mobilization rates due to such anthropogenic water quality changes. Flooding with surface water during the summer, in particular, can be very harmful for species-rich fens and may result in the rapid growth of tall grasses, such as *Phalaris arundinacea* or *Glyceria maxima*. Periodic local desiccation of the sediment during the summer, which is a natural phenomenon in fens, is important in sulfate-rich wetlands because this determines the ability of the soil to retain phosphates. The oxidation of FeS_x again increases the reactive Fe^{3+} concentration, stimulating P-binding in the top layer (Lucassen et al. 2004a).

The role of increased NO_3^- levels in the groundwater due to intensive agricultural fertilization is complex. Nitrate may increase the redox potential but also leads to a higher binding capacity of P due to FeS oxidation. A high input of nitrate inhibits SO_4^{2-} reduction because it is a more favorable electron acceptor and also the oxidation of FeS_x will increase the sulfate levels. So, high nitrate levels in the groundwater may counteract the negative effects of high sulfate concentrations. On the one hand, NO_3^- leaching into the groundwater increases the risk of eutrophication of groundwater discharge areas by mobilizing SO_4^{2-}, in FeS_x-containing soils or aquifers. On the other hand, high concentrations of NO_3^- may prevent P eutrophication (Lucassen et al. 2004b). As a rule of thumb, to avoid eutrophication due to internal release of nutrient in fens, one should not increase surface water tables above the potential groundwater table.

11.4 Seed Bank and Seed Dispersal

An increase in species richness over time following restoration depends on the environmental site conditions and on seed availability, germination, and survival. Fine-tuning of the environmental conditions for restoration is already a difficult task, but very often seed availability and also seed dispersal are restricting restoration prospects considerably (Bakker et al. 1996).

When degeneration processes have been taken place for a long time, seed banks might have been depleted and ways to introduction of locally extinct species have to be considered. Sometimes lack of seed availability can be improved by stimulating natural dispersal processes, such as flooding (Middleton 1999).

11.4.1 Seed Banks

Wet grassland species have a limited ability to regenerate by seed, because these species usually have only transient seed banks (Bekker et al. 1997, 2000) and seeds of rare and endangered species also have a comparatively low persistence (Fischer and Stöcklin 1999). In addition, seed dispersal distances are usually low (Bullock et al. 2002; Soons 2004) and thus it can be questioned whether seed persistence and dispersal significantly contribute to the restoration of species-rich wet grassland communities.

Jensen (1998) analyzed seed banks and the local seed rain of abandoned wet grasslands and found that the (anemochorous) seed rain of the investigated sites was indeed dominated by species of the aboveground vegetation, implying that long-distance seed dispersal by wind is a rather rare event in this habitat type. In contrast, the seed bank of late successional stages, in which most typical wet grassland species had disappeared in the vegetation, always contained some elements of the typical wet grassland flora. These results at least partly contradict the hypotheses of transient seed banks of most grassland species. Jensen's research group, therefore, carried out seed burial experiments to examine directly the seed mortality and longevity of approximately 45 species of the regional fen flora (Schütz 1997, 1998, 1999; Jensen 2004a, b; Brändel 2004a, b). These studies revealed that the mortality of artificially buried seeds of most species was low, that only a few species exhibited transient seed banks, and that the seeds of almost half of the investigated species were able to persist in the soil for at least five years (i.e., these seed banks were long-term and persistent). To explain such differences in seed longevity estimates, Jensen (2004b) compared the seed-longevity estimates for a set of 230 herbaceous wetland species from Northern Germany using different methods of seed bank analysis (e.g., burial experiments vs classic soil seed bank sampling) and various (indirect) indicators for seed bank classification (presence/absence in vegetation and seed bank, depth distribution in the soil; see Thompson et al. 1997). This analysis showed that burial experiments generally resulted in the highest longevity estimates. The criterion stated as "presence in the vegetation, but absence in the seed bank" is the criterion used by Thompson et al. (1997) to classify a species as "transient". However, many species classified as transients in this way had unrealistically low longevity estimates. From these results, it can be concluded that many typical wet grassland species are able to persist at least some years as viable seeds in the soil. We may conclude that much more experimental evidence is needed to develop realistic criteria for estimating seed longevity of soil seed banks.

11.4.2 Seed Dispersal

Water is an important dispersal vector and a number of studies have quantified hydrochorous seed dispersal in wetlands, such as clear streams in an alpine zone (Bill et al. 1999), forested temperate wetlands (Middleton 2000), boreal rivers (Andersson and Nilsson 2002), and a small lowland river (Boedeltje et al. 2003). All these studies revealed that running water might disperse seeds over quite long distances. Vogt et al. (2004) analyzed the hydrochorous seed dispersal in a fen in Northern Germany (Eider Valley) by means of seed traps. This two-year study revealed very large temporal and spatial variation in both quantity (seed number) and quality (species composition) of the hydrochorous seed transport. Overall, seeds of almost 200 different taxa were found in the traps, which is over 60 % of the regional species pool. Seed transport was dominated by (common) species which were abundant along the river shores, but rare and regionally endangered species were also present, albeit only in low quantities. Wind dispersal may also be important in these wetlands and human effects on landscapes, including habitat fragmentation and eutrophication, may affect the colonization capacity of species (Soons et al. 2004).

11.5 Fen Restoration: An Example From Hungary

11.5.1 Introduction

Restoration provides a way to offset some habitat loss for the rare species of fens and other wetland habitats (Middleton 1999; Grootjans et al. 2002; Lamers et al. 2002; Wheeler et al. 2002), particularly because more than 50 % of the world's wetlands have been lost (Munro 1988). Few specific estimates of loss for fens have been made in North America, but in some regions of the United States fen loss exceeds 40 % (Bedford and Godwin 2003). In Europe, less than 1 % of the original mire area has remained in most countries (Joosten and Clarke 2002), although some countries have some floodplain wetlands that are nearly natural (e.g., Turkey, Romania, France; Wenger et al. 1990). In eastern European countries, some 10–50 % of the former mire area still remains. In some Scandinavian countries, European Russia, and Romania, the mires are the best preserved (>50 % of the mires remaining). On a world scale, mire losses are highest in Europe due to its high population pressure on nature and the climatic suitability for agriculture and forestry. In Hungary, in particular, wetland losses are large. It is estimated that 97 % of the natural wetlands have been drained and converted to agricultural fields (Lájer 1998); and many of the remaining wetlands are damaged. While the restoration of fens may be a way to offset these worldwide losses of fens and related biodiversity loss, few examples exist of successful fen restoration, so this example from Hungary is of interest to our discussion.

11.5.2 Destruction and Restoration of a Fen System in Hungary

Fen restoration is less commonly attempted than some other types of restoration, possibly because of the difficulties of reestablishing the hydrology associated with groundwater. One attempt to restore fens has been made in the Hanság, a wetland that was once the largest fen system in the Carpathian basin of Hungary (55 000 ha; Szenkendi 1938). Before the nineteenth century, several rivers from the Alps flowed into the Danube valley, disappeared into the huge floating fen system, and re-appeared at the outflow. Local people used the biological resources of the fen for fishing, hunting, collecting eggs of water birds, hay-making, and reed-harvesting. However, human settlements and agricultural fields near the fen were continuously threatened by inundation, so attempts were made to drain the Hanság, beginning in the nineteenth century. Despite drainage and disturbance, the Hanság retained large patches of natural vegetation in the wet meadows and forests; and the vegetation of aquatic communities survived in drainage channels. Because of its natural value, the Hanság was designated a protected natural area in 1976 and in 1994 as the Fertő–Hanság National Park. It is within this park that the restoration project was conducted.

To restore the wetlands, managers of the Fertő–Hanság National Park reflooded 400 ha of poor quality grassland, with the support of the Hungarian and Dutch governments. To re-engineer the restoration sites, dikes were built around the three separate wetlands. Gravity transported water through sluices from the river Rábca and the channel Kismetszés (Fig. 11.8). The first and second units were flooded in spring 2001 and the third unit in autumn 2001. The water-level has stayed constant at 0–100 cm above the soil surface. While a floating fen was not restored by this procedure, a wetland with an open-water habitat was created for water-birds and fen plants. Invasive plant species such as *Solidago gigantea* were reduced in the restored wetland.

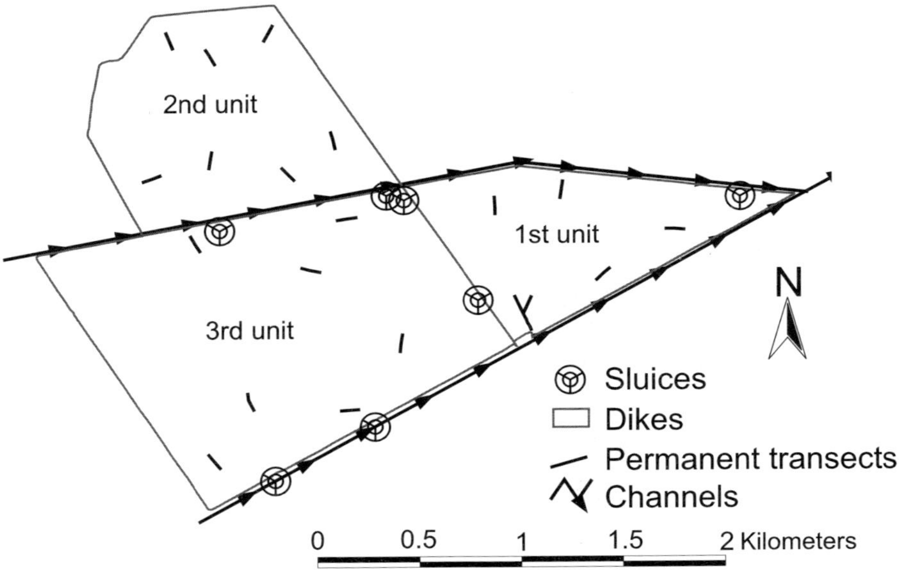

Fig. 11.8 Map of the restoration area, showing the location of permanent transects, in the Hanság fens of the Fertő–Hanság National Park, Hungary (Margóczi, unpublished data).

11.5.3 Monitoring and Evaluation of the Created Wetland

Monitoring of the restoration site was conducted in two ways, including field sampling and remote sensing. Field surveys of the vegetation were conducted by recording percent cover within 5 × 5 m phytosociological relevés along 21 permanent transects, 100 m long each, during 2001–2003. To create maps of vegetation change over time, fine resolution aerial orthophotos were compared every third year following the hydrologic restoration and a GIS database was developed. A rough vegetation map was made before the flooding began. The main vegetation types before flooding included fen meadows with large sedges (57 % *Carex riparia*, *C. acutiformis*), wet meadows (28 % *Alopecurus pratensis*, *Festuca arundinacea*), and reeds (5 % *Phragmites australis*, *Glyceria maxima*; Takács and Margóczi 2002). Important shifts in vegetation type were apparent during the first three years of flooding (Fig. 11.9). During the second year of flooding, upland vegetation disappeared; the area of the *Typha*, aquatic plants, and open water increased, while the area dominated by sedges decreased.

Natural wetland communities have redeveloped in the Ferto–Hanság National Park and are an attractive breeding and feeding area for birds. However, the hydrology is considerably different from the original fen, so the natural plant communities undoubtedly differ from those communities present

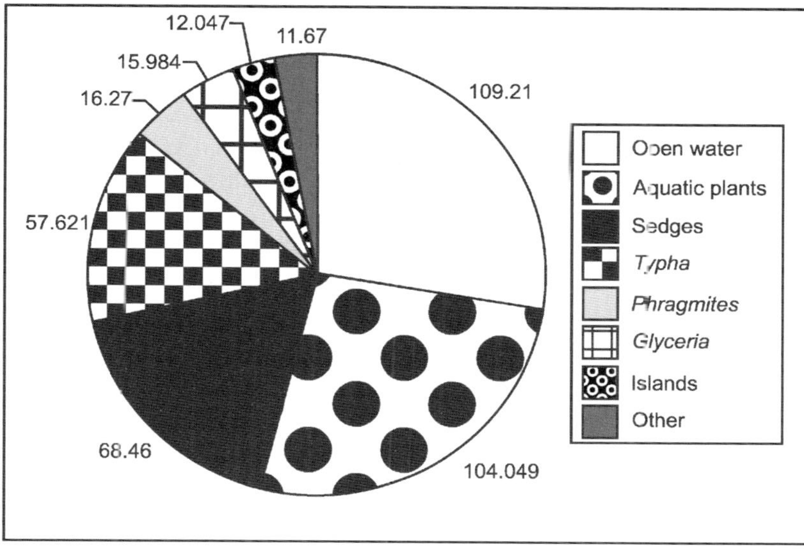

Fig. 11.9 Area (ha) of the main vegetation types in the flooded units (Units 1–3) in the Hanság fens, Hungary, in 2003. Area sizes were computed using raster statistics from the GIS database (Margóczi, unpublished data)

before the drainage of the ancient fen. While fen vegetation is not expected to re-develop, the Hanság National Park will continue to monitor the wetland restoration. Among the challenges that the managers of this national park face is that the restoration site is surrounded by agricultural lands, so that a considerable amount of water management and soil conservation in the regional landscape may be necessary to save the natural values of these wetlands.

11.6 Concluding Remarks

In times without much human influence, fen development was remarkably stable and vegetation types existed for centuries until they were replaced by meadows and alder wood. Nowadays fens cannot be maintained without management such as mowing or grazing. Nevertheless biochemical mechanisms exist that can retard a rapid succession to shrubs and forest. Such mechanisms are governed by subtle hydrological systems that keep the fen in a good condition.

A prerequisite for successful nature management is the creation of suitable habitat conditions (nutrient status, hydrology). There will be no adequate fen restoration without a proper assessment of the hydrological functioning of the fen system. It is very important that stagnation of surface water in the summer is prevented in restoration projects dealing with desiccated fens, in particular when the groundwater or surface water is rich in sulfate. The best option, which is not always available, is to stimulate the discharge of unpolluted (sulfate- and nitrate-poor) groundwater in such a way that this groundwater can flow through the system, leaving iron behind to contain the phosphates in the top soil.

Management measures are needed to counterbalance, for instance, the increased annual nitrogen input from atmospheric deposition. This may be accomplished by annual hay-harvesting, but it should be realized that hay-harvesting may induce a shift in the type of nutrient limitation, which also may affect biodiversity (Olde Venterink et al. 2003).

From studies on soil seed banks and seed dispersal, we may conclude that, until now, both seed persistence in the soil and dispersal distances of wet grassland species have been rather underestimated. Both types of seed sources can be manipulated by nature management, both in conservation and restoration projects. For successful restoration, both temporal (time since alteration of the habitat) and spatial (degree of fragmentation/isolation) aspects have to be taken into account.

We may expect that, in the near future, large areas on low-lying peat soils in large part of Europe and North America can no longer be maintained as intensely used agricultural areas. This offers good opportunities for the development of eutrophic marshes, as in our example of the Hungarian wetland restoration. The reflooding of former agricultural areas usually results in very eutrophic soil conditions (Richert et al. 2000), in particular when the surface water is rich in sulfates and the soil has been iron-depleted due to long-term drainage. Under such conditions persistent marsh species, such as *Glyceria maxima*, *Typha* spp, and *Phalaris arundinacea* can expand rapidly, but these species do not usually form peat, although they can rapidly form highly productive marsh vegetation, suitable to sustain large populations of waterfowl. To restore the peat-forming function of the mire, the vegetation should consist of marsh species with less easily degradable tissue (*Phragmites australis*, tall *Carex* spp; Richert et al. 2000). Restoring all of the fen qualities and functions will not be possible in many cases where restoring the hydrological system is not yet an option. Setting clear targets which can meet the opportunities presented, combined with a realistic strategy for future possibilities, may prevent much disappointment. A well explained modest result sometimes does more good for the public support than a scientifically sound project that lacks community or administrative support.

References

Adema EB, Grootjans AP, Petersen J, Grijpstra J (2002) Alternative stable states in a wet calcareous dune slack in The Netherlands. J Veg Sci 13:107–114

Andersson E, Nilsson C (2002) Temporal variation in the drift of plant litter and propagules in a small boreal river. Freshwater Biol 47:1674–1684

Bakker JP, Poschlod P, Strykstra RJ, Bekker RM, Thompson K (1996) Seed banks and seed dispersal: important topics in restoration ecology. Acta Bot Neer 45:461–490

Bedford BL, Godwin KS (2003) Fens of the United States: distribution, characteristics, and scientific connection versus legal isolation. Wetlands 23:608–629

Bekker RM, Verweij GL, Smith REN, Reine R, Bakker JP, Sneider S (1997) Soil seed banks in European grasslands: does land use affect regeneration perspectives? J Appl Ecol 34:1293–1310

Bekker RM, Verweij GL, Bakker JP, Fresco LFM (2000) Soil seed bank dynamics in hayfield succession. J Ecol 88:594–607

Bill HC, Poschlod P, Reich M, Plachter H (1999) Experiments and observations on seed dispersal by running water in an alpine floodplain. Bull Geobot Inst ETH 65:13–28

Bobbink R, Roelofs JGM (1995) Nitrogen critical loads for natural and semi-natural ecosystems: The empirical approach. Water Air Soil Pollut 85:2413–2418

Boedeltje G, Bakker JP, Bekker RM, Van Groenendael JM, Soesbergen M (2003) Plant dispersal in a low-land stream in relation to occurrence and three specific life-history traits of the species in the species pool. J Ecol 91:855–866

Brändel M (2004a) The role of temperature in the regulation of dormancy and germination of two related summer-annual mudflat species. Aquat Bot 79:15–32

Brändel M (2004b) Dormancy and germination of heteromorphic achenes of *Bidens frondosa*. Flora 199:228–233

Bullock JM, Moy IL, Pywell RF, Coulson SJ, Nolan AM, Caswell H (2002) Plant dispersal and colonization processes at local and landscape scales. In: Bullock JM, Kenward RE, Hails RS (eds) Dispersal ecology. Blackwell Science, London, pp 297–302

Ellenberg H (1986) Vegetation Mitteleuropas mit den Alpen. Ulmer, Stuttgart

Ellstrand NC, Elam DR (1993) Population genetic consequences of small population size: implication for plant conservation. Annu Rev Ecol Syst 234:217–242

Erisman JW, Draaijers GPJ (1995) Atmospheric deposition in relation to acidification and eutrophication. Elsevier, Amsterdam

Fischer M, Stöcklin J (1997) Local extinctions of plants in remnants of extensively used calcareous grasslands 1950–1985. Conserv Biol 11:727–737

Fischer M, Stöcklin J (1999) Plants with longer-lived seeds have lower local extinction rates in grassland remnants. Oecologia 120:539–543

Galatowitsch SM, Valk AG van der (1996) The vegetation of restored and natural prairie wetlands. Ecol Appl 6:102–112

Galatowitsch SM, Anderson NO, Ascher PD (1999) Invasiveness in wetland plants in temperate North America. Wetlands 14:733–755

García A (1992) Conserving the species-rich meadows of Europe. Agric Ecosyst Environ 40:219–232

Grime JP (1979) Plant strategies and vegetation processes. Wiley, Chichester

Grootjans AP, Diggelen R van (1995) Assessing the restoration prospects of degraded fens. In: Wheeler BD, Shaw SC, Foyt WJ, Robertson RA (eds) Restoration of temperate wetlands. Wiley, Chichester, pp 73–90

Grootjans AP, Van Wirdum G, Kemmers RF, Van Diggelen R (1996) Ecohydrology in The Netherlands; principles of an application driven interdiscipline. Acta Bot Neerl 45:491–516

Grootjans AP, Bakker JP, Jansen AJM, Kemmers RM (2002) Restoration of brook valley meadows in The Netherlands. Hydrobiologia 478:149–170
Grootjans AP, Alserda A, Bekker RM, Janáková M, Kemmers RF, Madaras M, Stanova V, Ripka J, Van Delft B, Wolejko L (2005) Calcareous spring mires in Slovakia; jewels in the crown of the mire kingdom. In: Steiner G (ed) Moore von Sibirien bis Feuerland. Biologiezentrum, Linz–Dornach, pp 97–116
Jansen AJM, Grootjans AP, Jalink M (2000) Hydrology of Dutch *Cirsio–Molinietum* meadows: prospects for restoration. Appl Veg Sci 3:51–64
Jensen K (1998) Species composition of soil seed bank and seed rain of abandoned wet meadows and their relation to aboveground vegetation. Flora 139:345–359
Jensen K (2004a) Dormancy patterns, germination ecology, and seed-bank type of twenty temperate fen grassland species. Wetlands 24:152–166
Jensen K (2004b) Longevity of soil seed banks of fen species of Northwest Germany; a review (in German). Ber Reinhold Tuexen Ges 16:17–28
Joosten H, Clarke D (2002) Wise use of peatlands. International Mire Conservation Group/International Peat Society, Jyväskylä
Koerselman W, Bakker SA, Blom M (1990) Nitrogen, phosphorus and potassium budgets for two small fens surrounded by heavily fertilized pastures. J Ecol 78:428–442
Komer SC (1994) Geochemistry and hydrology of a calcareous fen within the Savage Fen wetland complex, Minnesota, USA. Geochem Cosmochem Acta 58:3353–3367
Kotowski W (2002) Fen communities; ecological mechanisms and conservation strategies. PhD thesis, University of Groningen, Groningen
Lájer K (1998) Introduction to the vegetation ecology of the Hungarian fens (in Hungarian). Tilia 6:84–238
Lamers LPM, Tomassen HBM, Roelofs JGM (1998) Sulphate-induced eutrophication and phytotoxicity in freshwater wetlands. Environ Sci Technol 32:199–205
Lamers LPM, Smolders AJP, Roelofs JGM (2002) The restoration of fens in the Netherlands. Hydrobiologia 478:107–130
Loos S, Schipper A (2003) Hydrological and hydrochemical system analysis of Ob valley mire (Western Siberia) (in Dutch with English summary). Landschap 20:33–37
Lucassen ECHET, Crommenacker J van de, Roelofs JGM (2004a) Effects of stagnating sulphate-rich water on the mobility of phosphorus in freshwater wetlands: a field experiment. Arch Hydrobiol 160:117–131
Lucassen ECHET, Smolders AJP, Van der Salm AL, Roelofs JGM (2004b) High groundwater nitrate concentrations inhibit eutrophication of sulphate-rich freshwater wetlands. Biochemistry 67:249–267
Middleton BA (1999) Wetland restoration, flood pulsing and disturbance dynamics. Wiley, New York
Middleton BA (2000) Hydrochory, seed banks, and regeneration dynamics along the landscape boundaries of a forestered wetland. Plant Ecol 146:169–184
Middleton BA (2002a) Nonequilibrium dynamics of sedge meadows grazed by cattle in southern Wisconsin. Plant Ecol 161:89–110
Middleton BA (2002b) Winter burning and the reduction of *Cornus sericea* in sedge meadows in southern Wisconsin. Restor Ecol 10:1–8
Mitsch WJ, Dorge CL, Wiemhoff JR (1979) Ecosystem dynamics and phosphorus budget of an alluvial cypress swamp in southern Illinois. Ecology 60:1116–1124
Muller S, Dutoit T, Alard D, Grevilliot F (1998) Restoration and rehabilitation of species-rich grassland ecosystems in France: a review. Restor Ecol 6:94–101
Munro DA (1988) Wetlands conservation – world wide. In: Zelazny J, Feierabend JS (eds) Increasing our wetland resources. National Wildlife Federation, Washington, D.C., pp 20–23

Olde Venterink H, Vliet RE van der, Wassen MJ (2001) Nutrient limitation along a productivity gradient in wet meadows. Plant Soil 234:171–179

Olde Venterink H, Pieterse NM, Belgers JDM, Wassen MJ, Ruiter PC de (2002) N, P and K budgets along nutrient availability and productivity gradients in wetlands. Ecol Appl 12:1010–1026

Olde Venterink H, Wassen MJ, Verkroost AWM, Ruiter PC de (2003) Species richness-productivity patterns differ between N-, P- and K-limited wetlands. Ecology 84:2191–2199

P'Yavchenko NI (1976) Interrelations of forest and bog. Amerind Publishing, New Delhi

Ranwell DS (1959) Newborough Warren, Anglesay, I. The dune system and dune slack habitat. J Ecol 47:571–601

Richert M, Dietrich O, Koppisch D, Roth S (2000) The influence of rewetting on vegetation development and decomposition in a degraded fen. Restor Ecol 8:185–196

Schot P, Dekker S, Poot A (2004) The dynamic form of rainwater lenses in drained fens. J Hydrol 293:74–84

Schrautzer J, Asshoff M, Müller F (1996) Restoration strategies for wet grasslands in northern Germany. Ecol Eng 7:255–278

Schult M (2002) The development of percolation mires; an example the Trebel valley mire near Kirch-Baggendorf in northern Germany (in German). MSc thesis, Botanical Institute, University of Greifswald, Greifswald

Schütz W (1997) Primary dormancy and annual dormancy cycles in seeds of six temperate wetland sedges. Aquat Bot 59:75–85

Schütz W (1998) Seed dormancy cycles and germination phenologies in sedges (*Carex*) from various habitats. Wetlands 18:288–297

Schütz W (1999) Germination responses of temperate *Carex*-species to diurnally fluctuating temperatures – a comparative study. Flora 194:21–32

Schwaar J (1980) Sind die hygro- und xerokliner Phasen der Hochmoorbildung und bestimmte Phasen de Niedermoorbildung synchrone Vorgange gleicher Ursache? In: Wilmanns O, Tüxen R (eds) Epharmonie. Ber Int Symp Int Ver Vegetationsk 1980:95–116

Smolders AJP, Nijboer RC, Roelofs JGM (1995) Prevention of sulphide accumulation and phosphate mobilization by the addition of iron (II) chloride to a reduced sediment: an enclosure experiment. Freshwater Biol 34:559–568

Soons MB, Nathan R, Katul GG (2004) Human effects on long-distance wind disperal and colonization by grassland plants. Ecology 85:3069-3079

Succow M (1988) Landschaftökologische moorkunde (Landscape ecological mire research). Fischer, Jena

Succow M, Joosten H (eds) (2001) Landschafsökologische moorkunde, 2nd edn. Schweizerbart'sche Verlagsbuchhandlung, Stuttgart

Szekendi F (1938) History of the drainage attempts of the Hanság and Fertő (in Hungarian). Spec Diss Fac Philos Reg Hung Univ Elisabethinae Quinqeecclesiensis 126

Takács G, Margóczi K (2002) Biodiversity monitoring of the habitat reconstruction in South-Hansag (in Hungarian). Ferto-Hanság National Park, Sarrod

Tallowin JRB, Smith REN (2001) Restoration of a *Cirsio-Molinietum* fen meadow on an agricultural improved pasture. Restor Ecol 9:167–178

Taylor PD, Fahrig L, Henein G, Merriam G (1993) Connectivity is a vital element of landscape structure. Oikos 68:571–573

Thompson K, Bakker JP, Bekker RM (1997) The soil seed banks of north west Europe: methodology, density and longevity. Cambridge University, Cambridge

Van Andel J, Arondsen J (2005) Restoration ecology; the new frontier. Blackwell, Oxford

Van Breemen N, Buurman P (2002) Soil formation, 2nd edn. Kluwer, Dordrecht

Van Diggelen R (1998) Moving gradients: assessing restoration prospects of degraded brookvalleys. PhD thesis, University of Groningen, Groningen

Van Duren IC, Pegtel DM (2000) Nutrient limitations in wet, drained and rewetted fen meadows: evaluation of methods and results. Plant Soil 220:35–47

Van Duren IC, Van Andel J (1997) Nutrient deficiency in disturbed, drained and rewetted peat soils tested with *Holcus lanatus*. Acta Bot Neerl 46:377–386

Van Duren L, Bakker JP, Fresco LFM (1981) From intensively agricultural practices to hay-making without fertilization. Vegetatio 47:241–258

Verhoeven JTA, Koerselman W, Meuleman AFM (1996) Nitrogen- or phosphorus-limited growth in herbaceous, wet vegetation: relations with atmospheric inputs and management regimes. Trends Evol Ecol 11:494–497

Vogt K, Rasran L, Jensen K (2004) Water-borne seed transport and seed deposition during flooding in a small river-valley in Northern Germany. Flora 1999:377–388

Wassen MJ, Joosten JHJ (1996) In search of a hydrological explanation for vegetation changes along a fen gradient in the Biebrza Upper Basin (Poland). Vegetatio 124:191–209

Wassen MJ, Diggelen R van, Verhoeven JTA, Wolejko L (1996) A comparison of fens in natural and artificial landscapes. Vegetatio 126:5–26

Wenger EL, Zinke A, Gutzweiler K-A (1990) Present situation of the European floodplain forests. For Ecol Manage 33/34:5–12

Werner KJ, Zedler JB (2002) How sedge meadow soils, microtopography, and vegetation respond to sedimentation. Wetlands 22:451–466

Wheeler BD, Shaw SC, Fojt W, Robertson RA (1995) Restoration of temperate wetlands. Wiley, Chichester

Wheeler BD, Money RP, Shaw SC (2002) Freshwater wetlands. In: Perrow MR, Davy AJ (eds) Handbook of ecological restoration, volume 2: restoration in practice. Cambridge University, Cambridge, pp 325–354

Wolejko L, Aggenbach C, Diggelen R van, Grootjans AP (1994) Vegetation and hydrology in a spring mire complex in western Pomerania, Poland. Proc R Acad Sci 97:219–245

Woo I, Zedler JB (2002) Can nutrients alone shift a sedge meadow towards dominance by the invasive *Typha* x *glauca*? Wetlands 22:509–521

12 Social Learning in Wetland Development

Erik Van Slobbe, E.Dick Morris, Niels Röling, Reinder Torenbeek, Koen Broker, and Hero Heering

12.1 Introduction

An important issue in wetland planning and design is the most appropriate relationship between expert knowledge and public participation. How is it possible to organise a meaningful input from the public and what does this mean for the role of specialists? The insight that stakeholders, for instance local farmers, must be involved in the development of wetlands is widely accepted. But the extent of this involvement appears to be a matter for discussion. Some argue that only sound research can be the basis for the analysis of the actual situation and for setting objectives in planning and design. They claim that the most important intention of wetland development is to improve their ecological value. A consequence of this paradigm is that the involvement of stakeholders would be limited to including them in a 'value-free' planning process that is to lead to achieving a given objective. At most, stakeholders might be consulted in order to check whether data are well interpreted and processed; and they may eventually be given a role in wetland management.

An alternative to this specialist-dominated planning approach is social learning. In rural resource management, social learning "reflects the idea that the shared learning of interdependent stakeholders is a key mechanism for arriving at more desirable futures" (Leeuwis and Pyburn 2002, p. 11). Röling puts it as follows: "Social learning is about the interactive way of getting things done in theatres with actors who are interdependent with respect to some contested natural resource or ecological service. The interactive way of getting things done is based on conflict resolution, negotiated agreement, shared learning, convergence of goals, theories, and systems of monitoring, and concerted action. It stands in sharp contrast to instrumental use of technologies to control nature for assumed human purposes. It also stands in sharp contrast to economics which ascribes people with reasons (in assuming

Ecological Studies, Vol. 191
R. Bobbink, B. Beltman, J.T.A. Verhoeven, and D.F. Whigham (Eds.)
Wetlands: Functioning, Biodiversity Conservation, and Restoration

that humans are hard wired to optimise utility). Given that our predicaments are increasingly anthropogenic, technology and market increasingly fail to solve our problems, social learning is a hope giving approach for designing the future" (Leeuwis and Pyburn 2002, p. 11–12).

This chapter develops the concept of social learning. It first proposes that wetland development is not a simple undertaking in which linear planning paradigms are likely to be effective. Second, it presents different perspectives on the role of expert knowledge in planning. After this discussion of existing planning paradigms, the concept of social learning is developed as an alternative. Care must be taken however not to replace one paradigm with another equally one-sided paradigm. Therefore, social learning is positioned as one of four co-ordination mechanisms that often occur in a mix. After this positioning, the concept of social learning is further developed in terms of elements of changing behaviour and the changing understanding of stakeholders in wetland development. We conclude by attempting to answer the initial question about the relation between expert knowledge and public participation in wetland development.

12.2 Socio-Technical Characteristics of Wetlands

Understanding wetlands requires us to integrate different disciplines, such as hydrology, ecology and biology, with knowledge about social relations among inhabitants or stakeholders and knowledge about the management and development of institutions. In addition, knowledge based on local experience makes a crucial contribution. Changes in understanding have an impact on the wetland, but the converse is also true: changes in the wetland can influence perceptions. In order to illustrate this complex inter-connectedness, a case study from The Netherlands (the Wieden and Weerribben wetland complex) is used as an example to highlight some key socio-technical characteristics of wetlands (Torenbeek and Riegman 2004).

The Wieden and Weerribben constitute a 2000-ha highly modified wetland complex situated in the central Netherlands. This network of canals, lakes and emergent wetlands is part of a huge peat bog that has been managed for centuries. In previous centuries, peat was exploited for energy, resulting in peat removal and the formation of many water bodies. The general hydrological characteristics of the system are shown in Fig. 12.1, but this does not do justice to the actual complexity of the network of canals and lakes. Polders drain their excess waters into this network. River water draining from elevated sandy plateaus flows into the system too, bringing urban waste with it. The canal and lake system itself is drained through pumping stations into the large central IJssel Lake. The water level of Wieden and Weerribben is carefully managed so that water levels vary within a range of 1 cm. Before the

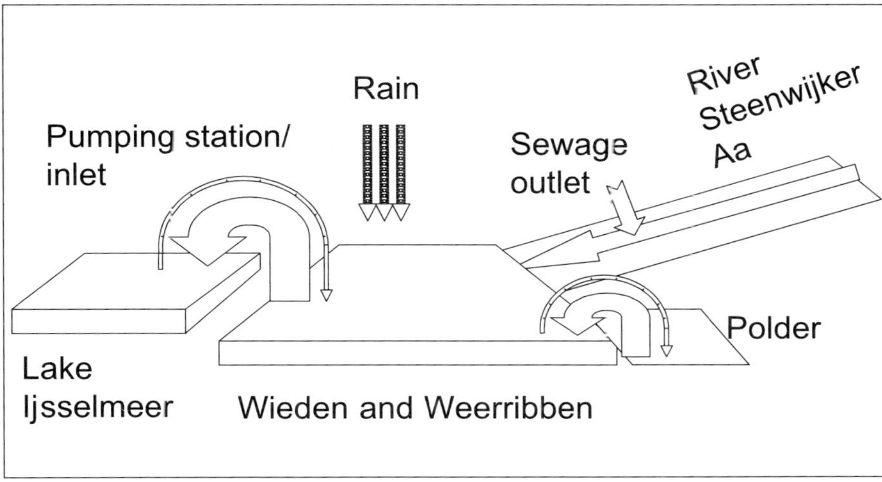

Fig. 12.1 Scheme of the Wieden and Weerribben water system

1990s, water levels were maintained to suit the needs of agriculture. In winter, water levels were kept low in order to create a basin to capture drainage water from the agricultural polders (Fig. 12.2). In summer, levels were raised to allow water to flow back into the polders for irrigation. But not only agriculture was affected by the water level. In the lakes, a small but substantial amount of reed (*Phragmites australis*) was harvested for thatch, an industry that provided employment to local farmers. Also, recreation is an important source of income; and the local villages and houses built on wetland soils depend for their foundation on stable groundwater tables. In the1990s, the area was declared a National Landscape Park, because of its high value for wildlife. So, in addition to the already existing purposes, water is now also being managed to suit the needs of nature. The second diagram (Fig. 12.2) shows the development of the management of water levels across the seasons from 1996 (when summer and winter levels were optimal for agriculture) to a possible future optimum for nature. In future, the management of the water levels is likely to be left to the natural dynamics of rainfall and drought. This in turn is likely to lead to substantial decreases of summer water levels of up to 25 cm.

This proposed management of water levels has led to controversy among stakeholders. Farmers resist the changes, but other interests also claim disadvantages. House-owners fear fluctuations in groundwater, recreation needs high water tables in summer and reed growers need low water in winter. The water board, which is responsible for managing the water levels, is cautiously experimenting with new water levels (the 'now' line in Fig. 12.2). There are many reasons for this cautious approach. In the first place, since 1992, the

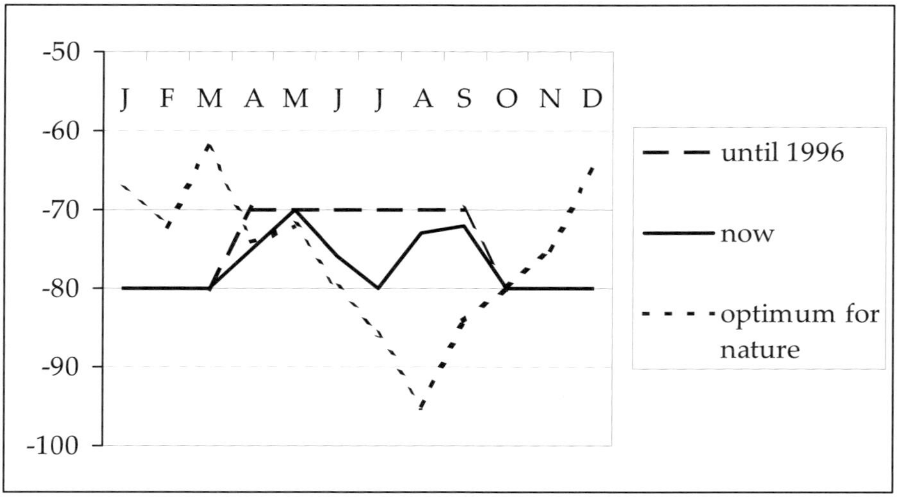

Fig. 12.2 Water levels in the Wieden and Weerribben before 1996, nowadays and levels considered best for nature (expressed in centimetres below the 'Nieuw Amsterdams Peil', the standard used throughout the Netherlands)

water board has been a democratic organisation, in which all interests are represented. Before that year, water boards used to be dominated by farmers' interest. The designation of a National Park arose from national policy and did not meet with much local sympathy. Thus, it exposed the water board to two forces: national policy and both regional and local opposition. As nature conservation interests slowly gain power, opinions and perspectives in the water board are also slowly changing. Another element is that the impact of changing water levels on hydro-ecological processes was not known. Much research has since been done, leading to more insight but also to more discussion among ecologists. Some advocate the restoration of (average low) water levels to improve conditions for natural peat formation. Others resist this because they see more value in (higher) water levels that favour particular plant species.

The decision-making process on water levels is thus influenced by many forces. At the time this manuscript was completed, no final decision had been made. The water board does not know how to proceed and continues to carry out more research and to produce more plans. The example from the Wieden and Weerribben also draws attention to the three characteristics presented below, which seem common for wetlands the world over (SLIM 2004a).

- *Interdependence and competing claims*: The use of natural resources through one type of human activity affects ecological processes in ways that interact with other peoples' uses of natural resources, both across geographic and ecosystem boundaries and timescales. Different stakeholders

in the Wieden and Weerribben (i.e., farmers, nature conservation organisations, reed producers) and the tourist industry all need different water levels; and the goals of water management are contested because what suits one is seen as a disadvantage for the other.
- *Complexity*: Natural resources are under the influence of a complex of enmeshed natural, technical and social processes, including changes in public policy, organisations and a diversity of stakeholders, each with their own perceptions. The Wieden and Weerribben case shows feedback loops that make the system complex. The water board managing the waters in Wieden and Weerribben is changing from a 'farmer-dominated' board to an organisation which considers multiple interests. Representatives of farmers, nature organisations, house owners, tourism and industry are all present in the board. But, at the same time, it is under pressure from the National Landscape policy and ecologists do not agree on the desired management of water levels. The result is a complex management and decision-taking system.
- *Uncertainty*: Despite extensive research, stakeholders do not know yet how changing water levels will affect their interests. The influence of new phenomena (such as climate change) or new regulations (such as the EU Water Framework Directive) is also unknown. For example, as uncertain weather events lead to unpredictable water levels downstream, water level management goals might shift from nature conservation to water retention and flood prevention.

The question is how planners and designers of wetlands can effectively deal with these characteristics? In order to answer this question, the following section will examine the role of (expert) knowledge in the management of wetlands.

12.3 Different Perspectives on Planning

A useful typology of paradigms of ecosystem management is a classification based on two dimensions: (1) an ontological dimension with reductionism and holism as poles and (2) an epistemological dimension with subjectivism (or constructivism) and objectivism (or positivism) as poles. This typology is based on Miller's (1983) observations on the behaviour of his colleagues in battling the Spruce Budworm outbreak in New Brunswick in Canada; and it was later adapted by Bawden (1997), Röling (2000) and SLIM (Morris et al. 2003).

The right-hand side of Fig. 12.3 represents an objectivist perspective. It assumes that the world can be known and understood, on condition that the right data and analytical tools are available. Planning based on this paradigm

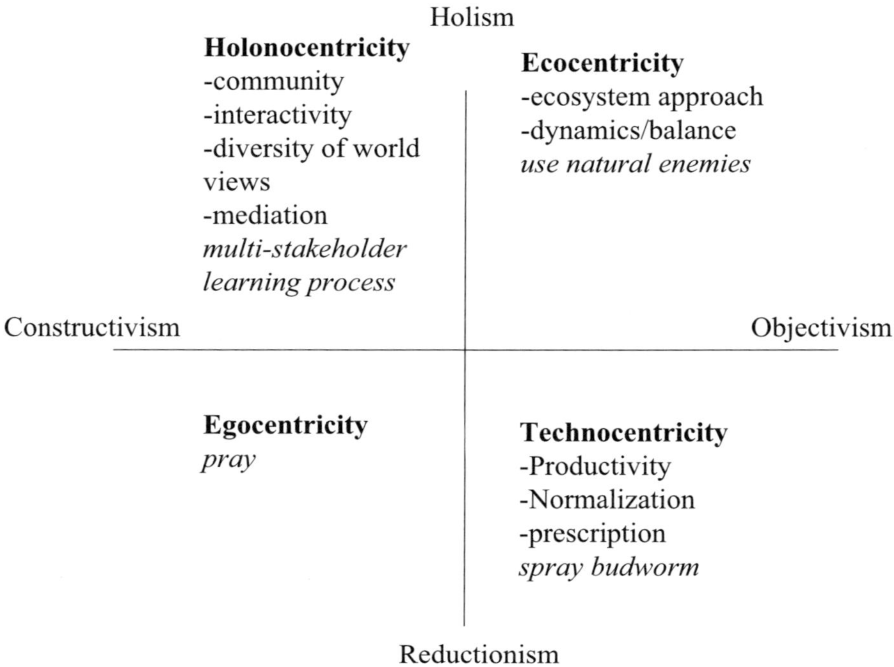

Fig. 12.3 Four paradigms for natural resource management. *Italics* indicate the behaviours of experts observed by Miller (1983, 1985) during a spruce budworm outbreak in Canada. Based on Bawden (1997), Röling (2000) and SLIM (Morris et al. 2003)

assumes value-free and objective analysis of situations. Because of its 'value-free' character, experts can assess actual situations of wetlands and formulate visions of desired futures. The challenge is to use the right data and analytical tools. Goals can be assumed given and the challenge is goal achievement. Public participation is seen as a source of information to be used as one of many inputs. Complex situations can be managed by reducing them to separate elements and by analysing these one by one. Within the *techno-centric quadrant*, design objectives for wetlands are assumed given and expressed in terms of productivity, or the occurrence of specific species in the wetlands. Scientists in the *eco-centric quadrant* consider ecosystems more in terms of sets of interrelated processes. Patterns and characteristics of these processes are studied; and natural processes that can be modelled by scientists determine the outcomes. An example is the new European Union Water Framework (European Union 2000). The Water Framework Directive assumes that there are 'natural ecosystems' that have some objective validity – in their own right – independent of human valuations of them. Environmental objectives are derived from so-called 'reference conditions' that are derived from what are assumed to be undisturbed locations. These 'value-free' references are then processed into ecological objectives.

The constructivist left-hand side of the diagram assumes that the planning and design of wetlands is the result of a concerted effort of stakeholders with different perspectives and competing claims on the wetland. Goals are not given but contested. The construction of objectives is not based on an objective reality, but is the result of negotiation and the discussing and sharing of different views. Views usually change during the planning and design process because people learn and develop their insights. In the left-hand side of the diagram, the position of expert knowledge changes radically compared to the one in the right-hand side. Experts are no longer considered as the centre where all information comes together and where it is processed in a 'value-free' way, but as stakeholders in a negotiation process. The platform on which the analysis, planning, design and construction of objectives takes place is now somewhere in between many groups and parties, each with a different perspective on what is necessary. Research suggests that the management of wetlands, typically characterised by interdependence and competing claims, complexity and uncertainty, requires a *holocentric* approach to planning and design. In order to better understand such an approach, we position it as one of several governing principles or co-ordination mechanisms used by policy makers and planners.

12.4 Social Learning as Part of a Mix of Governance Approaches

Governance serves to co-ordinate individual and collective behaviour. The commonly recognised co-ordination mechanisms are hierarchy, market and information. Each is associated with characteristic policy tools that make it operational (Powell 1991; SLIM 2004b; Table 12.1). Conventional development processes of wetlands draw on the first three governing principles: hierarchy, market and information. Attempts to resolve resource dilemmas through a combination of these tools have met some problems:

- They draw heavily on expert knowledge, but expert knowledge cannot resolve conflicts, deal with competing claims or create concerted action.
- The desire to resolve a resource dilemma must be shared by the stakeholders.
- Regulation brings the problems and costs of non-compliance; markets fail to redress the externalisation of costs; and information campaigns serve mainly to raise awareness.
- Sectoral mandates and approaches do not generate integrated outcomes.

Table 12.1 Governance principles (Powell 1991; SLIM 2004b)

Hierarchy	Market	Information	Social learning
Regulation	Rational choice	Public relations	Create spaces for learning
Technical measures	Fiscal policy	Education campaigns	Facilitate process
Laws	Liberalisation	Raising awareness	Multi-stakeholder process

Research (SLIM 2004a, 2004b) demonstrates the efficacy in the water sector of a fourth governance mechanism, based on interactive learning processes among area-based stakeholders, and the co-creation of knowledge and objectives brought about by joint experimentation and facilitated interaction. Transformation over time in behaviours and relationships leads to new kinds of concerted action. The policy tools that characterise this approach include: participatory processes, facilitation of concerted action, collective experimentation and investment in social spaces for learning (SLIM 2004b).

Care has to be taken however that the fourth principle is not seen as a panacea to all kind of processes. Social learning adds to existing principles, but does not replace them. Finding the right mix between the four principles is the challenge that developers of wetlands have to meet. The right mix depends on the local situation, as experience from Sri Lanka shows (Broker and Bentinck 2004).

Analyses of different wetland-management projects in Sri Lanka reveal that there is too strong a reliance on bottom-up processes. A common claim is that, when communities are trained and provided with alternative livelihood opportunities, environmental degradation decreases. In practice, however, this is not the case. Experience shows that a bottom-up approach can lead to a general lack of enforcement of environmental laws and regulations that had earlier been imposed in top-down fashion. This lack of enforcement results from insufficient will and/or capacity of responsible agencies and police, inefficiency of the legal system and loopholes in the legal framework. Political patronage of offenders is also a disincentive. The inter-agency National Wetland Steering Committee, established to overcome these problems, has been useful for policy formulation and co-ordination, but it has not been sufficient to effectively protect wetlands at the community level.

An example is provided by local polluting industries. Preliminary success in reducing industrial pollution can be achieved by facilitating dialogue among local communities, governmental agencies and industrialists and by providing assistance in implementing cleaner production. This concept was introduced during the Integrated Resources Management Programme for the catchment of the Negombo Lagoon. Industries were approached and asked to co-operate to apply cleaner production. Although most industries were willing to co-operate, there was no possibility to take legal action against those that refused to co-operate or did not want to implement action to produce in a cleaner way. Clearly, such free riders do not motivate industries that are investing in cleaner production, thus the "The Tragedy of the Commons" makes itself felt.

In these conditions, the threat of legal action can be a stimulus for non-compliant stakeholders to respect environmental standards. A return to top-down approaches is not advocated, however, because stakeholder consultation and participatory planning and implementation are essential to do meaningful work with communities. It is nevertheless essential that government-funded projects do not solely focus on local learning processes but also stimulate the relevant authorities to enforce environmental laws.

12.5 Social Learning in Wetland Development

Participatory practices, such as consulting civil society organisations, citizens and end-users, reflect a desire to go beyond simple objectivist approaches. Knowledge from non-expert participants can and often does contribute to different approaches, new relations, richer debates and eventually better plans and designs. Within this mindset, expert knowledge constitutes but one source of pertinent input into the process of defining the issue and building successful outcomes. Local and changing forms of knowledge, concerns and constraints also feed into the equation. The process entails changes in practices and behaviours as well as changes in perceptions and understanding. The development of wetlands regulation and management is thus a flexible process, which is adaptive to changing patterns of understanding and behaviour.

Interactive learning requires explicit investment of time to allow interaction, creation of social spaces where people can interact in an organised way and design of experiments around material objects from which all can learn. The experience is that such investment costs are highest at the beginning of the process, while the benefits come later. With a more coercive approach, the costs are low at the beginning but mount later as court cases, bad feelings, lack of trust in government and open conflict lead to unworkable solutions (SLIM 2004b).

An example of an adaptive process during which different stakeholders initiated a process to cope with over-use of groundwater and falling groundwater tables is the Benelux middle area (Jiggins 2003, 2005).

The Benelux is a partnership between the governments of Belgium, The Netherlands and Luxemburg. The Benelux middle area includes the southern provinces of the Netherlands and the northern provinces of Belgium. The middle of the Benelux harbours an extensive area of sloping, free-draining and nutrient-poor sandy soils. It is the centre of a very intensive and profitable agriculture, consisting of pig production, asparagus growing, intensive dairying, glasshouse production and broad-acre horticulture. The withdrawal of water for overhead sprinkler irrigation and infiltration of nutrients threatens groundwater reservoirs. Nature conservation areas are desiccating and suffering from eutrophication. In order to stop the over-use of ground water, the provincial Government in North Brabant in 1992 proclaimed severe restrictions on sprinkler irrigation. This led to massive farmer protests. Leaders of the regional farmers union realised that farmers' interests would not be served by continuing obstruction and protest and they accepted that agriculture was contributing to the environmental problems.

Together with provincial administrations and water boards, they initiated a process to improve the groundwater situation. The first phase of the project eventually involved 3000 farmers and glasshouse owners as active participants in 30 experimental areas; and maybe three times as many adopted one or more of the available measures. The broad objective was to strengthen the role of the agrarian sector in water conservation. Farmers and market gardeners were given a central role in implementation. In the Netherlands context, this represented a break with the historical trend of increasingly handing over water management to professionals and specialist water management agencies. The project espoused a facilitated social learning process as its main operating principle in implementation, at both the farm level and at the level of the management of the water system as a whole. The project had two, complementary, pillars:

- Water management at enterprise level, encompassing a range of measures designed to improve the efficiency of water use in farming operations. This included experimentation with different varieties and types of crops, alternative cultivation methods, reduced fertilisation, mineral budgeting, simulation modelling to improve strategic decision-making, advisory programmes and metered use of overhead sprinkler irrigation.
- Water conservation, encompassing physical measures designed to call a halt to the mid-season soil water deficit, principally by holding water longer in field ditches by means of weirs or by means such as raising the bed of the ditch and by managing seepage in wind blown sand dunes in order to improve the infiltration of water to nature areas. Monitoring by farmers of water levels by means of water gauges also formed part of these activities.

The ultimate proof of the success or failure of the project is in the practice. The internalised experience reported by the project partners and the wider set of stakeholders who were interviewed suggest that the project, in a very short time, brought about significant changes in attitude, understanding and practice not only at farm level but also at system management and policy-making levels. An important outcome is the shared understanding that, for effective concerted action to become possible, a process approach needs to become embedded in the routine procedures and behaviours of a multitude of actors.

The example describes a social learning process initiated by a farmers' organisation. The driving force of this process was severe regulation of overhead sprinkler irrigation by provincial authorities. A social learning process, such as the one described in the example, can be seen as a system of five interconnected elements shown in Fig. 12.4 (SLIM 2004a).

1. Any given resource problem exists in a historically based social, cultural and institutional setting which frames the issue. This wide set of historical circumstances helps explain what is at stake in any issue and enables it and its causes to be understood more fully. The situation in the Benelux middle area project is influenced by a history of small and poor farmers struggling to develop poor sandy soils. These farmers, although now well-to-do, still clearly remember past hardships and much of their actions and their relations with provincial authorities are governed by this history.
2. Stake-holding expresses the idea that individuals or groups actively construct and promote their stakes in relation to those of others. Through social interaction, new stakes can emerge which to transform both the

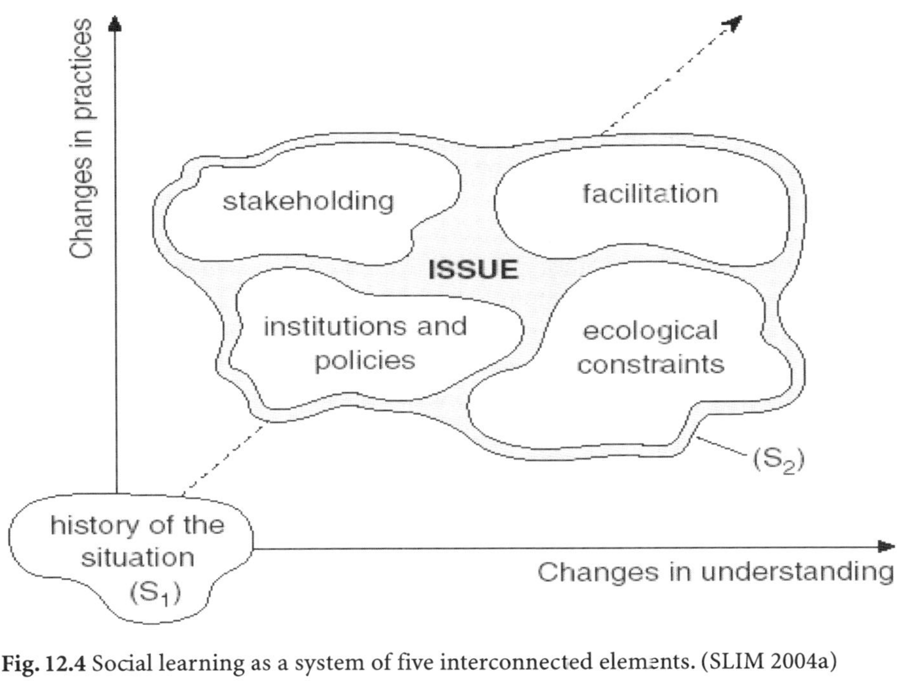

Fig. 12.4 Social learning as a system of five interconnected elements. (SLIM 2004a)

issue and the relationships among stakeholders. It is in the processes of debating, discussing, questioning and acting that stakes and stakeholders are defined, the roles and actions of stake holding emerge and the stakes of concern begin to be expressed in the context of wider communities and networks of other stakeholders. At first, only the groundwater situation in the Benelux middle area project was at stake. But through better relations, now other policy issues (such as nature conservation) are entering the arena.

3. Ecological constraints are a set of identifiable and quantifiable factors that are perceived to influence ecosystem functions. Scientific knowledge is often accorded primacy in defining ecological constraints that reflect the experience and understanding of researchers and experts. Just like the scientists or experts, the individual users also build their own understanding of the ecosystem and their roles in it, through their relationship with it within a given system of interest. Rather than a purely objective description, the term 'ecological constraints' can thus be used to define an observer's understanding of the relationship between people and their biophysical environment. In the Benelux case, professional water managers, through being forced to work with farmers to install weirs in field-level ditches, discovered to their amazement that farmers have very sound hydrological knowledge of their areas.
4. Facilitation in relation to natural resource management is understood as a combination of skills, activities and tools used by a facilitator to support and guide learning processes among multiple interdependent stakeholders. Its main role is to bring about systemic change in complex situations for achieving concerted action. Facilitation is about the process management of concerted action, at individual and collective scales. It helps stakeholders involved in deliberations to better understand 'what they are doing' and 'why they are doing what they do', in order to act more purposefully. In the Benelux middle area project, the farmers' organisation took up the facilitator's role. By taking this initiative, trust grew among the provincial, water board and municipal authorities and the farmers.
5. Institutions play a significant role in shaping the management of resources and in providing the norms and values that underpin and inform many policy decisions and management practices. The increasing complexity and diversity of environmental problems, in part, arises from the increased number of often-conflicting institutional factors which affect how resources are managed. This generates much uncertainty, particularly on how the different institutional elements are expected to fit together into a coherent whole. In the Benelux water conservation project, one of the interesting institutional factors that emerged was the difference between the Netherlands provinces of Brabant and Limburg in terms of their differential approach to hierarchy and power. In Brabant, it is somehow much easier to trust an egalitarian interactive process to generate outcomes at

the macro level while, in Limburg, a tradition of trust in hierarchy and deference to central power leads to search for ways in which stakeholder agreements can be made legally binding and be subjected to guarantees.

12.6 Conclusions

Wetlands constitute arenas in which interdependent stakeholders make competing claims on the natural resources but need each other to design concerted action that ensures sustainable management of the wetlands for all. Hydrology, ecology and management expertise provides only part of the understanding of a system that is complex and marked by a high degree of uncertainty. The Wieden and Weerribben case in this article shows that not only 'hard sciences' but also history and social and political relations among stakeholders create a relevant context for wetland development.

In such conditions, taking an objectivist expert position on wetland development has little impact. This observation does not diminish the fact that, in the current practice of wetland management, the role of expert knowledge in the formulation of plans and designs is still dominant. The idea persists that an objective 'value-free' plan, based on a more-or-less holistic conception of the situation and sound data analysis, will lead to effective implementation. The European Union Water Frame Work methodology is an example of this planning paradigm. It takes a more holistic view than previously was the case, but is still based on a purely positivist ontology (Morris et al 2003).

Research shows that the sustainable management of wetlands is the result of a process that involves several interconnected elements: (a) taking into account the history of a situation, (b) involving stakeholders, (c) paying explicit attention to institutions and (d) constructing of a common perspective on ecological constraints. Facilitation of action and debate among stakeholder groups and individuals with different perspectives is essential and replaces the imposition of blueprint plans. Of course, input from experts is still necessary, along with input from non-specialist knowledge. However, the role of experts is that of stakeholders and participants in a learning process. Experts must therefore be prepared to leave behind fixed value-free scientific positions and instead use their scientific knowledge in a flexible way to enrich the learning process. A good example of a development process presented in this article is the "Benelux Midden Gebied" project. In this project, explicit space was created for learning.

The Sri Lanka example shows that too much emphasis on participatory practices or learning and underestimation of regulatory instruments leads to inefficient management of wetlands. Social learning alone is not a simple alternative to existing planning and design practices. All co-ordination mechanisms, regulation, stimulation and information, as well as social learning are needed to form an effective governance mix for wetland management. Local circumstances dictate the specific configuration of the mix in each individual situation.

Acknowledgments This chapter reflects discussions in the session "Planning and design of wetland restoration in multi-stakeholder environments" during the Intecol conference on Wetlands in Utrecht (25–30 July 2004). Some case presentations are used in this chapter as examples. The theory on social learning presented in this chapter is based on results of a European Union-funded research program called SLIM (Social learning for the integrated management and sustainable use of water at catchment scale; http://slim. open.ac.uk).

References

Bawden R (1997) Learning to persist. In: Stowell FA, Ison RL, Armson R, Holloway J, Jackson S, McRobb S (eds) Systems for sustainability: people, organizations and environment. Plenum, New York

Broker KCA, Bentinck JV (2004) Management of wetlands in Sri Lanka. A strategy for combining community participation with effective protection. Arcadis, Arnhem (unpublished report)

European Union (2000) Directive 2000/60/EC of the European Parliament and of the Council of 23 October 2000 establishing a framework for community action in the field of water policy. European Commission, Brussels

Jiggins J (2003) Key informant study. (Final report, interreg project water management in the Central Benelux area) Wageningen University, Wageningen. Available at: www.waterconservering.nl

Leeuwis C, Pyburn R (2002) Wheelbarrows full of frogs. Social learning in rural resource management. Koninklijke van Gorcum, Assen

Miller A (1983) The influence of personal biases on environmental problem-solving. J Environ Manage 17:133–142

Miller A (1985) Technological thinking: its impact on environmental management. Environ Manage 9:179–190

Morris D, Roggero PP, Steyaert P, Van Slobbe E, Watson D (2003) Ecological constraints and management practices in social learning for integrated catchment management and sustainable use of water. Open University, Milton Keynes. Available at http://slim.open.ac.uk

Powell W (1991) Neither market nor hierarchy: network forms of organisation. In: Thompson G, Frances J, Levavcic R, Mitchell J (ed) Markets, hierarchies and networks. The co-ordination of social life. Sage, London

Röling N (2000) Gateway to the global garden: beta/gamma science for dealing with ecological rationality. (Eighth annual Hopper lecture, 24 October 2000) University of Guelph, Guelph

SLIM (2004a) Social learning as a policy approach for sustainable use of water. A field tested-framework for observing, reflecting and enabling slim (social learning for the integrated management and sustainable use of water at catchment scale). Open University, Milton Keynes. Available at http://slim.open.ac.uk

SLIM (2004b) The role of learning processes in integrated catchment management and the sustainable use of water. Open University, Milton Keynes. Available at http://slim.open.ac.uk

Torenbeek R, Riegman R (2004) Eutrophication control in the peat bog wetland "Wieden & Weerribben" (unpublished report). Arcadis, Apeldoorn

13 Eco-Hydrological Functioning of the Biebrza Wetlands: Lessons for the Conservation and Restoration of Deteriorated Wetlands

Martin J. Wassen, Tomasz Okruszko, Ignacy Kardel, Jarek Chormanski, Dorota Swiatek, Waldemar Mioduszewski, Wladimir Bleuten, Erik P. Querner, Mohssine El Kahloun, Okke Batelaan, and Patrick Meire

13.1 Introduction

Human activities have led to the loss of a large proportion of biodiversity in riverine wetlands in western Europe (Van Urk 1984; Cirujano et al. 1996). In the second half of the twentieth century, many floodplains, fens, and riparian woodlands were cultivated for agricultural purposes. In addition, the remaining riverine wetlands lost species due to the impact of human activities (Rich and Woodruff 1996; McCollin et al. 2000). Recently, policy has become more focused on conservation of the remaining wetlands and on rehabilitation of disturbed rivers and floodplains (Jongman 1998). The management and rehabilitation of wetlands is difficult without adequate knowledge of the hydrological and ecological processes responsible for the functioning and biodiversity of undisturbed wetlands.

The last decade has seen an increasing interest in using eco-hydrological knowledge in environmental management. Eco-hydrology studies the two-way linkage between hydrological processes and plant growth (Baird and Wilby 1999) and more specifically is a landscape ecological specialization aiming at a better understanding of hydrological factors determining the development of wet ecosystems. It includes the study of the origin, flow, and quality of groundwater and surface water and their ecological implications for wetlands (Wassen and Grootjans 1996) and takes into account the functional interrelations between hydrology and biota at the catchment scale (Zalewski 2000). Eco-hydrological knowledge is essential for successfully restoring wetlands. Research focusing on general relationships between hydrology and

Ecological Studies, Vol. 191
R. Bobbink, B. Beltman, J.T.A. Verhoeven, and D.F. Whigham (Eds.)
Wetlands: Functioning, Biodiversity Conservation, and Restoration

wetland ecosystems, however, is hampered by the fact that many of the ecosystems studied are influenced strongly by a combination of man-induced disturbances, e.g. hydrological changes, increased levels of atmospheric nitrogen deposition, and presence of contaminants. For instance, studies of quite undisturbed floodplains reveals that flood duration, water tables, and soil wetting are primary factors regulating riparian vegetation abundance (Cellot et al. 1984; Stromberg 1993a; Large et al. 1994), while in catchments heavily influenced by man these relationships are less clear, since pollution of the river water and physical changes such as regulation, normalization, and constructing embankments have a dominant effect on vegetation (Girel 1994; Trémolières et al. 1994).

Hence, there is a strong need for relatively undisturbed reference ecosystems. Such reference wetlands may help to determine target communities (cf. Bakker and Berendse 1999) and to define the environmental conditions necessary for establishment of these target communities in restoration projects. Relationships between ecological variables including productivity, species richness and abundance of threatened species and for instance depth of the groundwater table in a relatively undisturbed area can be used in a space-for-time substitution to predict effects of raising the groundwater table in a drained area (Stromberg et al. 1996; Toner and Keddy 1997; Stromberg 2001). The lowland Biebrza valley in northeastern Poland contains fairly undisturbed floodplain marshes and fens. The river has not been regulated nor embanked, the valley is characterized by a relatively low level of contamination, and hydrology is not dominated by man. Atmospheric deposition is less than 10 kg N ha^{-1} $year^{-1}$, which is low for European standards compared to the 30–80 kg N ha^{-1} $year^{-1}$ in western Europe (Holland et al. 1999). Thus, the Biebrza valley may offer a reference site for comparable floodplain and fen ecosystems elsewhere in Europe (Wassen et al. 2002).

In this chapter, we summarize eco-hydrological research results from the Biebrza wetlands. After introducing some general features of the Biebrza wetlands, we present studies on hydrology and plant ecology giving an overview of groundwater and surface water patterns, processes, and relations with plant communities. From these studies, we will be able to learn about the importance of water flow and water quality for species composition. The spatial patterns are different for different parts of the valley. Further coupling between productivity, nutrient limitation, and species composition of marsh and fen vegetation is reported. We discuss the observed patterns and processes in the context of other European river marginal fens and floodplains and we discuss how information from Biebrza can be used for the conservation and restoration of deteriorated wetlands.

With respect to the used methods, we refer to previously published papers as much as possible. Methods new to this paper are briefly described when appropriate.

13.2 General Characteristics of the Biebrza Valley

13.2.1 Introduction

Biebrza is an almost natural lowland river of intermediate size (156 km long, average annual discharge in the lower course ca. 30 $m^3 s^{-1}$), running through a valley of about 1000 km^2 in northeastern Poland (22°30'–23°60' E, 53°30'–53°75' N). The valley contains non-drained floodplains, marshes, and fens and is surrounded by a post-glacial landscape with ice-pushed hills, moraines, and outwash plains (Succow and Jeschke 1986; Okruszko 1990). The altitude of the valley ranges from ca. 95 m to 130 m above mean sea level; and the catchment area of ca. 7000 km^2 has a maximum altitude of ca. 220 m (Byczkowski and Kicinski 1984). The average precipitation is 583 mm, of which 244 mm falls in summer. The average annual temperature is 6.8 °C; and the growing season is ca. 200 days (Kossowska-Cezak 1984). The almost natural character of the Biebrza peatlands is reflected in a regular pattern of plant communities (Palczynski 1984). Parts of the valley are mown or are grazed by cattle and the rest is grazed and browsed by wildlife such as elk (*Alces alces*) and roe deer (*Capreolus capreolus*). The river is not regulated and it has many oxbow lakes, which are overgrowing. Extensive areas are flooded by the river in spring. Further away from the river, groundwater-fed rich fens cover large areas. On the moraines, agriculture is practised at a low intensity. The dunes in and around the valley are mainly covered by woods. Population pressure is low. Occasionally, ditches and canals are present, especially in the Middle Basin (Fig. 13.1).

13.2.2 Geomorphology, Lithology, and Geo-Hydrology

The Biebrza valley is surrounded on the east, south, and west by morainic plateaux of the last but one glaciation (Saalien). In the north, it contacts the Elk Lake District in the northwest and the Augustow Outwash Plain in the northeast (Fig. 13.1). Both were formed during the last glaciation (Weichselien). During the whole Pleistocene, the successive ice-sheet advances and retreats were associated with repeated erosive and accumulative processes. As a result, highly diversified surfaces of individual beds and great lithologic variability are present in the Pleistocene formations (Pajnowska et al. 1984). There are boulder clay strata of varying thickness and deep gravel–sand–silt series. The quaternary deposits range from ca. 130 m to 210 m in thickness.

Within the valley, the following lower-order morphological units can be distinguished: the Upper, the Middle, and the Lower Basin (Fig. 13.1).

The Upper Basin is ca. 34 km long and 6–7 km wide. The Upper Basin consists of a 5–10 m trough cut in the morainic upland. The valley is com-

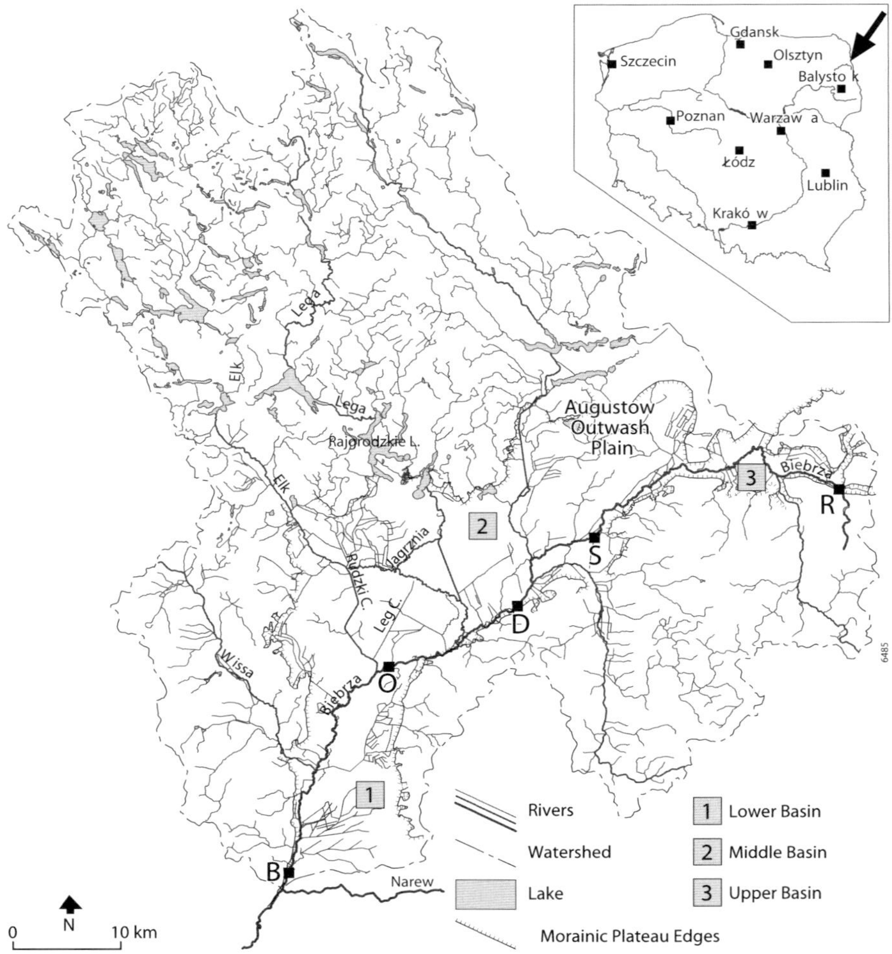

Fig. 13.1 The Biebrza catchment, location in Poland, and general topographic features. Characters refer to the gauge stations in Table 13.1 (*R* Rogozyn, *S* Sztabin, *D* Dolistowo, *O* Osowiec, *B* Burzyn)

pletely peat-covered, with peat deposits of 3–7 m thickness partly underlain by calcareous gyttja. Together, they overlie a sandy-gravel bed, to a depth of 10 m. In the north, the valley joins without any distinct elevation difference with the rather flat sandy plain of the Augustow outwash extending 40 × 12 km. In the south, the valley is delimited by erosive edges of the morainic plateau that rises gradually from 125–135 m to 160 m or even 200 m a.s.l. (Zurek 1984).

The Middle Basin is ca. 33 km long and up to 23 km wide. Peats are less thick than in the Upper Basin and form only 1–3 m beds. They are underlain by sands, with gravel series at the bottom of the northern part and silts and clays in the southern part. To the north and west, the valley passes into slightly higher (1–4 m) outwash levels that are in turn delimited by plateau edges. The valley contacts directly with the 10–14 m slope of the morainic plateau in the east and southeast only. The plateau rises to 140–150 m a.s.l. The Middle Basin is distinguished by extensive sandy tracts, transformed in many places by aeolian processes to dunes partly covered by peat (Zurek 1984).

The Lower Basin is 25 km long and up to 15 km wide. The river runs at the right-hand side of the valley. The valley floor slopes down to 104 m a.s.l. at the confluence with the Narew. The peat is usually not more than 2 m thick and is underlain by thick sandy gravel beds locally covered by loam and silt deposits. Small dunes form islands several metres high, submerged in the peat of the surrounding flat floodplain. Numerous oxbows are situated in a 1–2 km muddy zone along the river. The Lower Basin is separated from the downstream Wizna Basin by a fluvial cone stretching out along and deposited by the Narew River. This cone makes the passage from the Biebrza Lower Basin to the Wizna Basin very narrow (<1 km). To the west and east, the valley is bordered by morainic plateaux. Steep high slopes accompany the valley in the southwest, whereas the edges are long and gentle in the remaining parts (Zurek 1984).

Recurring glaciations produced a great lithologic variability, resulting in a diverse pattern of aquitards, semi-permeable layers and aquifers. The Biebrza valley is fed: (a) from below, by water under pressure coming from confined aquifers by diffuse upward groundwater flow or through hydrogeologic windows, and (b) from the sides, by subsurface phreatic groundwater flows or by groundwater flow coming from semi-confined aquifers, both directed from the watershed towards the valley (Pajnowska et al. 1984).

13.2.3 Vegetation

The valley harbors a well preserved, almost natural peatland, with a variety of fen and river marginal floodplain types. These differ in species richness and productivity and are arranged in regular zones, which run both from upstream in the downstream direction as well as parallel to the river (Palczynski 1984). The floodplain contains highly productive tall sedge, reed, and grass communities, whereas outside the reach of the river floods, low-productive rich fen and moderately rich fen communities occur. In the Middle Basin, drainage has led to extensive areas of fen meadows. Plant species include boreal and continental species such as *Betula humilis*, *Trichophorum alpinum*, *Dianthus superbus*, *Pedicularis sceptrum-carolinum*, and *Saxifraga hirculis*, but more than 90 % of plant species present at Biebrza also belong to the Western European flora (Bootsma 2000).

13.2.4 Birds and Mammals

Birdlife includes 157 breeding species, of which 21 species are considered as threatened in Europe (Dyrcz et al. 1985). The Biebrza wetlands are famous for their breeding populations of great snipe (*Gallinago media*, 300–400 displaying males) and aquatic warbler (*Acrocephalus paludicola*, ca. 3000 pairs; Dyrcz and Zdunek 1993). In spring and autumn, large numbers of rough geese and ducks rest and feed in the wetlands. Furthermore, mammals like wolf (*Canis lupus*, ca. ten individuals), elk (*Alces alces*, 500–700 individuals), otter (*Lutra lutra*, ca. 50 individuals), and European beaver (*Castor fiber*, ca. 300 individuals) are present.

13.3 Hydrology of the Biebrza Valley

13.3.1 Surface Water: Hydrography and Hydrology

The headwaters are located at an altitude of 162 m a.s.l.; the confluence with the Narew River is at 102.5 m. The width of the riverbed ranges from 10 m to ca. 80 m. The major part of the drainage system (75 %) is to the north of the river, in the Elk and Augustow lakeland. The free surface slopes are variable. The averages over several years range from 0.07 % in the Middle Basin to almost 3.0 % near to the Biebrzas origin. The average slope from the origin to the mouth is 0.36 %. The fluctuation in the free water surface is large. Over several years, these fluctuations range from up to 1.5 m close to the origin to about 3.0 m at the lower course. Because of such large variations, vast areas along the river are flooded, particularly in spring.

The Biebrza and its tributaries are primarily draining rivers. Numerous lakes are also part of the Biebrza Basin drainage system. They lie outside the valley, on the upland, especially in the northern part of the watershed (Fig. 13.1). Other significant waterways in the area are six canals built over the past 150 years in or northward of the Middle Basin. Hydrologic conditions of the Biebrza river are described by characteristic discharges (mean discharge, mean low discharge, mean high discharge) calculated for six gauges (Table 13.1). Mean discharge in the Lower Basin is ca. seven times the discharge of the Upper Basin (respectively, 30.6 $m^3\ s^{-1}$ and 4.61 $m^3\ s^{-1}$). This ratio resembles the ratio of the discharge areas (ca. 1:8), which implies that the specific discharges of the Upper and Lower Basins are comparable and amount to 5.45 $dm^3\ s^{-1}\ km^{-2}$ and 4.43 $dm^3\ s^{-1}\ km^{-2}$.

The acidity of the river water is neutral to slightly alkaline and the water is quite rich in nutrients. No large differences in water quality were observed going downstream: both nutrient concentrations and specific nutrient loads are in the same order of magnitude as the Upper and Lower Basins (Table 13.2).

Three numerical models have been developed that describe the river flow in the Upper, Middles and Lower Basins. These models either work independently or can be merged in a single unit. They include river network models

Table 13.1 Discharge characteristic of the Biebrza River. *SNQ* Mean low discharge, *SNq* mean low specific runoff, *SQ* mean discharge, *Sq* mean specific runoff, *SWQ* mean high discharge, *SWq* mean high specific runoff. For locations of gauge stations, see Fig. 13.1

Biebrza basin	Gauge station	Area (km^2)	SNQ ($m^3\ s^{-1}$)	SNq $\left(\frac{dm^3}{s \cdot km^2}\right)$	SQ ($m^3\ s^{-1}$)	Sq $\left(\frac{dm^3}{s \cdot km^2}\right)$	SWQ ($m^3\ s^{-1}$)	SWq $\left(\frac{dm^3}{s \cdot km^2}\right)$
Upper	Rogozyn	102.8	0.17	1.69	0.62	6.02	3.23	31.4
	Sztabin	846.0	1.23	1.45	4.61	5.45	31.6	37.4
Middle	Dolistowo	3065.1	3.83	1.25	15.2	4.96	53.8	17.6
	Osowiec	4365.1	6.48	1.48	20.1	4.60	81.9	18.8
Lower	Burzyn	6900.4	10.6	1.53	30.6	4.43	138.0	21.1

Table 13.2 Water quality and nutrient loads of the Biebrza River. Winter sampling was carried out in April 1992 and 1993 during spring floods ($n=2$), summer sampling in July 1987, 1990, 1991, 1992, and 1993 ($n=5$). For average discharge data for winter (including spring floods) and summer, see Table 13.1 and Byczkowski and Kicinski (1984). See Fig. 13.1 for location of the sampling and discharge measurement stations (Sztabin, Burzyn). See Wassen et al. (1990) for sampling and analyzing methods. *EC* Electro-conductivity

	Upper Basin (Sztabin)		Lower Basin (Burzyn)	
	Winter	Summer	Winter	Summer
pH	7.7	8.0	8.2	7.8
EC (μS cm^{-1})	410	388	410	398
Ca (mg l^{-1})	87	73	79	75
N (mg l^{-1})	1.15	0.46	1.12	0.59
P (mg l^{-1})	0.23	0.13	0.19	0.17
K (mg l^{-1})	2.0	1.5	3.2	3.0
Nutrient loading rates				
N loading rate (t [0.5 year]$^{-1}$)	110.4	22.7	652.8	225.2
(t $year^{-1}$)	133	878		
P loading rate (t [0.5 year]$^{-1}$)	22.0	6.4	111.0	63.4
(t $year^{-1}$)	28	174		
K loading rate (t [0.5 year]$^{-1}$)	192.4	72.6	1867.2	1145.0
(t $year^{-1}$)	265	3012		
Specific nutrient load				
N load/catchment area (kg km^{-2} [0.5 year]$^{-1}$)	130.4	26.8	94.6	32.6
P load/catchment area (kg km^{-2} [0.5 year]$^{-1}$)	26.0	7.4	16.0	9.2
K load/catchment area (kg km^{-2} [0.5 year]$^{-1}$)	227.4	85.5	270.6	166.0

of instream flow in Biebrza and its main tributaries. For the Lower and Middle Basins, the floodplain is also included in the models. The geometry of the floodplain has been reproduced using cross-sections obtained from a digital elevation model (DEM). DEM was also used in order to compare the model flood extension with the flooding range obtained by analysis of satellite images for specific records. For details, see Kubrak and Okruszko (2000), Swiatek et al. (2002), and Verhoeven et al. (2004). The models were used for the analysis of floods (see Section 13.3.3) and assessment of the effect of water management strategies on the discharge and flooding regime of Biebrza. There is a good match between modelled flood zones and the natural vegetation patterns. The existing hydraulic structures in the Middle Basin appear to have a large impact on the hydrology of the northern part of the Middle Basin but seem to have an almost negligible effect for flooding features in the Lower Basin.

13.3.2 Groundwater

In the Upper Biebrza Basin, groundwater plays a key role in the contribution of water to the valley; it seeps out in the wetlands all over the valley bottom. The fens of the Biebrza Upper Basin are typical soligenous fens, i.e. fens with relatively little restriction of water outflow but kept wet by constancy of water supply, mostly groundwater discharging into the fen. A MODFLOW groundwater model has been set up to predict the area and magnitude of the upward flux of groundwater discharge (Batelaan and Kuntohadi 2002). Strong upward fluxes occur in the river bed, in the northern part of the basin in the lower part of the Augustow outwash, and also along the morainic edges (Fig. 13.2). In general, these locations correspond well with the occurrence of peat as derived from peat thickness mappings.

Hydrological modelling of the Middle and Lower Biebrza shows a set of superimposed flow systems of various orders. The size and the distribution of the local systems depend on the hydraulic conductivity of the peat soils and the underlying sandy soil, but also on the changing precipitation and evapotranspiration patterns within a year (Mioduszewski and Wassen 2000; Mioduszewski and Querner 2002).

13.3.3 Flooding

In the Upper Basin, flooding is restricted to a narrow belt of approximately 10–20 m along the river. This is because the river discharges are relatively small (Table 13.1) and the peat body is loosely structured, since there is no sediment deposited. The peat, 3–7 m thick, rises and shrinks with water level fluctuations, preventing it from being flooded by the river. In the Middle and

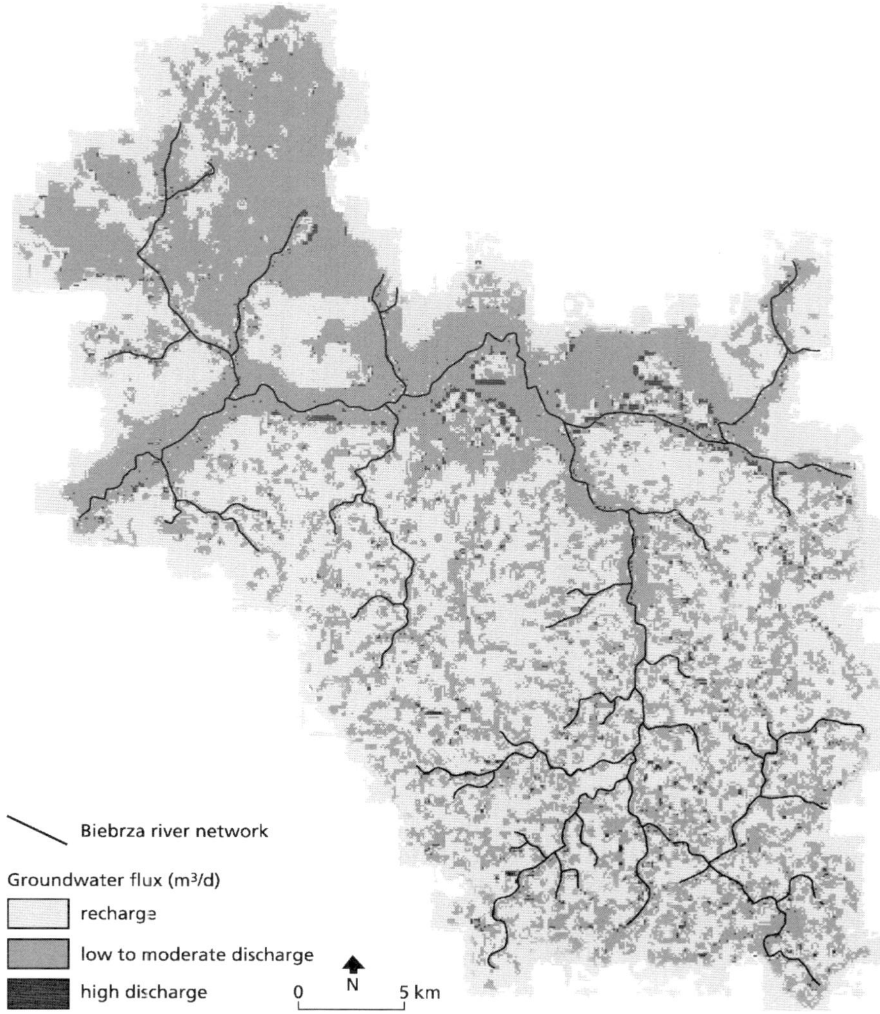

Fig. 13.2 Upward fluxes of groundwater simulated for the Upper Biebrza Basin using MODFLOW (after Batelaan and Kuntohadi 2002)

Lower Basin, river water levels exceed bankful conditions during a significant number of days in most years (Fig. 13.3), leading to almost annual spring floods. Especially in the Lower Basin, the inundated zone is very wide in the flat valley in which the river floodplain and the topogenous fen (i.e. flat fens kept wet by retention of water) merge. This inundation water may derive from three different sources: the atmosphere (rain water/melting snow), river water, and groundwater. Chormanski (2003) was able to distinguish these water sources during spring inundations, using a combination of hydrological

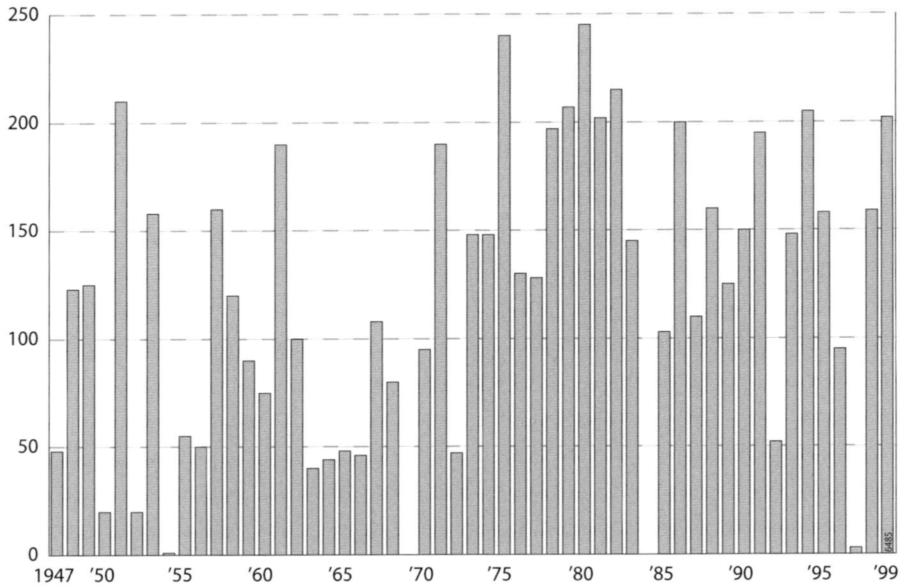

Fig. 13.3 Flood duration in the Lower Biebrza Basin: number of days when river water levels exceed bankful for the period 1947–1999 (calculated for Burzyn gauge station; Okruszko et al. 2003)

modelling, remote sensing, GIS, and statistical analysis of a large water chemistry data base. The area flooded by the river in spring 2002 determined in this way covered 89 km^2 (Fig. 13.4). The total inundated area was 214 km^2, of which 125 km^2 was inundated with rainwater/meltwater and groundwater discharging to the surface (Chormanski 2003).

13.3.4 Drainage

Drainage for grassland farming undertaken in the past has resulted in moist conditions, especially in the Middle Basin. The result is an accelerated mineralization of peat in fen meadows (Okruszko 1990) and, when hay cutting is abandoned, a succession towards vegetation types dominated by scrubs and trees, leading to a loss of biodiversity. To counterbalance these negative effects, the Biebrza National Park (which was founded in 1993) aims to restore the original hydrological regime. For this purpose, we developed dynamical models for assessing the effects of drainage by large canals in the Middle Basin (Mioduzewski and Wassen 2000) and shallow ditches in the Lower Basin (Mioduszewski and Querner 2002). Recently, several re-wetting measures have been realized in the Middle Basin (raising the local drainage basis

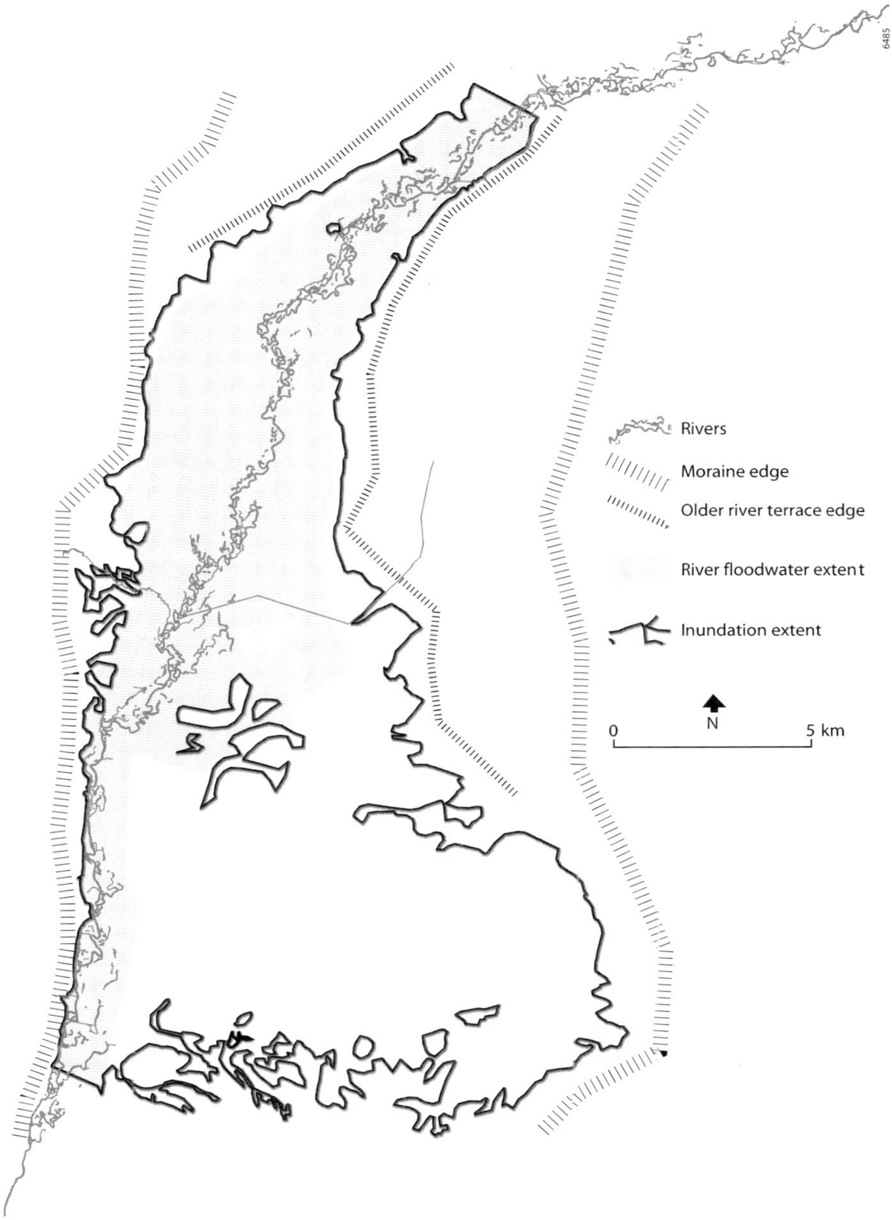

Fig. 13.4 Inundation extent in 2002 and extent of river flood zone in 2002 for the Biebrza Lower Basin (after Chormanski 2003)

by the construction of submerged weirs in canals, re-opening of former river sections, closing small ditches and canals, and filling-in canals). For the Lower Basin, we showed that re-wetting by complete closing of the shallow drainage systems (canals and ditches) results in a water level rise of 0.2–0.5 m and a significant increase in the width of the area fed by groundwater discharge. In large areas of groundwater discharge, the vertical fluxes increased by more than 0.5 mm day^{-1} during late spring and early summer (Fig. 13.5). Based on these results, the Biebrza National Park will realize re-wetting measures in the Lower Basin in the near future.

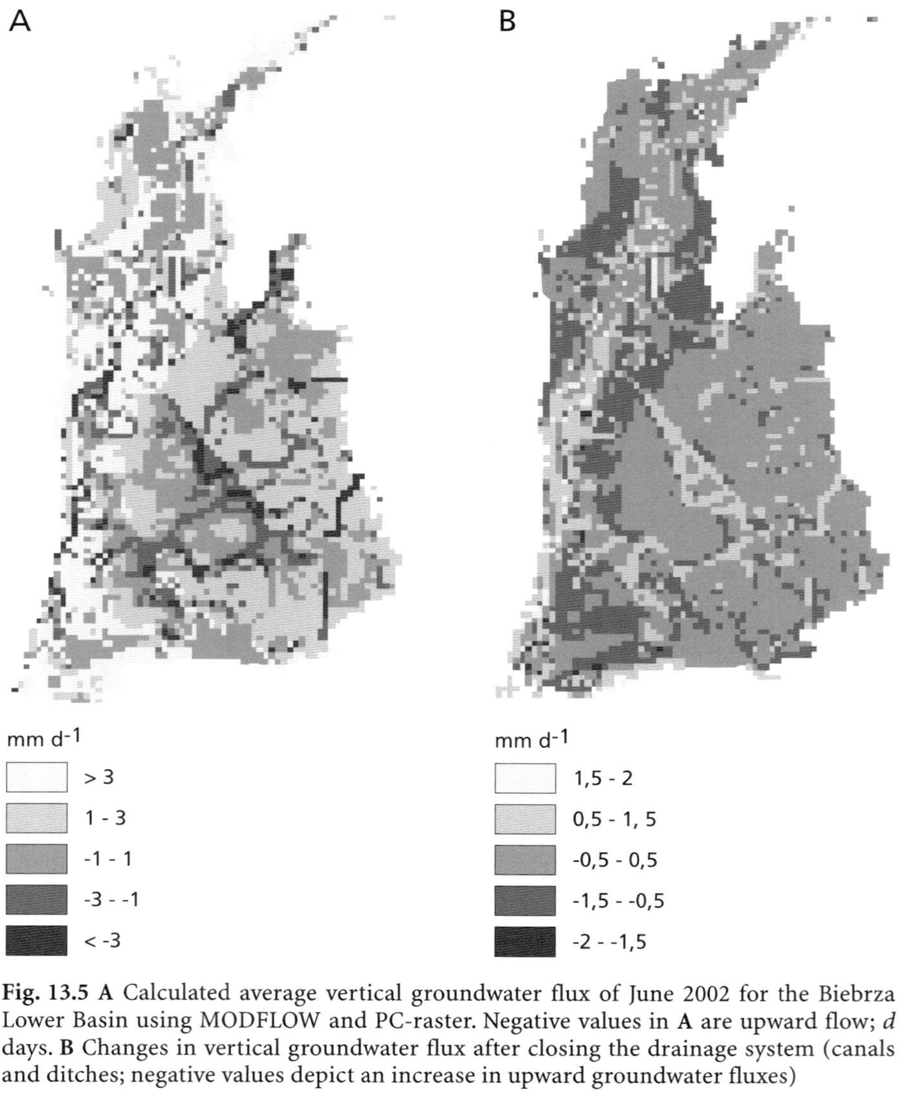

Fig. 13.5 **A** Calculated average vertical groundwater flux of June 2002 for the Biebrza Lower Basin using MODFLOW and PC-raster. Negative values in **A** are upward flow; *d* days. **B** Changes in vertical groundwater flux after closing the drainage system (canals and ditches; negative values depict an increase in upward groundwater fluxes)

13.4 Relation Between Hydrology and Vegetation Zoning

The original zoning of plant communities has disappeared in the Middle Basin, due to drainage works. The characteristic zoning in the Upper and Lower Basin is still present and can be summarized as follows: (a) floodplain marshes along the river (absent in the Upper Basin), (b) rich fens in the occasionally flooded belt further away from the river, (3) moderately rich fen outside the reach of river floods, and finally again (d) rich fen along the valley edges. These vegetation gradients are smooth.

The floodplain contains highly productive marshes which belong syntaxonomically to *Glycerietum maximae*, *Caricetum gracilis* and *C. elatae* (Palczynski 1984). These are tall sedge, grass, and herb vegetations, relatively poor in species (Table 13.3). Typical associations of the occasionally flooded belt are *C. caespitosae* and *Peucedano-Caricetum appropinquatae*.

Moderately rich fen is found in an intermediate belt, outside the reach of the seasonal river floods on places where the calcareous groundwater from the moraines does not reach the fen surface. It is fed mainly by rainwater (Wassen et al. 1990) and belongs syntaxonomically to *Betuletum humilis*, with affinity to *C. rostrato-diandrae*. It is a thin dwarf-shrub vegetation with low sedges and occasionally some *Sphagnum* hummocks (*S. recurvum*, *S. squarrosum*, *S. palustre*). The moss layer has a fairly high standing crop (Table 13.3).

In a belt along the moraines, but also further away from the moraines, provided that the calcareous groundwater still reaches the fen surface (Wassen et al. 1992), low sedge-rich fen types are abundant. In the Biebrza valley, several species-rich associations of this fen type (the *Caricion diandrae*) are present (Palczynski 1984). These are low-productivity sedge and herb vegetation with a well developed moss layer of Hypnaceae.

Water levels show a large seasonal dynamics in the floodplain (Table 13.3). In dry summers, as for instance the summer of 1992, the water levels in the floodplain even dropped >0.5 m below the peat surface (see also Palczynski and Stepa 1991). The floodplain is relatively nutrient-rich; potassium concentrations are especially higher than in the fens. The rich-fen type has the lowest phosphorus concentration in both peat water (in summer) and in peat. The peat of the moderate rich fen has a large phosphorus content. Phosphorus availability is larger than in the rich fens, which is probably due to the dissolution of precipitated calcium phosphates by infiltrating rainwater (Wassen et al. 1990).

Floodplain and rich fen are calcareous and have a near-neutral pH; the moderate rich fen is slightly acidic. Both inorganic the nitrogen concentration in the peat water in summer and the total nitrogen content in the peat do not differ much between the distinguished wetland ecosystems (Table 13.3).

Table 13.3 Average values of groundwater, peat, and vegetation variables of three characteristic wetland ecosystems in the Biebrza river valley. Values were measured at the peat surface. Spring sampling was carried out in April 1992 during spring floods (wet spring conditions), summer sampling in July 1987 (average summer conditions).Superscript letters (*a*, *b*, *c*) refer to significant differences between average values (different letters indicate significant differences; Tukey test, $P<0.05$). See Wassen et al. (1990) for sampling and analyzing methods. Limitation refers to nutrient limitation as determined by fertilization experiments (Wassen et al. 1998; and unpublished data). *EC* Electro-conductivity, *n* number of samples, *s* summer, *w* winter. *none* In a second fertilization experiment in the floodplain, we measured no significant growth response upon nutrient addition, indicating that plant growth was not limited by nutrients (Wassen et al. 1998)

Wetland ecosystem		Floodplain marsh	Rich fen	Moderately "rich fen
Groundwater	(*n*, w/s)	(5/6)	(16/26)	(7/11)
Level (cm surface)	Winter	-54.8 ± 6.8^{b}	-3.5 ± 6.2^{a}	0.6 ± 1.3^{a}
	Summer	4.0 ± 6.5^{ab}	2.2 ± 3.8^{a}	6.1 ± 4.2^{b}
pH	Winter	8.12 ± 0.14^{a}	7.10 ± 0.21^{b}	6.58 ± 0.19^{c}
	Summer	6.59 ± 0.36^{a}	6.51 ± 0.40^{a}	6.01 ± 0.16^{b}
EC ($\mu S\ cm^{-1}$)	Winter	377 ± 1^{a}	233 ± 43^{b}	180 ± 71^{b}
	Summer	413 ± 110^{a}	392 ± 106^{a}	248 ± 14^{b}
Ca ($mg\ l^{-1}$)	Winter	74 ± 1^{a}	44 ± 10^{b}	29 ± 11^{c}
	Summer	102 ± 14^{a}	56 ± 14^{b}	24 ± 15^{c}
N ($mg\ l^{-1}$)	Winter	0.39 ± 0.08^{a}	0.15 ± 0.06^{b}	0.17 ± 0.04^{b}
	Summer	1.72 ± 1.11^{ab}	1.43 ± 0.65^{a}	1.04 ± 0.39^{b}
H_2PO_4 ($mg\ l^{-1}$)	Winter	0.21 ± 0.10^{a}	0.13 ± 0.13^{a}	0.17 ± 0.08^{a}
	Summer	0.19 ± 0.16^{a}	0.01 ± 0.04^{b}	0.27 ± 0.36^{a}
K ($mg\ l^{-1}$)	Winter	2.95 ± 0.09^{a}	0.42 ± 0.28^{b}	0.55 ± 0.34^{b}
	Summer	0.88 ± 0.32^{a}	0.55 ± 0.50^{b}	0.29 ± 0.23^{c}
Peat (summer, *n*)		(3)	(11)	(5)
Organic matter content (% dry weight)		53 ± 43^{a}	85 ± 5^{a}	79 ± 11^{a}
P_{lime}		4.9 ± 1.1^{ab}	4.5 ± 0.2^{a}	3.9 ± 0.4^{b}
C/N		16 ± 0^{a}	20 ± 3^{a}	20 ± 3^{a}
N_t (% dry weight)		1.92 ± 1.62^{a}	2.59 ± 0.46^{a}	2.32 ± 0.65^{a}
PO_4 (% dry weight)		0.14 ± 0.07^{ab}	0.07 ± 0.03^{a}	0.70 ± 0.64^{b}
Vegetation (*n*)		(5)	(11)	(5)
Standing crop				
Phanerogams ($g\ m^{-2}$)		983 ± 315^{a}	248 ± 46^{c}	345 ± 396^{b}
Cryptogams ($g\ m^{-2}$)		0	226 ± 93^{a}	314 ± 167^{a}
Limitation		N (*n*=1), none	N (*n*=2), P (*n*=1)	N (*n*=1)
Number of species ($n\ [10\ m]^{-2}$)		12 ± 5^{b}	28 ± 7^{a}	22 ± 4^{a}

Figure 13.6 integrates eco-hydrological patterns in two typical cross-sections, one for the Upper Basin and one for the Lower Basin. In Biebrza, electro-conductivity (EC_{25}) was a suitable parameter to distinguish rainwater from groundwater and river water, the latter two having higher solute concentrations. We measured EC_{25} values at every 10-cm depth and at horizontal intervals ranging from a few meters to 50 m, depending on the differences in values between adjacent sites. For this, we used a probe allowing for measurement of EC_{25} values at every desirable depth in water-saturated peat (Van Wirdum 1984). EC_{25} patterns were validated with water quality analyses of samples taken from surface water and from piezometers with filters at several depths (see Wassen et al. 1992). The groundwater flow pattern in the Biebrza Upper Basin shows groundwater discharge at the foot of the moraine and slight groundwater flow fed by infiltrating rainwater further away from the upland (Fig. 13.6A). EC_{25} values are fairly homogeneous over large distances in both cross-sections, showing that the gradient in water quality from the valley edge to the river is fairly gradual. There is little dilution at the surface, except for the shallow rainwater lens in the Upper Basin, where groundwater is recharged by inflitrating rainwater. In the Upper Basin, there is no river flooding (Fig. 13.6A). The larger part of the peatland in the Lower Basin is flooded annually by the river (see also Okruszko 1990). In the Lower Basin (Fig. 13.6B), groundwater discharges originate from the dune complex, which borders the peatland here. These dunes consist of leached sand (Wassen et al. 1996), resulting in low EC_{25} values in the groundwater coming from these dunes. In the direction of the river, EC_{25} values are increasing, which is ascribed to the river floods and exchange of solutes from sedimentary loam (Wassen et al. 1992). In both the Upper and Lower Basins, a number of mesotrophic species (species of moderate nutrient-rich conditions) are exclusively restricted in their distribution to the groundwater discharge zone: *Carex diandra* and *C. lasiocarpa* (depicted in Fig 13.6A) are representative for a larger number of species, e.g. *Parnassia palustris*, *C. lepidocarpa*, *C. panicea*, and *Pedicularis palustris*. A number of species which tolerate slightly more nutrient-rich conditions are present along almost the entire cross-section: *Menyanthes trifoliata*, *Caltha palustris*, *Equisetum fluviatile*. Some species are only present in the mineral-poor rainwater infiltration belt in the Upper Basin: *Betula humilis*, *Drosera rotundifolia*, *Oxycoccus palustris*, *Ledum palustre*. Eutrophic species (species of nutrient-rich habitats) tolerant for inundation are only present in the Upper Basin along the river bank (*Carex cespitosa*), whereas in the Lower Basin, these are arranged in wide zones according to decreasing flood tolerance (respectively *Glyceria maxima*, *C. acuta*, *C. elata*; Fig. 13.6B).

From a dynamic 3-D model based on the MODFLOW code (Janssen 2000; Batelaan and Kuntohadi 2002) the percent contribution of water fluxes to the rooted peat top-layers were estimated for some vegetation communities (Table 13.4). In addition to rain water, the vegetation zones close to the river

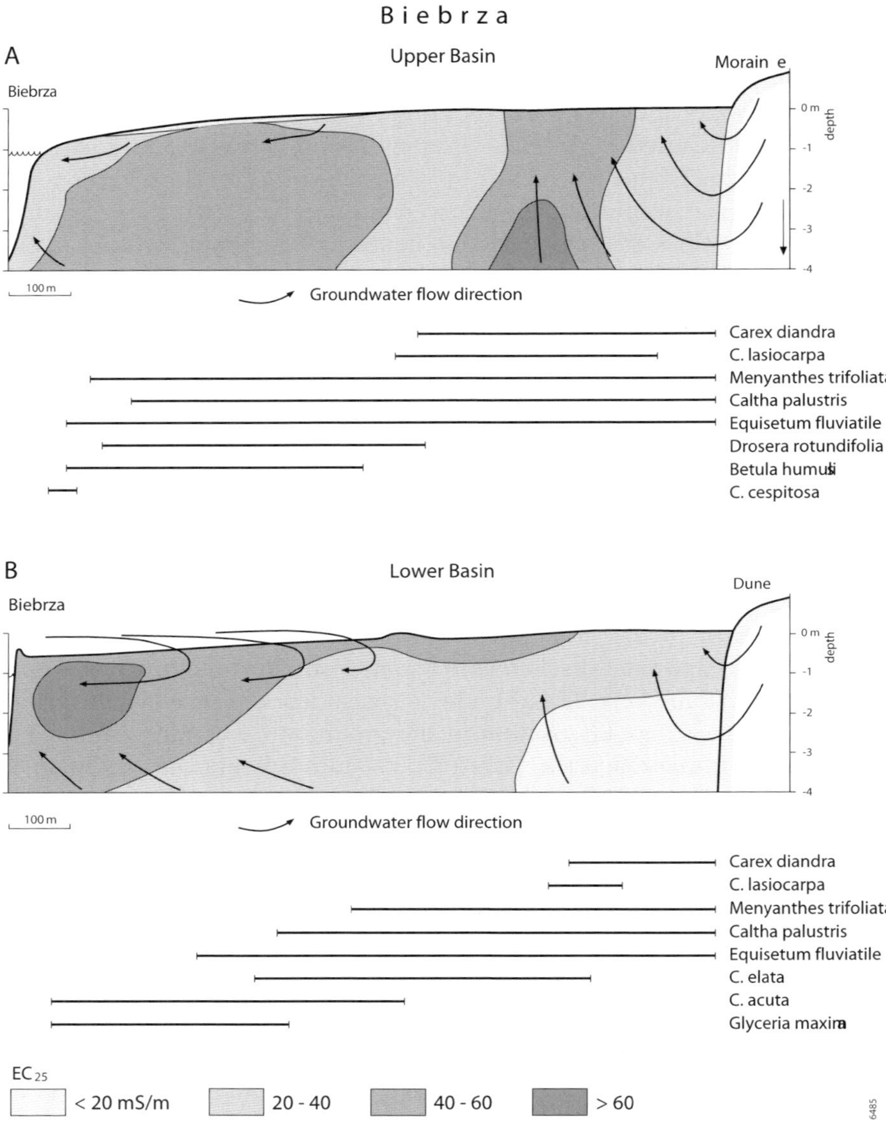

Fig. 13.6 Cross-sections of the Biebrza valley (**A** Upper Basin, **B** Lower Basin) from the upland (*right*) to the river (*left*). *Arrows* show schematic groundwater flow pattern (simulated with FLOWNET), *gray shades* show electro-conductivity (EC_{25}) of the groundwater, *horizontal lines* below the cross-sections show approximate distribution of some characteristic fen plant species (after Wassen et al. 1992, 1996)

Table 13.4 Attribution of different water sources to major vegetation types in the Biebrza wetlands

Vegetation type	Attribution (%) of water fluxes to the root zone[a]		
	Ground-water	River water	Overland flow
Reedland (*Scirpo-Phragmitetum*)	0	23	0
Tall grasses (*Glycerietum maximae*)	1	12	2
Tall sedges (*Caricetum acutae*)	7	8	0
Tall sedges (*C. elatae*)	29	0	0
Sedge grassland (*Calamagrostietum strictae* type a[b])	37	0	0
Sedge grassland (*C. strictae* type b[b])	46	0	0
Moss sedgeland (*Caricetum limoso-diandrae*)	76	0	0
Scrub–moss–sedgeland (*Betuletum humilis*)	45	0	0

[a] The rest of the water flux consists of precipitation water (after Janssen 2000; Batelaan, unpublished data).
[b] Refers to two types of *Calamagrostietum strictae* vegetation (see Wassen et al. 2002).

(*Scirpo-Phragmitetum*, *Glycerietum maximae*) receive substantial amounts of river water in flood periods, while zones at a greater distance from the river (*Caricetum elatae*, *Calamagrostietun strictae*) are supplied with groundwater and rainwater. Especially, the rich-fen type *Caricetum limoso-diandrae* is dominated by discharging groundwater, while in contrast, the moderately rich-fen type *Betuletum humilis* receives mainly rainwater.

13.5 Productivity and Nutrient Limitation of Marsh and Fen Vegetation

Species richness–productivity relationships for threatened species in wetlands show hump-shaped patterns, with a narrow hump at low productivity (Moore et al. 1989, Wheeler and Shaw 1991; Olde Venterink et al. 2003). The narrow hump indicates that these species are sensitive to productivity increases, and hence to increased availabilities of nutrients which limit plant growth. Nitrogen, phosphorus, and potassium availability limit biomass production in most wetlands. Although many freshwater wetlands seem to be limited by nitrogen, phospohorus limitation can also occur in fens with high inputs of iron-, aluminium- and/or calcium-rich waters (Boyer and Wheeler 1989; Boeye and Verheyen 1994; Wassen et al. 1992) and in nitrogen-enriched fens where annual mowing depletes the soil phosphorus pool relatively faster than the nitrogen pool (Verhoeven and Schmitz 1991). The Biebrza floodplain marshes are highly productive and growth is limited by nitrogen, if limited by nutrients at all (Table 13.3). The groundwater-fed rich fens were in two cases limited by nitrogen and in one case growth was phosphorus-limited, whereas the moderate rich fen was nitrogen-limited.

In the Biebrza Upper Basin, productivity and nutrient availability were analyzed along a gradient from rich fen to moderately rich fen (see Fig. 13.6A). The traditional mowing regime has been abandoned here for some decades and Schmidt et al. (2000) observed a invasion of *Betula pubescens* in the rich fen. El-Kahloun (2004) investigated possible shifts in nutrient limitation in this fen by analyzing peat and water and, in 2003, repeating measurements of nitrogen/phosphorus ratios in vascular plant tissue, as done earlier by Wassen et al. in 1992 (published by Wassen et al. 1995, 1998). The main objective of this comparison was to analyze whether the observed changes in vegetation composition (especially the invasion of *Betula pubescens*) during the past ten years were related to changing nutrient availabilities.

It appeared that, in 2003, water tables and phosphorus concentrations in water and peat were lower than in 1992 in both fen types, whereas pH values were higher than in 1992 (Table 13.5). Nitrogen and phosphorus concentrations in above-ground plant material declined. Nitrogen/phosphorus ratios indicated nitrogen limitation in 1992 (for critical values, see Güsewell and Koerselman 2002; Olde Venterink et al. 2003). In 2003, nitrogen/phosphorus ratios were higher than in 1992, both in the rich fen and in the moderately rich fen. In the rich fen, the nitrogen/phosphorus ratios of 2003 clearly indicate phosphorus-limited growth (Table 13.5).

An explanation for this shift in limitation from nitrogen to phosphorus limitation could be an increase in external eutrophication. However, from the actual lower nutrient concentrations in groundwater in comparison with data from 1992, we can infer that no external eutrophication through the groundwater has occurred, which is in line with earlier observations by Wassen et al. (1990, 1992). We have found no indications that atmospheric nitrogen deposition has increased. We therefore conclude that external eutrophication is limited and has not increased since the study of Wassen et al. (1990). Present groundwater levels are lower compared with the levels measured in the period from 1987 to 1993 (Wassen et al. 1996). We suggest a lowering of the groundwater table as a possible cause for the lowered availability of phosphorus and the shift from nitrogen to phosphorus limitation in the rich fen. Lower water tables may have enhanced co-precipitation of phosphorus with aluminum and iron hydroxides under oxic conditions (Patrick and Khaled 1974; Boeye and Verheyen 1994). The development of trees may have lowered the groundwater table by increasing evapotranspiration (Wassen and Joosten 1996). This hypothesis is supported by observations by El-Kahloun et al. (2003). They observed a very low water table in summer when evapotranspiration rates were high, while water levels in spring were close to the mire surface, as in 1992. These observations suggest not only that the removal of biomass by hay-

Table 13.5 Water tables, peat water quality, nutrient content in peat and in standing crop, and above-ground biomass production in two major fen types of the Biebrza Upper Basin, in 1992 and 2003. Values are average ±SD (number of observations given in parentheses). Water levels were measured relative to the peat surface. Standing crop was harvested at the height of the growing season (July). See Wassen et al. (1995, 1998) and El-Kahloun et al. (2003, 2005) for sampling and analyzing methods

	Rich fen		Moderately rich fen	
	1992	2003	1992	2003
Water table (cm above peat surface)	April: 0±3 (51) July: −33±6 (17)	April: −7±1.7 (3) July: −55±1 (3)	April: 2±4 (51) July: −24±4 (17)	April: 1.3±5.0 (3) July: −50±2 (3)
Peat water quality (mg l^{-1})				
PO_4	0.31±0.26 (3)	0.02±0.02 (4)	7.09±6.53 (3)	0.01±0.0 (4)
NH_4	0.34±0.24 (4)	0.89±0.06 (4)	0.41±0.13 (3)	0.48±0.5 (4)
NO_3	0.39±0.51 (4)	0.22±0.3 (4)	2.28±3.94 (3)	0.00±0.0 (4)
Ca	75.0±3.0 (4)	89.2±19.9 (4)	58.0±13.0 (3)	98.9±32.6 (4)
pH	6.17±0.09 (4)	6.95±0.4 (4)	6.52±0.20 (3)	7.01±0.1 (4)
Nutrients in peat (% dry weight)				
N_{tot}	1.84±0.28 (6)	2.05±0.28 (4)	2.74±0.69 (5)	2.89±0.62 (4)
P_{tot}	0.08±0.01 (6)	0.07±0.03 (4)	0.49±0.23 (5)	0.10±0.04 (4)
Nutrients in above-ground biomass (mg g^{-1} dry weight)				
N	14.3±2.3 (5)	10.61±1.5 (4)	16.7±1.0 (5)	9.45±0.4 (4)
P	1.06±0.23 (5)	0.43±0.1 (4)	3.43±0.4 (5)	0.82±0.4 (4)
N/P	13.7±2.37 (5)	24.16±1.3 (4)	4.86±0.15 (5)	12.81±5.9(4)
Above-ground biomass (g dry weight m^{-2})	534±198 (9)	496.4±69 (4)	660±453 (9)	394±105 (4)

making influences nutrient availabilities directly by exporting nutrients but also that abandonment of this management practice may exert an indirect influence on nutrient availability via evapotranspiration and water tables. We did not find large differences in productivity, neither in space nor in time. Still, the type of nutrient limitation has changed and may possibly affect the species composition of the fens. In 2003, five threatened species (*Viola persicifolia, Linum catharticum, Dactylorhiza incarnata, Carex flava, Briza media*) which were not present in the species list of Wassen et al. (1992) were found along the cross-section. These species are common in phosphorus-limited sites (Wassen et al. 2005) and likely possess adaptations enabling them to obtain the limited phosphorus resources.

13.6 Discussion and Conclusions

We found a number of clear hydrological and ecological patterns in the Biebrza valley.

Hydrology in the Upper Basin was very different from the Lower Basin. In the Upper Basin groundwater discharge was dominant, whereas in the Lower Basin inundation was the predominant phenomenon. The hydro-chemistry of the mire water was governed by three principal sources of water: precipitation, groundwater seeping to the surface, and river floods. Vegetation zoning and productivity were related to the dominance of these water types in the mires and floodplains. In the river flood zone, nutrient availability and productivity was governed by the river, which provided the vegetation with nutrients dissolved in river water as well as attached to sedimentary loam and silt. The relationship between nutrient availability, productivity, and species composition outside the river flood zone in the Upper Basin was more complicated and its dynamics seemed to be related to succession and management. Generally speaking, the present situation in the Biebrza Upper Basin and in the Lower Basin, outside the reach of the river floods, allows for the study of processes in so-called through-flow fens (Durchströmungsmoore; cf. Succow 1988) without the disturbing effects of raised atmospheric nitrogen deposition, heavy drainage or eutrophication of groundwater (see Wassen and Joosten 1996). The same holds for the floodplains in the Biebrza Lower Basin, which in contrast to most European river marginal floodplains, are still flooded over vast areas. Even more fascinating, the floodplain interacts with a vast topogenous mire inundated in spring by atmospheric water and groundwater. Here we can learn how the interaction between river water, discharging groundwater, and atmospheric water (Chormanski 2003) establishes a biomass gradient with peat-forming vegetation and marsh vegetation arranged in a zoning pattern parallel to the river. Water fluxes, water level dynamics, water chemistry, and mineralization rates seem key factors determining these vegetation gradients (De Mars et al. 1997; Wassen et al. 2003).

In Western Europe and North America human, interference in hydrological systems led to changes in the distribution and size of water systems, causing fragmentation which in turn led to changes in the distribution of plant species and communities (Van Diggelen et al. 1991; Mitsch and Gosselink 1993; Stromberg 1993b; Barendregt et al. 1995; Wassen et al. 1996). Hydrological processes such as groundwater discharge and rainwater and surface water recharging the groundwater are of course still active in disturbed landscapes, but the place where they occur, their intensity, and their water quality has changed (Giller and Wheeler 1988; Wassen et al. 2003). For instance, deterioration of ecosystems in the Vecht river plain is partly due to poor water quality (Barendregt 1993) and drainage (De Mars 1996) but the fragmented distribution of ecosystems in the Vecht river plain is caused by hydrological fragmentation; and restoration should thus try to re-enforce larger regional water systems. The fairly undisturbed hydrology at Biebrza allowed us to assess how vegetation was fed by different water sources. Moreover, it became apparent that, where large regional water systems meet, gradual spatial differences in water fluxes facilitate the hydrological conditions for large-scale vegetation gradients. Such gradients cannot exist in a hydrologically fragmented landscape. Bootsma et al. (2000) compared the Drentse Aa valley with Biebrza; and Bootsma and Wassen (1996) compared three large fen wetland areas in the Netherlands with Biebrza. They concluded that the Drentse Aa valley was not as severely disturbed as the other Dutch areas. The eutrophication problem was especially severe in the Vecht river plain, whereas acidification of fens was an especially serious problem in the Wieden and Weerribben.

In using Biebrza as a reference for other areas, we should consider that Biebrza is a lowland river of intermediate size (average annual discharge in the upper course ca. 5 $m^3 s^{-1}$ and in the lower course ca. 30 $m^3 s^{-1}$) running through a flat valley filled with peatlands and marshes and surrounded by a post-glacial landscape with ice-pushed hills, moraines, and outwash plains. Its origin, geomorphology, and topography in northeastern Poland is similar to that in northern Germany, parts of Denmark and the United Kingdom, and the northern half of the Netherlands. The present-day climate shows differences: the above-mentioned western European countries have an Atlantic climate with mild winters and a precipitation surplus of ca. 300 mm $year^{-1}$, whereas the sub-continental climate of northeastern Poland shows colder winters and an annual precipitation surplus of only 150 mm. However, these climatic differences do not appear to have a predominant influence on the flora. Although the flora of Biebrza contains continental and sub-boreal elements, 96 % of the plant species present in the Biebrza valley also belong to the Dutch flora (Bootsma 2000). Thus, we conclude that Biebrza can be used as a reference example for lowland rivers, floodplains, and fens in western and central Europe, such as as the rivers Peene and Ems in Germany, Drentse Aa, Vechtstreek, Wieden and Weerribben in the Netherlands, Zwarte Beek in Belgium, and the Norfolk Broads in Britain (see also Van Wirdum 1991; Grootjans and Van Diggelen 1995; Wheeler and Shaw 1995).

We should also realize that a far-reaching objective will automatically add uncertainty to knowledge from a reference area applied in the area that is to be restored. Wassen (2005) discerns the following objectives of reference studies: (a) to discover the relationship between natural key processes and ecosystem functioning, (b) to estimate the degree of degradation of areas to be restored, (c) to set targets for nature conservation and to define conditions necessary for (re-)establishment of target species and communities, and (d) to design restoration measures in restoration projects. We conclude that the first two objectives are feasible by using information from a reference wetland. However, when the aim of the reference study is to set targets for nature conservation, or to define environmental standards, or specific restoration measures necessary for (re-)establishment of target species and communities, we also have to deal with added uncertainty related to recovery processes. This leads us to the fundamental question: Can we predict recovery? Knowledge from reference areas in itself does not enable us to predict the potentials for recovery of disturbed and stressed ecosystems, since recovery does not necessarily follow the same path as deterioration and also the rate of recovery may differ from the rate of deterioration. Hysteresis occurs in both abiotic and biotic processes. Drainage may lead to irreversible decomposition of peat (Okruszko 1995). Also, the redox status of peat may irreversibly change following drainage (De Mars and Wassen 1999) and the nutrient dynamics is strongly impacted by drying and re-wetting (Olde Venterink et al. 2002). Restoration projects aiming at counterbalancing acidification show that pH and base saturation of the soil recover very slowly (Beltman et al. 2001). Also, high atmospheric deposition and poor water quality may prevent full recovery of nutrient-poor conditions (Bakker and Berendse 1999). For this reason, Van Diggelen (1998) is more optimistic about the restoration prospects of eutrophic floodplains than those of mesotrophic fens and fen meadows. In this respect, it is also important to note that re-establishment of the desired species is hampered if they are absent in the actual vegetation and the seed-bank. Regeneration of the vegetation depends in such cases on dispersion possibilities, which are unfavorable for many wetland species in the present-day fragmented landscape (Bakker et al. 1996; Poschlod and Bonn 1998; Van Diggelen 1998). Additionally, above- and belowground communities influence each other through a variety of direct and indirect interactions (Wardle 2002). Time lags needed by soil organisms to respond to change lead to different selection pressures, for example as exerted by above- and belowground herbivores and pathogens (Van der Putten et al. 2001). De Deyn et al. (2003) showed that a strong linkage exists between succession in vegetation and soil community composition. This linkage is often overlooked in restoration projects in which either only soil fertility is reduced or aboveground vertebrate grazers are introduced. The slow recovery of belowground communities might be a factor responsible for the disappointing recovery of ecosystems after restoration measures were applied to former agricultural fields.

What these examples (in which time lags, feedbacks, unpredictable recovery processes, and hysteresis occur) demonstrate is that they are all potential reasons for slow or incomplete recovery. The conclusion should be that information from reference areas is very useful to demonstrate the general potential for restoration or rehabilitation, but it has to be treated with care in order not to raise too high an expectation for the outcome of restoration projects (Wassen 2005). We can never guarantee success fora resource manager or a landscape planner in his restoration attempt, when using solely the knowledge obtained in reference areas. Research in reference areas undoubtedly has invaluable importance for conservation and restoration. Therefore if we fail to protect or if we should lose these unspoilt areas, we are automatically cut off from a source which enables us to gain knowledge and information needed for restoration elsewhere.

Acknowledgements We thank Vincent Roodenburg for correcting the English and Ton Markus for drawing the figures.

References

Baird AJ, Wilby RL (1999) Eco-hydrology. Routledge, London

Bakker JP, Berendse F (1999) Constraints in the restoration of ecological diversity in grassland and heathland communities. Trends Ecol Evol 14:63–68

Bakker JP, Poschlod P, Strykstra RJ, Bekker RM, Thompson K (1996) Seed banks and seed dispersal: important topics in restoration ecology. Acta Bot Neerl 45:461–490

Barendregt A (1993) Hydro-ecology of the Dutch polder landscape. PhD thesis, Utrecht University, Utrecht

Barendregt A, Wassen MJ, Schot PP (1995) Hydrological systems beyond a nature reserve, the major problem in wetland conservation of Naardermeer (the Netherlands). Biol Conserv 72:393–405

Batelaan O, Kuntohadi T (2002) Development and application of a groundwater model for the Upper Biebrza River basin. Ann Warsaw Agric Univ SGGW Land Reclam 33:57–69

Beltman B, Van den Broek T, Barendregt A, Bootsma MC, Grootjans AP (2001) Rehabilitation of acidified and eutrophied fens in The Netherlands: effects of hydrologic manipulation and liming. Ecol Eng 17:21–31

Boeye D, Verheyen RF (1994) The relation between vegetation and soil chemistry gradients in a groundwater discharge fen. J Veg Sci 5:553–560

Bootsma MC (2000) Stress and recovery in wetland ecosystems. PhD thesis, Utrecht University, Utrecht

Bootsma MC, Wassen MJ (1996) Water quality of fen vegetation types in three European lowland mires. Vegetatio 127:173–189

Bootsma MC, Wassen MJ, Jansen AJM (2000) The Biebrza-valley as an ecological reference for Dutch stream valleys (in Dutch with English summary). Landschap 17:113–130

Boyer MLH, Wheeler BD (1989) Vegetation patterns in spring-fed calcareous fens: Calcite precipitation and constraints on fertility. J Ecol 77:597–609

Byczkowski A, Kicinski T (1984) Surface waters in the Biebrza drainage basin. Pol Ecol Stud 10:271–299

Cellot B, Dole-Olivier MJ, Bornette G, Patou G (1984) Temporal and spatial environmental variability in the Upper Rhone river and its floodplain. Freshwater Biol 31:311–325

Chormanski J (2003) Methodology of the flood extent determination. PhD thesis Warsaw Agricultural University, Warsaw

Cirujano S, Casado C, Bernues M, Camargo JA (1996) Ecological study of Las Tablas de Daimiel National Park (Ciudad Real, central Spain): differences in water physicochemistry and vegetation between 1974 and 1989. Biol Conserv 75:211–215

De Deyn GB, Raaijmakers CE, Zoomer HR, Berg MP, De Ruiter PC, Verhoef HA, Bezemer TM, Van der Putten WH (2003) Soil invertebrate fauna enhances grassland succession and diversity. Nature 422:711–713

De Mars H (1996) Chemical and physical dynamics of fen hydro-ecology. PhD thesis, Utrecht University, Utrecht

De Mars H, Wassen MJ (1999) Redox potentials in relation to water levels in different mire types in the Netherlands and Poland. Plant Ecol 140:41–51

De Mars H, Wassen MJ, Olde Venterink H (1997) Flooding and groundwater dynamics in fens in eastern Poland. J Veg Sci 8:319–328

Dyrcz A, Zdunek W (1993) Breeding statistics of the Aquatic warbler on the Biebrza marshes, NE Poland. J Ornithol 134:317–323

Dyrcz A, Okulewicz J, Witkowski J (1985) Bird communities on natural eutrophic fen mires in the Biebrza river valley, N.E. Poland. Vogelwarte 33:26–52

El-Kahloun M (2004) Vegetation dynamics in P-limited rich fens. PhD thesis, Antwerp University, Antwerp

El-Kahloun M, Verhagen B, Van Haesebroeck V, Verhagen B (2003) Differential recovery of above- and belowground fen vegetation following fertilization. J Veg Sci 14:451–458

El-Kahloun M, Gerard M, Meire P (2005) Phosphorus and nitrogen cycling in fen vegetation along different trophic conditions in the Biebrza valley, Poland. Ecohydrol Hydrobiol 5:68–79

Giller KE, Wheeler BD (1988) Acidification and succession in a floodplain mire in the Norfolk Broadland, U.K. J Ecol 76:849–866

Girel J (1994) Old distribution procedure of both open water and matter fluxes in floodplains of Western Europe: impact on present vegetation. Environ Manage 18:203–221

Grootjans AP, Van Diggelen R (1995) Assessing the restoration prospects of degraded fens. In: Wheeler BD, Shaw SC, Fojt WJ, Robertson RA (eds) Restoration of temperate wetlands. Wiley, Chichester, pp 73–90

Güsewell S, Koerselman W (2002) Variation in nitrogen and phosphorus concentrations of wetland plants. Perspect Plant Ecol 5:37–61

Holland EA, Dentener FJ, Braswell BH, Sulzman JM (1999) Contemporary and preindustrial global reactive nitrogen budgets. Biogeochemistry 46:7–43

Janssen A (2000) Modeling groundwater flow in a floodplain in the lower Biebrza catchment with the MODFLOW three dimensional finite differences code. (Report of the Department of Physical Geography) Utrecht University, Utrecht

Jongman RGH (1998) Rivers: key elements in European ecological networks. In: Nienhuis PH, Leuven RSEW, Ragas AMJ (eds) New concepts for sustainable management of river basins. Backhuys, Leiden, pp 53–67

Kossowska-Cezak U (1984) Climate of the Biebrza ice-marginal valley. Pol Ecol Stud 10:253–270

Kubrak J, Okruszko T (2000) Hydraulic model of the surface water system for the central Biebrza basin. In: Mioduszewski W, Wassen MJ (eds) Some aspects of water management in the Valley of the Biebrza River. Institute of Land Reclamation and Grassland Farming, Falenty, pp 83–90

Large ARG, Prach K, Bickerton, MA, Wade PM (1994) Alteration of patch boundaries on the floodplain of the regulated river Trent, U.K. Regul Rivers 9:71–78

McCollin D, Moore L, Sparks T (2000) The flora of a cultural landcape: environmental determinants of changes revealed using archival sources. Biol Conserv 92:249–263

Mioduszewski W, Querner EP (eds) (2002) Hydrological system analysis in the valley of Biebrza River. Institute of Land Reclamation and Grassland Farming, Falenty

Mioduszewski W, Wassen MJ (eds) (2000) Some aspects of water management in the valley of the Biebrza river. Institute of Land Reclamation and Grassland Farming, Falenty

Mitsch WJ, Gosselink JG (1993) Wetlands, 2nd edn. Van Nostrand Reinhold, New York

Moore DRJ, Keddy PA, Gaudet CL, Wisheu IC (1989) Conservation of wetlands: do infertile wetlands deserve a higher priority. Biol Conserv 47:203–217

Okruszko H (1990) Wetlands of the Biebrza valley; their value and future management. Polish Academy of Sciences, Warsaw
Okruszko H (1995) Influence of hydrological differentiation of fens and their transformation after dehydration and on possibilities for restoration. In: Wheeler BD, Shaw SC, Fojt WJ, Robertson RA (eds) Restoration of temperate wetlands. Wiley, Chichester, pp 113–119
Okruszko T, Wasilewicz M, Dembek W, Rycharski M, Matuszkiewicz A (2003) Analysis of the changes in water, soil and plant conditions in Bagno Lawki in the Lower Biebrza Basin. (Report 3-1) Institute of Land Reclamation and Grassland Farming, Falenty, pp 107-128
Olde Venterink H, Davidsson TE, Kiehl K, Leonardson L (2002) The impact of drying and re-wetting on N, P and K dynamics in a wetland soil. Plant Soil 243:119–130
Olde Venterink H, Wassen MJ, Verkroost AWM, De Ruiter PC (2003) Species richness-productivity patterns differ between N, P and K limited wetlands. Ecology 84:2191–2199
Pajnowska H, Pozniak R, Wienclaw E (1984) Groundwater of the Biebrza valley. Pol Ecol Stud 10:301–311
Palczynski A (1984) Natural differentiation of plant communities in relation to hydrological conditions of the Biebrza valley. Pol Ecol Stud 10:347–385
Palczynski A, Stepa T (1991) Biomass production in main plant associations of the Biebrza valley with respect to soil conditions. Pol Ecol Stud 17:53–62
Patrick WH, Khaled RA (1974) Phosphate release and sorption by soil and sediments: effect of anaerobic and anaerobic conditions. Science 186:53–55
Poschlod P, Bonn S (1998) Changing dispersal perspectives in the central European landscape since the last ice age: an explanation for the actual decrease of plant species richness in different habitats? Acta Bot Neerl 47:27–44
Rich TCG, Woodruff ER (1996) Changes in vascular plant floras of England and Scotland between 1930–1960 and 1987–1988: the BSBI monitoring scheme. Biol Conserv 75:217–229
Schmidt A, Piórkowski H, Bartoszuk H (2000) Remote sensing techniques and geographic information systems for wetland conservation and management: monitoring scrub encroachment in Biebrza National Park. Report Alterra, Wageningen
Stromberg JC (1993a) Instream flow models for mixed deciduous riparian vegetation within a semiarid region. Regul Rivers 8:225–235
Stromberg JC (1993b) Riparian mesquite forests: a review of their ecology, threaths, and recovery potential. J Ariz Nev Acad Sci 27:111–124
Stromberg JC (2001) Biotic integrity of *Platanus wrightii* riparian forests in Arizona: first approximation. Forest Ecol Manage 142:251–266
Stromberg JC, Tiller R, Richter B (1996) Effects of groundwater decline on riparian vegetation of semiarid regions: the San Pedro, Arizona. Ecol Appl 6:113–131
Succow M (1988) Landschaftsökologische Moorkunde. Borntraeger, Berlin
Succow M, Jeschke L (1986) Moore in der Landschaft. Urania, Leipzig
Swiatek D, Okruszko T, Chormanski J (2002) Surface water system analysis. In: Mioduszewski W, Querner EP (eds) Hydrological system analysis in the Valley of Biebrza River. Institute of Land Reclamation and Grassland Farming, Falenty, pp 39–80
Toner M, Keddy P (1997) River hydrology and riparian wetlands: a predictive model for the ecological assembly. Ecol Appl 7:236–246
Trémolières M, Roeck U, Klein JP, Carbiener R (1994) The exchange processes between river and groundwater on the central Alsace floodplain (eastern France): the case of a river with functional floodplain. Hydrobiologia 273:19–36
Van der Putten WH, Vet LEM, Harvey JA, Wäckers LA (2001) Linking above- and belowground multitrophic interactions of plants, herbivores, pathogens and their antagonists. Trends Ecol Evol 16:547–554

Van Diggelen R (1998) Moving gradients; assessing restoration prospects of degraded brook valleys. PhD thesis, University of Groningen, Groningen

Van Diggelen R, Molenaar W, Casparie WA, Grootjans AP (1991) Paläoökologische Untersuchungen als Hilfe der Landschaftanalyse im Gorecht-gebiet. Telma 21:57–73

Van Urk G (1984) Lower Rhine–Meuse. In: Whitton BA (ed) Ecology of European rivers. Blackwell, Oxford, pp 437–468

Van Wirdum G (1984) Development of techniques for ecohydrological research. (Annual report 1983) Research Institute for Nature Management, Leersum

Van Wirdum G (1991) Vegetation and hydrology of floating rich fens. PhD thesis, University of Amsterdam

Verhoeven JTA, Schmitz MB (1991) Control of plant growth by nitrogen and phosphorus in mesotrophic fens. Biogeochemistry 12:135–148

Verhoeven R, Banasik R, Swiatek D, Chormanski J, Okruszko T (2004) Surface water modeling of the Biebrza River network. In: River flow. Balkema, Rotterdam, pp 1057–1063

Wardle DA (2002) Communities and ecosystems: linking the aboveground and belowground components. Princeton University, Princeton

Wassen MJ (2005) The use of reference areas in the conservation and restoration of riverine wetlands. Ecohydrol Hydrobiol 5:41–49

Wassen MJ, Grootjans AP (1996) Ecohydrology: an interdisciplinary approach for wetland management and restoration. Vegetatio 126:1–4

Wassen MJ, Joosten JHJ (1996) In search of a hydrological explanation for vegetation changes along a fen gradient in the Biebrza Upper Basin (Poland). Vegetatio 124:191–209

Wassen MJ, Barendregt A, Palczynski A, De Smidt JT, De Mars H (1990) The relationship between fen vegetation gradients, groundwater flow and flooding in an undrained valley mire at Biebrza, Poland. J Ecol 78:1106–1122

Wassen MJ, Barendregt A, Palczynski A, De Smidt JT, De Mars H (1992) Hydro-ecological analysis of the Biebrza mire, Poland. Wetland Ecol Manage 2:119–134

Wassen MJ, Olde Venterink H, De Swart EOAM (1995) Nutrient concentrations in mire vegetation as a measure of nutrient limitation in mire ecosystems. J Veg Sci 6:5–16

Wassen MJ, Van Diggelen, R, Wolejko L, Verhoeven JTA (1996) A comparison of fens in natural and artificial landscapes. Vegetatio 126:5–26

Wassen MJ, Van der Vliet RE, Verhoeven JTA (1998) Nutrient limitation in the Biebrza fens and floodplain (Poland). Acta Bot Neerl 47:241–253

Wassen MJ, Bleuten W, Bootsma MC (2002) Linking hydrology to ecology; Biebrza as a geographical reference. Ann Warsaw Agric Univ 33:27–34

Wassen MJ, Peeters WHM, Olde Venterink H (2003) Patterns in vegetation, hydrology, and nutrient availability in an undisturbed river floodplain in Poland. Plant Ecol 165:27–43

Wassen MJ, Olde Venterink H, Lapshina ED, Tanneberger F (2005) Endangered plants persist under phosphorus limitation. Nature 437:547–550

Wheeler BD, Shaw SC (1991) Above-ground crop mass and species richness of the principal types of herbaceous rich-fen vegetation of lowland England and Wales. J Ecol 79:285–301

Wheeler BD, Shaw SC (1995) A focus on fens – Controls on the composition of fen vegetation in relation to restoration. In: Wheeler BD, Shaw SC, Fojt WJ, Robertson RA (eds) Restoration of temperate wetlands. Wiley, Chichester, pp 49–72

Zalewski M (2000) Eco-hydrology – the scientific background to use ecosystem properties as management tools toward sustainability of water resources. Ecol Eng 16:1–8

Zurek S (1984) Relief, geologic structure and hydrography of the Biebrza ice-marginal valley. Pol Ecol Stud 10:239–251

Subject Index

A
acid sulphate soils 112
acidification 253
adaptations
– physiological 2
– morphological 2
adventitious roots 3
aerenchyma 2, 15, 16
agricultural development of wetlands 247, 258, 261
agriculture 171, 175, 180, 181, 191, 193
Allee effects 78
allelopathy 3
alternative stable states 149
Amazon River 170, 174, 175
1-aminocyclopropane-1-carboxylic acid (ACC) 28
anaerobic respiration 19–21
Andes mountains 183
anoxia 2, 19–22, 27, 29–31
aquatic ape theory 94
Aswan high dam 105
– impact on sardine industry 105
atmospheric deposition 248, 249, 256

B
barriers 62, 64, 82, 83
biomass 126
birds 180–182

C
carbon dioxide 17, 18, 22–27, 30, 31
catchment 6, 10, 285, 286, 288
chalk deposition 253
chemical ecology 42
– allelopathy 42, 52
clean development mechanism (Kyoto Protocol) 170
clear versus turbid state 38, 42, 48, 52
clear water 149, 151, 154, 165
climate change 101, 109, 170, 175, 186, 191, 193, 253
coastal 8
co-creation 276
co-existence 78
competing claims 272
competition 63, 71, 76, 78
complexity 273
– composition 74
Congo River 174
connectivity 74, 152, 174
constructivism 274, 275
consumers 71
Convention on Biological Diversity 95, 169, 176–178, 189
crassulacean acid metabolism (CAM) 23
crayfish 49, 52

D
dams 170–172, 175, 191
Danube River 183
decision making 272
deforestation 175, 180, 181
dispersal 62, 237, 238
dispersion 306
disturbance 71, 74, 75, 77, 82

diversity
– loss of 226, 227
– genetic 241, 242
drainage 294
dredging 181
dykes (levees, bunds) 181

E
ecohydrological 10
ecohydrology 285
ecological roulette 73
ecoregions 6, 172–176, 189, 192, 193
ecosystem approach 6, 95, 97, 176, 186
– principles 98
– operational guidance 99
– source book 98
– societal choice 99
ecosystem
– characteristics 68
– engineers 5, 63
– invaded 79
– management 94
– new paradigm 97
– reference 286
ecotourism 181
elongation growth 21, 26–31
endogenous cycles 84
environmental flows 174, 191
environmental impact assessment 177
epiphyton 3, 4, 45–49, 52
estuaries 7, 118
ethylene 3
eutrophication 226, 253–258
– internal 235
– external 234
exotic species 248, 261
expansions 27
expert system 164
extinction 61, 63, 81

F
facilitation 7, 74, 279, 280
fen 10, 11, 247, 286, 287, 289, 293, 297, 298, 301, 302, 304–306
– alkaline fen 249
– agricultural abandonment 254, 257, 264
– calcareous spring mire 249, 251
– community structure 252, 254, 262
– dune slack 249, 250
– definition 247
– distribution 248, 249
– drainage 250
– erosion 250, 253
– eutrophication 251
– fen meadow 247
– floating 261
– grazing 254, 264
– history of loss/change 247, 257, 258, 260–264
– landuse change 248
– management 254, 263, 264
– mire systems 250
– mowing 264
– productivity 254–256
– species richness 247, 250, 254, 255
– spring mire 252
– woody invasion 251, 254, 263, 264
fermentation 2, 20, 21, 30
fire 254
fish 131, 132, 172, 180–182, 192
floods 180–182, 190, 191
flooding 293
flooding tolerance in plants 15–31
floodplains 172, 180, 181, 289, 304, 306
Florida 205–208, 219
food webs 4, 150
fouling 62
fragmentation 247, 260, 264, 304, 305

G
gas diffusion 17–19
genetic diversity 64, 78
geographic information systems (GIS) 172, 175
glaciers 170
governance
– approaches 275
– principles 276
grazing 4, 154
groundwater 9, 10, 285, 292, 296–299, 302, 304
grubbing 4
Gulf of Mexico, hypoxia 103

H
habitat modification 63
hay removal 254
herbaceous 9

herbivory 63
- vertebrate 4
high marsh 126
Holocene 203
holonocentricity 274
horizontal migration
hurricane Katrina 94
hybridization 64
hydroelectricity (-power) 170, 171, 191, 193
hydrology 8, 249–253
- alteration of hydrology 250–252, 256
- flooding 259
- flow through system 249
- groundwater 249, 250, 252–253
- local water table drop 250
- landscape level change 253, 257
- surface water table 256–258
hydroperiod, shifts in 226, 234–237, 242
hyponastic curvature 27, 28
hypoxia 20–22

I
impacts
- human 2
- invasion 61, 63–67
indigenous peoples 177
inoculation, pressure 74
INTECOL 2
integrated river basin management (IRBM) 176, 182, 189, 190, 193
integrated water resources management (IWRM) 171, 186
interdependence 272
inundation 293, 295
invasibility 70, 71, 75, 76
invasion
- biological 4
- failure 81
- history 68, 83
- impact 61, 63–67
- pathways 66
- resistance 71, 72
- stages, steps 68, 79–81
- success 68, 69, 74
invasive species 128, 249, 261 (see also exotic species)
- characteristics 68, 69, 79
- examples of 226, 231–233, 242
- control of 233, 240, 241
invasiveness 70

K
key processes 305
keystone species 63

L
lakes, shallow 4, 7, 37, 49, 50
land tenure 176
leaf heterophylly 24, 25
Lotka-Volterra–equilibrium 72
Louisiana 209–213, 219
low marsh 126

M
macrophytes 3
- freely-floating 38, 42, 51
managed realignment 109
management 6–9, 165, 166, 271, 273, 278, 280
- adaptive 197–200, 205, 206, 209, 213, 214, 217, 219, 220
mangrove 200, 204, 216–219
marshes, floodplain 286, 287, 297
Mediterranean Wetlands Initiative (MedWet) 183–185
Mekong Delta 112
Mekong River 170, 172, 174, 185
Mesopotamian marshes 103
- drainage 105
- hydraulic civilisation 104
- importance 104
- Madan 104
- restoration 105–107
- satellite image 106
mineralization 251
model
-PCLake 159–161, 163–165
-BASIS 159, 163, 164
models
- abiotic resistance 74
- biotic resistance 70–74, 80
- fluctuating resource availability 75, 76
- comparative historical approach 83
- invaded ecosystem approach 70–73
- invading species approach 68–70
- invasional meltdown 71, 73, 74
- key-lock approach 68
- niche-based framework 77, 78
- steps and stages 79–81

MODFLOW 299
multidimensional 158

N

natural resource 1
naturalization, Darwin's hypothesis 79
Netherlands, wetlands of 229
niche
- breadth 77, 78
- overlap 77, 78
Niger River 174
Nile Delta 107
nitrogen 233, 235
- deposition 10, 286, 304
non-government organizations 169, 175, 176, 184, 185, 187, 190, 193
nutrient
- concentrations 158, 290
- cycling 4, 134–138, 254–258
- limitation 298, 301, 303
- loading 290, 291

O

objectivist perspective 273
outbreeding, depression 64
oxygen deficiency 18–20, 26
- sensing 20
- adaptation 19–22

P

palynology 251, 252
parasitism 63, 78, 84
participation 269
pathogens 63, 65, 84
peat 270, 288, 298, 303, 306
peatlands 1, 100, 289, 305
- carbon flux 100
- functioning 101
- SE Asia peat swamps 101
- tropical 100
- medicinal plants 102
perspectives on planning 273
pH 201, 204
phenology 129
phosphorous 233, 235
phosphorus 151, 154, 155, 297, 302
- cycling 137
photosynthesis 18–20, 22–25, 31
plant hormones
- abscisic acid 24, 27–29, 31
- auxin 26–29, 31
- ethylene 18, 24, 26–29, 31
- gibberellins 24, 26–29, 31
Pocosins 108
pollution 179, 182, 190
post-anoxic stress 22
poverty 171, 180–182, 191–193
predation 63, 64, 77, 78, 84
process 197, 198, 200, 205, 206, 209–215, 217, 219, 220
- productivity 76
propagule pressure 69, 77, 85
protected areas 169, 176, 178, 182, 185, 187–191

Q

quadrant
- techno-centric 274
- eco-centric 274

R

Ramsar Convention 1, 94, 176–178, 182–189
reactive oxygen species (ROS) 19, 22
recreation 271
Red List Species 249
redundancy 78
reference 304–306
- system 10
refuge 40–42
resource availibility 71, 75, 76
- limitation 77
restoration 257–264
- goals 230, 243
- monitoring 262, 263
- strategy 239–242
- sulfate-rich water 264
rice (*Oryza sativa*) 16, 19, 21, 22, 26, 27
river, lowland 287, 305
riverine wetlands 285
runoff 170, 174

S

saltmarsh 200, 203, 204, 207, 208, 214, 216, 219
saturation, niches 72
scales 150

Schnorkel effect 26
sea level rise 109
seasonality 128–130
sedge meadow 247
sedimentation 136, 182, 248
seed
– dispersal 247, 248, 258–260
– bank 130, 237–239, 242, 259, 260, 306
– hydrochory 260
– persistence/longevity 259
– rain 259
– wind dispersal 260
seeding 238–242
snails 39, 45, 47, 49, 52
social learning 6, 269–281
soils 201, 204–206
– acid sufate 210, 219
speciation 64
species 64
– alien 62, 82
– endemic 5
– exotic 62
– introduced 5, 62, 80–81
– invasive 62, 82
– native 64, 83
– non-indigenous 62
– non-native 62, 83
– pest 80, 81
– richness 70, 71, 76, 80
stable states, alternative 149, 155
stress 71
submergence, escaping 3
succession 251, 252, 259, 294, 306
sustainable development 94
swamps 128

T

taxon
– diversity 157
– richness 158, 163
techno-centric design 274
tidal creeks 124
tidal freshwater wetlands 7, 117–148
– animals 131–134
– European vegetation 124–126
– historical development 121, 122
– North American vegetation 126–128
– restoration 138–141
travertine formation 253
treaty 176–179, 193
trophic
– interactions 3
– levels 7
– state 74
tsunami 94
turbid state 149, 151, 154

U

urbanization 247
– runoff 248

V

value free planning 269
vegetation zonation 123
virus diseases 65

W

Water Framework Directive 99, 166, 176, 178, 179, 189, 273, 274
water levels 271, 272
water plants, submerged 151
water quality 122
wetland
– biogeochemistry 111
– creation 10, 198, 210, 211, 216
– development 269, 277
– economics 107, 109
– ecotones 93
– goods and services 107, 108
– hydrological change 111
– interface 93, 95
– loss 105,108, 205, 209, 212, 213
– restoration 197, 198, 200, 209–212, 214, 216, 219
– socio-technical characteristics 270
– values 101, 109
wildlife 131
Wisconsin, wetlands of 228
World Commission on Dam 177
World Wide Fund for Nature (WWF) 169, 171, 172, 174, 175, 180, 184, 187–193

Z

zooplankton 3, 40–42

Printing: Krips bv, Meppel, The Netherlands
Binding: Stürtz, Würzburg, Germany